Study Guide and Solutions Manual

Leo J. Malone
Saint Louis University

to accompany

Basic Concepts of Chemistry

Ninth Edition

Leo J. Malone
Saint Louis University

Theodore O. Dolter
Southwestern Illinois College

In collaboration with **Steven Gentemann** *Southwestern Illinois College*

WILEY

John Wiley & Sons, Inc.

ISBN 978-1118156438

Printed in the United States of America

SKY10051655_072423

Table of Contents

1

Chemistry and Measurements

Review of Part A *The Numbers Used in Chemistry*

Review of Part B *The Units Used in Chemistry*

Chapter Summary Assessment

Answers to Assessments of Objectives

Answers and Solutions to Green Text Problems

Review of Part A *The Numbers Used in Chemistry*

OBJECTIVES AND DETAILED TABLE OF CONTENTS

1-1   The Numerical Value of a Measurement

OBJECTIVE *Determine the number of significant figures and the degree of uncertainty in a measurement.*

1-2   Significant Figures in Chemical Calculations

OBJECTIVE *Perform arithmetic operations and percent error calculations, rounding the answer to the appropriate number of significant figures.*

1-2.1 The Mass of a Compound
1-2.2 Molecular Weight and the Rules for Addition and Subtraction and Rounding Off
1-2.3 The Rules for Multiplication and Division
1-2.4 Calculating Percent Error

1-3   Expressing the Large and Small Numbers Used in Chemistry

OBJECTIVE *Perform arithmetic operations involving scientific notation.*

1-3.1 Chemistry and Scientific Notation
1-3.2 Mathematical Manipulation of Scientific Notation

SUMMARY OF PART A

Our journey into the science of chemistry begins with the definitions **chemistry** and the **matter** of which it is composed. Most of this text will center on the basic forms of matter that we see about us, which are the **elements** and the chemical combinations of elements known as **compounds**. In the submicroscopic world we can visualize **atoms** and **molecules**, which are the particles of which elements and compounds are composed.

A discussion of chemistry can't go far, however, without application of quantitative laws. Since quantitative laws are based on **measurements**, it is necessary for us to establish how to appropriately express a measurement and eventually how to carry out mathematical operations on specific measurements. A measurement is composed of a numerical value and a **unit**. In this first section of the review, we will concentrate on properly expressing the numerical value of the measurement.

Accuracy reflects how close a measurement is to the true value. Measurements are also expressed to various degrees of **precision**, which are expressed by the number of **significant figures** in the measurement. Accuracy also reflects its degree of **uncertainty**. Since a zero may or may not be a significant figure, it is necessary to be familiar with the certain rules. For example, the following numbers represent measurements. The zeros that are significant are underlined.

$$13.\underline{0}7 \quad 1\underline{0}.7 \quad 0.3\underline{0} \quad 1\underline{0}.3\underline{0}1 \quad 1030 \quad \underline{20}.\underline{0} \quad 0.02\underline{0}$$

You should also know that there are different rules concerning the proper form for answers for addition and subtraction as opposed to those for multiplication and division. A common mistake is to apply the rules on the number of significant figures to addition and subtraction when, in fact, this is not important. For example, in the addition example that follows, the answer is rounded off to the same number of decimal places as the measurement with the fewest number. In the multiplication example, the answer is rounded off to the same number of significant figures as the multiplier with the fewest number of significant figures.

$$
\begin{array}{l}
0.063 \\
+\ 1.02 \quad \leftarrow \left\{ \begin{array}{l}\text{term with least} \\ \text{number of} \\ \text{decimal places}\end{array}\right. \\
\underline{+\ 0.003} \quad \left\{ \begin{array}{l}\text{answer to two} \\ \text{decimal places} \\ \text{(rounded off)}\end{array}\right. \\
\ \ 1.086\ =\ \underline{1.09} \quad \leftarrow
\end{array}
$$

number of significant figures

$$
\begin{array}{cccc}
(2) & (4) & (4) & (2) \\
0.23 \times 14.68 \times 7.882 = 26.6127 = \underline{27}
\end{array}
$$

multiplier that limits answer

In laboratory situations, it is often required to evaluate a measurement as to how it compares to the actual known value. This is usually expressed in terms of **percent error**. This is calculated as follows:

% error = [(known value – measured value)/known value] x 100%

Atoms and molecules are so small that their weights require very small numbers, and it requires a very large number of them to deal with in the visible world. Exponential notation is introduced as a necessary tool to express these numbers without writing a lot of zeros. In **scientific notation** the number is written with one digit to the left of the decimal point in the **coefficient** multiplied by 10 raised to a power. The power is known as the **exponent** of 10. In later chapters, we will assume that the student is not only familiar with scientific notation, but also with the mathematics of manipulating numbers expressed in this fashion. Refer to Appendix C in the text for additional help in this area if needed.

ASSESSMENT OF OBJECTIVES

A-1 Multiple Choice (There is only one correct answer for each question.)

___ 1. A container is known to have a volume of one quart. Which of the following measurements of the volume is the most precise?

(a) 1.0 qts   (b) 0.98 qts   (c) 1.052 qts   (d) 0.9 qts

___ 2. How many significant figures are in the measurement, 0.03010 cm?

(a) 6   (b) 5   (c) 4   (d) 3   (e) 2

___ 3. When 78.5 cm^2 is divided by 2.654 cm, the answer is:

(a) 29.577995 cm   (b) 29.5 cm   (c) 30 cm

(d) 29.6 cm   (e) 29.58 cm

___ 4. When the numbers 0.508, 25.0 and 1.188 are added, the answer is:

(a) 26.7   (b) 26.6   (c) 27   (d) 27.0   (e) 26.696

___ 5. A sample of gold is known to weigh 1.088 oz. A student measures the weight as 1.05 oz. What is the percent error?

(a) 0.037%   (b) 4%   (c) 3.7%   (d) 3.8%

___ 6. The number 0.0070×10^{-8} expressed in scientific notation is:

(a) 7×10^{-11} (b) 7×10^{-5} (c) 7.0×10^{-11}

(d) 7.0×10^{-10} (e) 7.0×10^{-5}

A-2 Significant Figures

1. How many significant figures are in each of the following numbers?

0.02 _____   0.50 _____   120.0 _____

103 _____   0.209 _____   1200 _____

2. Assume that the following numbers represent measurements. Express the answer for the calculation to the proper number of significant figures or decimal places.

(a)   0.0375
         0.12
     + 0.001
        0.1585 = _____

(b)   9700
     - 180
     8520 = _____

(c) $\dfrac{14.222 \times 0.3010}{0.200} = 21.40411 = $ _____

(d) $(0.6)^2 = 0.36 = $ _____   (e) $(6.0)^2 = 36 = $ _____

A-3 Scientific Notation

1. Express the following numbers in scientific notation.

(a) $0.0042 = $ _____   (b) $47900 = $ _____

(c) $0.00040 = $ _____   (d) $0.037 \times 10^{-6} = $ _____

(e) $1560 \times 10^{-8} = $ _____

2. Express the following as normal numbers without scientific notation.

        (a) 4.83×10^{-3} = _____   (b) 3.77×10^{4} = _____

        (c) 0.091×10^{-2} = _____   (d) 915×10^{4} = _____

3. Carry out the following calculations. Express the answer to the proper number of significant figures or decimal places.

        (a) $(14.6 \times 10^{-3}) + (0.141) + (237 \times 10^{-5})$ _____

        (b) $(18.3 \times 10^{-6}) \times (0.00320 \times 10^{-3})$ _____

        (c) $\dfrac{392 \times 10^{-22}}{0.022 \times 10^{24}}$ _____

        (d) $\dfrac{(0.0631 \times 10^{6}) \times (1.009 \times 10^{8})}{(7.71 \times 10^{-4})}$ _____

Review of Part B *The Units Used in Chemistry*

OBJECTIVES AND DETAILED TABLE OF CONTENTS

1-4   The Units Used in Chemistry

OBJECTIVE *List several fundamental and derived units of measurement in the metric (SI) system.*

1-5   Conversion of Units by Dimensional Analysis

OBJECTIVE *Interconvert measurements by dimensional analysis.*

1-6   Measurement of Temperature

OBJECTIVE *Interconvert several key points on the Fahrenheit, Celsius and Kelvin temperature scales.*

> 1-6.1 Thermometer Scales
> 1-6.2 Relationships Between Scales

SUMMARY OF PART B

The units of measurement used in chemistry are those of the **metric system**. Metric units also serve as the basis of the SI system and are universally recognized units. The metric units use **meter** for length, **gram** for **mass (weight)** and **liter** for **volume**. These units employ prefixes such as **milli**, **centi**, and **kilo**. Table 1A, which follows, summarizes the more common relationships within and between the English and metric systems. The **English system** of measurement lacks systematic relationships between units. Most measurements in this system now relate to a specific value in the metric system. It should be noted that relationships within a system are exact, but between systems can be expressed to various numbers of significant figures. *Table 1A* presents relationships within both systems and between the two systems.

Table 1A

RELATIONSHIPS BETWEEN UNITS

	English	Metric	English-Metric
Length	1 ft = 12 in. 3 ft = 1 yd 1760 yd = 1 mile	$1\ m = 10^3\ mm$ $1\ m = 10^2\ cm$ $1\ m = 10^{-3}\ km$	2.54 cm = 1 in. (exact)
Volume	2 pt = 1 qt 4 qt = 1 gal 42 gal = 1 barrel	$1\ L = 10^3\ mL$ $1\ L = 10^{-3}\ kL$ $1\ mL = 1\ cm^3$	1.057 qt = 1 L
Mass	16 oz = 1 lb 2000 lb = 1 ton	$1\ g = 10^3\ mg$ $1\ g = 10^{-3}\ kg$	453.6 g = 1 lb

Many chemistry problems involve the conversion of a measurement in one system of measurement to that of another. **Dimensional analysis** is used for working problems of this nature and is introduced in this chapter. It is a powerful problem-solving tool when used consistently. The method can be illustrated by conversions within and between English and metric units. Basically, this method involves the inclusion of units in all operations as well as the numerical value of the measurement. In this fashion, the units lead us to the proper mathematical operation and thus to the correct answer. For a simple one-step conversion, the given measurement is multiplied by a **conversion factor** that transforms the original measurement to the desired measurement. For a conversion involving more than one step, a **unit map** is useful. Conversion factors are usually expressed as **unit factors**, which relate to *one* of a unit.

$$[\text{What's given}] \; x \; (\text{Conversion Factor}) = [\text{What's requested}]$$

The conversion factor is constructed from an equality or equivalency that relates the given and requested measurements. An example problem follows.

Example B-1 Converting Pounds to Grams

Convert 0.78 lb to grams.

PROCEDURE

A conversion factor is needed that relates lb to g. From Table 1A, this is 453.6 g = 1 lb. The conversion factor expresses the relationship in factor or fractional form.

$$\frac{453.6\,g}{lb} \quad \text{or} \quad \frac{1\,lb}{453.6\,g}$$

Choose the proper conversion factor that cancels the given unit (lb) and leaves the requested unit in the numerator (g).

SOLUTION

| (Given) | (Conversion Factor) | (Requested) |

$$0.78\ \cancel{lb} \quad x \quad \frac{453.6\,g}{\cancel{lb}} \quad = \quad \underline{350\,g}$$

Given unit cancels   Requested unit remains
in the numerator

Many problems, however, involve conversions where one simple relationship between given and requested units is not available. In these cases, one must travel from given to requested in two or more carefully planned steps. For each step in the calculation, a conversion factor is needed.

Example B-2 Converting Meters to Feet

Convert 61.7 m to ft.

PROCEDURE

The only relationship between English and metric systems given in Table 1A is between in. and cm. Thus, three steps and three corresponding conversions are needed. The unit map follows.

Given   **Requested**

      (1)   (2)   (3)

m $\longrightarrow$ **cm** $\longrightarrow$ **in** $\longrightarrow$ **ft**

SOLUTION

$$61.7 \text{ m} \times \underset{\substack{\uparrow \\ \text{converts} \\ \text{m to cm}}}{\frac{10^2 \text{ cm}}{\text{m}}} \times \underset{\substack{\uparrow \\ \text{converts} \\ \text{cm to in.}}}{\frac{1 \text{ in.}}{2.54 \text{ cm}}} \times \underset{\substack{\uparrow \\ \text{converts} \\ \text{in. to ft}}}{\frac{1 \text{ ft}}{12 \text{ in.}}} = \underline{202 \text{ ft}}$$

Temperature is a measure of the heat intensity of the substance. Temperature is measured with a device called a **thermometer**. The **Fahrenheit** [T(°F)], **Celsius** [T(°C)] and **Kelvin** [T(K)] temperature scales relate to each other as follows:

$$T(°F) = [T(°C) \times 1.8] + 32$$

$$T(°C) = [T(°F) - 32]/1.8 \qquad\qquad T(K) = [T(°C) + 273]$$

ASSESSMENT OF OBJECTIVES

B-1 Multiple Choice

____ 1. The metric prefix *milli* relates to the basic unit by:

    (a) 10^3 (b) 10^2 (c) 10^{-6} (d) 10^{-3} (e) 1 cm^3

____ 2. Which of the following relationships is false?

    (a) $1 \text{ mg} = 10^6 \text{ kg}$ (b) $100 \text{ cm} = 1 \text{ m}$
    (c) $10^3 \text{ L} = 1 \text{ kL}$ (d) $1 \text{ cg} = 10^{-2} \text{ g}$

____ 3. Which of the following is an exact relationship?

    (a) $1 \text{ yd} = 3 \text{ ft}$ (b) $1 \text{ lb} = 453.6 \text{ g}$
    (c) $2.205 \text{ lb} = 1 \text{ kg}$ (d) $1 \text{ km} = 0.6215 \text{ mile}$

____ 4. One milliliter is the same as:

    (a) 1 dm^3 (b) 1 m^2 (c) 1 mm^3 (d) 1 cm^2 (e) 1 cm^3

___ 5. Jupiter is a planet with a larger gravity than Earth. On Jupiter, a person:

(a) weighs less.   (b) has more mass.
(c) weighs more.   (d) has less mass.

___ 6. If the Fahrenheit temperature increases by 36 degrees, the Celsius temperature increases by:

(a) 65 degrees.   (b) 20 degrees.   (c) 52 degrees.
(d) 33 degrees.   (e) 15 degrees.

B-2 Matching

___ 1 m	(a) 2.205 kg	(f) 30.5 in.
___ 1 ft	(b) 0.4536 kg	(g) 1 m^3
___ 1 L	(c) 1 dm^3	(h) 10^6 mg
___ 1 kg	(d) 10^3 KL	(j) 30.5 cm
___ 1 lb	(e) 10^3 mm	(i) 453.6 kg

B-3 Conversion Factors

Write a relationship in factor form that accomplishes the following conversions.

e.g.: cm to in.   <u>1 in./2.54 cm</u>

1. cm to m   _____

2. oz to lb   _____

3. g to lb   _____

4. L to qt   _____

5. in. to cm   _____

B-4 Conversions

Make the following conversions. Use dimensional analysis.

1. 14,850 in. to mi

2. 958 mm to km

3. 495 in. to cm

4. 0.811 L to qt

5. 18.8 oz to g

6. 37.0 barrels to kL (42.0 gallons = 1 barrel)

7. 195 m to yd

8. 85 mi/hr to m/min

9. 127°F to °C and K

10. -78°C to °F

Chapter Summary Assessment

S-1 Matching

___ Accuracy

___ Exponent

___ Mass

___ Thermometer

___ Fahrenheit

___ Exact relationship

(a) A measured relationship between units in two different systems of measurement

(b) The attraction of gravity for a substance

(c) A device used to measure temperature

(d) Refers to how close a measurement is to the true value

(e) The number that is raised to a power

(f) A defined relationship between units in the same system of measurement

(g) The amount of matter in a substance

(h) A scale of temperature where the freezing point of water is defined as zero

(i) The power to which a number is raised

(j) Refers to how many significant figures are in a measurement

(k) A temperature scale with 180 divisions between the boiling point and freezing point of water

S-2 Problems

1. Add the following numbers. Express the answer to the proper decimal place.
   $$0.321 + 4.24 \times 10^{-2} + 4 \times 10^{-4}$$

2. What is 0.062×10^{8} divided by 0.17×10^{-3}? Express the answer to the proper number of significant figures.

3. Convert 0.273 kg to mg.

4. Convert 17.5 lb to g.

5. Convert 23.0 mm/sec to in./hr.

6. Convert 100°F to °C and to K.

Answers to Assessments of Objectives

A-1 Multiple Choice

1. **c**   The more precise measurement has the most significant figures.

2. **c**   The first two zeros are not significant.

3. **d**   29.6 cm

4. **a**   26.7. Round off 26.696 to 26.7.

5. **b**   4%. Round off to one significant figure.

6. **c** 7.0×10^{-11}

A-2 Significant Figures

1.

0.02 - 1	0.50 - 2	120.0 - 4
103 - 3	0.209 - 3	1200 - 2

2.

(a) 0.1585 = <u>0.16</u>   (b) 8520 = <u>8500</u>
(c) 21.40411 = <u>21.4</u>   (d) 0.36 = <u>0.4</u>
(e) 36 = <u>36</u>

A-3 Scientific Notation

1. (a) 4.2×10^{-3} (b) 4.79×10^5 (e) 1.56×10^{-5}
   (c) 4.0×10^{-4} (d) 3.7×10^{-8}

2. (a) 0.00483   (b) 37,700
   (c) 0.00091   (d) 9,150,000

3. (a) $14.6 \times 10^{-3} = $ 0.0146
                         +0.141
        $237 \times 10^{-5} = $ +0.00237
                         $\underline{0.15797} = \underline{0.158}$

   (b) $0.00320 \times 10^{-3} = 3.20 \times 10^{-6}$
        $(18.3 \times 10^{-6}) \times (3.20 \times 10^{-6}) = 58.56 \times 10^{-12} = \underline{5.86 \times 10^{-11}}$

   (c) $\dfrac{392 \times 10^{-22}}{0.022 \times 10^{24}} = \dfrac{3.92 \times 10^{-20}}{2.2 \times 10^{22}} = \underline{1.8 \times 10^{-42}}$

   (d) $\dfrac{(0.0631 \times 10^6) \times (1.009 \times 10^8)}{(7.71 \times 10^{-4})} =$

   $\dfrac{(0.0631 \times 1.009)}{7.71} \times \dfrac{(10^6 \times 10^8)}{10^{-4}} = 0.0082578 \times 10^{18} = \underline{8.26 \times 10^{15}}$

B-1 Multiple Choice

1. **d** 10^{-3} 5. **c**   weighs more

2. **a** $1\ mg = 10^6\ kg$ 6. **b**   20 degrees

3. **a**   1 yd = 3 ft $(36\ \cancel{F\ div.} \times \dfrac{100\ C\ div.}{180\ \cancel{F\ div.}} = 20\ C\ div.)$

4. **e** $1\ cm^3$

B-2 Matching

e $1\ m = 10^3\ mm$ **h** $1\ kg = 10^6\ mg$

j   1 ft = 30.5 cm   **b**   1 lb = 0.4536 kg

c $1\ L = 1\ dm^3$

B-3 Conversion Factors

1. $1\ m/10^2\ cm$ or $10^{-2}\ m/cm$

2. $1\ lb/16\ oz$

3. $1\ lb/453.6\ g$

4. $1.057\ qt/L$

5. $2.54\ cm/in.$

B-4 Conversions

1. **in.** $\longrightarrow$ **ft** $\longrightarrow$ **yd** $\longrightarrow$ **mi**

$14{,}850\ \cancel{in.}\ \times\ \dfrac{1\ \cancel{ft}}{12\ \cancel{in.}}\ \times\ \dfrac{1\ \cancel{yd}}{3\ \cancel{ft}}\ \times\ \dfrac{1\ mi}{1760\ \cancel{yd}}\ =\ \underline{0.2344\ mi}$

2. **mm** $\longrightarrow$ **m** $\longrightarrow$ **km**

$958\ \cancel{mm}\ \times\ \dfrac{1\ \cancel{m}}{10^3\ \cancel{mm}}\ \times\ \dfrac{10^{-3}\ km}{\cancel{m}}\ =\ 958 \times 10^{-6}\ km = \underline{9.58 \times 10^{-4}\ km}$

3. **in.** $\longrightarrow$ **cm**

$495\ \cancel{in.}\ \times\ \dfrac{2.54\ cm}{\cancel{in.}}\ =\ 1.26 \times 10^3\ cm$

4. **L** $\longrightarrow$ **qt**

$0.811\ \cancel{L}\ \times\ \dfrac{1.057\ qt}{\cancel{L}}\ =\ \underline{0.857\ qt}$

5. **oz** $\longrightarrow$ **lb** $\longrightarrow$ **g**

$18.8\ \cancel{oz}\ \times\ \dfrac{1\ \cancel{lb}}{16\ \cancel{oz}}\ \times\ \dfrac{453.6\ g}{\cancel{lb}}\ =\ \underline{533\ g}$

6. **barrels** $\longrightarrow$ **gal** $\longrightarrow$ **qt** $\longrightarrow$ **L** $\longrightarrow$ **kL**

$37.0\ \cancel{barrel}\ \times\ \dfrac{42\ \cancel{gal}}{\cancel{barrel}}\ \times\ \dfrac{4\ \cancel{qt}}{\cancel{gal}}\ \times\ \dfrac{1\ \cancel{L}}{1.057\ \cancel{qt}}\ \times\ \dfrac{10^{-3}\ kL}{\cancel{L}}\ =\ \underline{5.88\ kL}$

7. **m** $\longrightarrow$ **cm** $\longrightarrow$ **in.** $\longrightarrow$ **ft** $\longrightarrow$ **yd**

$195\ \cancel{m}\ \times\ \dfrac{10^2\ \cancel{cm}}{\cancel{m}}\ \times\ \dfrac{1\ \cancel{in.}}{2.54\ \cancel{cm}}\ \times\ \dfrac{1\ \cancel{ft}}{12\ \cancel{in.}}\ \times\ \dfrac{1\ yd}{3\ \cancel{ft}}\ =\ \underline{213\ yd}$

8. mi/hr $\longrightarrow$ km/hr $\longrightarrow$ m/hr $\longrightarrow$ m/min

$$\frac{85 \text{ mi}}{\text{hr}} \quad \text{x} \quad \frac{1.609 \text{ km}}{\text{mi}} \quad \text{x} \quad \frac{1 \text{ m}}{10^{-3} \text{ km}} \quad \text{x} \quad \frac{1 \text{ hr}}{60 \text{ min}} = 2.3 \times 10^3 \text{ m/min}$$

9. $T(^{\circ}C) = T(^{\circ}C) = [T(^{\circ}F) - 32]/1.8 = \frac{(127 - 32)}{1.8} = \frac{95}{1.8} = 53^{\circ}C$

   $T(K) = 53 + 273 = \underline{326 \text{ K}}$

10. $T(^{\circ}F) = 1.8[T(^{\circ}C)] + 32 = 1.8(-78.0) + 32 = -140 + 32 = \underline{-108^{\circ}F}$

S-1 Matching

d   **Accuracy** refers to how close a measurement is to the true value.

i   An **exponent** is the power to which a number is raised.

g   **Mass** is the amount of matter in a substance.

c   A **thermometer** is a device used to measure temperature.

k   **Fahrenheit** is a temperature scale with 180 divisions between the boiling and freezing points of water.

f   An **exact relationship** is a defined relationship between units in one system of measurement.

S-2 Problems

1.
$$
\begin{aligned}
& \phantom{4.24 \times 10^{-2} =\ } 0.321 \\
4.24 \times 10^{-2} = & 0.0424 \\
4 \times 10^{-4} \phantom{^{-2}} = & \underline{0.0004} \\
& 0.3638 = \underline{0.364}
\end{aligned}
$$

2. $\dfrac{0.0662 \times 10^8}{0.17 \times 10^{-3}} = \dfrac{6.62 \times 10^6}{1.7 \times 10^{-4}} = 3.9 \times 10^{10}$

       kg $\longrightarrow$ g $\longrightarrow$ mg

3. $0.273 \text{ kg} \quad \text{x} \quad \dfrac{1 \text{ g}}{10^{-3} \text{ kg}} \quad \text{x} \quad \dfrac{10^3 \text{ mg}}{\text{g}} = 2.73 \times 10^5 \text{ mg}$

$$ \text{lb} \longrightarrow \text{g} $$

4. $17.5 \; \cancel{\text{lb}} \; \times \; \dfrac{453.6 \; \text{g}}{\cancel{\text{lb}}} \; = 7.94 \times 10^3 \; \text{g}$

$$ \text{mm/sec} \longrightarrow \text{sec/min} \longrightarrow \text{min/hr} \longrightarrow \text{cm/nm} $$

5. $\dfrac{23.0 \; \cancel{\text{mm}}}{\cancel{\text{sec}}} \quad \times \quad \dfrac{60 \; \cancel{\text{sec}}}{\cancel{\text{min}}} \quad \times \quad \dfrac{60 \; \cancel{\text{min}}}{\text{hr}} \quad \times \quad \dfrac{1 \; \cancel{\text{cm}}}{10 \; \cancel{\text{mm}}}$

$$ \longrightarrow \text{in./cm} \longrightarrow \text{in/hr} $$

$\times \; \dfrac{1 \; \text{in.}}{2.54 \; \cancel{\text{cm}}} \; = \; \underline{3.26 \times 10^3 \; \text{in./hr}}$

6. $T(^{\circ}C) = \dfrac{100 - 32}{1.8} \; = 38^{\circ}C \quad T(K) = 273 + 38 = \underline{311 \; K}$

Answers and Solutions to Green Text Problems

1-1   (b) 74.212 gal. (It has the most significant figures.)

1-2   If a device used to produce a measurement is itself inaccurate, the measurement may be reproducible (precise) but not accurate. Examples are a ruler with the tip broken off or a weight scale with some dirt on it.

1-4   (a) three   (b) two   (c) three   (d) one   (e) four   (f) two   g) two   (h) three

1-6   (a) ±10   (b) ±0.1   (c) ±0.01   (d) ±0.01   (e) ±1   (f) ±0.001   (g) ±100
(h) ±0.00001

1-8   (a) 16.0   (b) 1.01   (c) 0.665   (d) 489   (e) 87,600   (f) 0.0272   (g) 301

1-10   (a) 0.250   (b) 0.800   (c) 1.67   (d) 1.17

1-12   (a) ±0.1   (b) ±1000   (c) ±1   (d) ±0.01

1-14   (a) 188   (b) 12.90   (c) 2300   (d) 48   (e) 0.84

1-16   37.9 qt

1-18   (a) 7.0   (b) 137   (c) 192   (d) 0.445   (e) 3.20   (f) 2.9

1-20   (a) two   (b) three   (c) two   (d) one

1-23   (a) 6.07   (b) 0.08   (c) 8.624   (d) 24   (e) 0.220   (f) 0.52

1-26   (a) (63) + 75.0 = <u>138</u>   (b) (45) x 25.6 = <u>1200</u>   (c) (2.7) x (10.52) = <u>28</u>

1-28   (13.5 – 11.7)/13.5 x 100% = <u>13%</u>

1-30   (29035 – 29002)/29035 x 100% = <u>0.11%</u>

1-32   (a) 1.57×10^2 (b) 1.57×10^{-1} (c) 3.00×10^{-2} (d) 4.0×10^7 (e) 3.49×10^{-2}
(f) 3.2×10^4 (g) 3.2×10^{10} (h) 7.71×10^{-4} (i) 2.34×10^3

1-34   (a) 9×10^7 (b) 8.7×10^7 (c) 8.70×10^7

1-36   (a) 0.000476   (b) 6550   (c) 0.0078   (d) 48,900   (e) 4.75   (f) 0.0000034

1-38   (a) 4.89×10^{-4} (b) 4.56×10^{-5} (c) 7.8×10^3 (d) 5.71×10^{-2}
(e) 4.975×10^8 (f) 3.0×10^{-4}

1-40   (b) < (f) < (g) < (d) < (a) < (e) < (c)

1-42   (a) 1.597×10^{-3} (b) 2.30×10^7 (c) 3.5×10^{-5} (d) 2.0×10^{14}

1-44   (a) 10^7 (b) $10^0 = 1$ (c) 10^{29} (d) 10^9

1-46   (a) 3.1×10^{10} (b) 2×10^9 (c) 4×10^{13} (d) 14   (e) 2.56×10^{-14}

1-48   (a) 2.0×10^{12} (b) 3.7×10^{16} (c) 6.0×10^2 (d) 2×10^{-12} (e) 1.9×10^8

1-50   (a) 1.225×10^7 (b) 9.00×10^{-12} (c) 3.0×10^{-24} (d) 9×10^4 (e) 1×10^{10}

1-51   (a) milliliter (mL)   (b) hectogram (hg)   (c) nanojoule (nJ)
(d) centimeter (cm)   (e) microgram (g)   (f) decipascal (dPa)

1-53   (a) 720 cm, 7.2 m, 7.2×10^{-3} km
(b) 5.64×10^4 mm, 5640 cm, 0.0564 km
(c) 2.50×10^5 mm, 2.50×10^4 cm, 250 m

1-54   (a) 8.9 g, 8.9×10^{-3} kg   (b) 2.57×10^4 mg, 0.0257 kg   (c) 1.25×10^6 mg, 1250 g

1-56   (a) 12 = 1 doz   (c) 3 ft = 1 yd   (e) 10^3 m = 1 km

1-58 (a) $\dfrac{1\text{ g}}{10^3\text{ mg}}$ (b) $\dfrac{1\text{ km}}{10^3\text{ m}}$ (c) $\dfrac{1\text{ L}}{100\text{ cL}}$ (d) $\dfrac{1\text{ m}}{10^3\text{ mm}}$, $\dfrac{1\text{ km}}{10^3\text{ m}}$

1-60 (a) $\dfrac{1\text{ ft}}{12\text{ in.}}$ (b) $\dfrac{2.54\text{ cm}}{\text{in.}}$ (c) $\dfrac{5280\text{ ft}}{\text{mi}}$ (d) $\dfrac{1.057\text{ qt}}{\text{L}}$ (e) $\dfrac{1\text{ qt}}{2\text{ pt}}\ \dfrac{1\text{ L}}{1.057\text{ qt}}$

1-62 (a) 47 L (b) 98 cm (c) 1.85 mi (d) 51.56 yd (e) 92 m (f) 10.27 bbl (g) 32 Gg

1-64 (a) $7.8 \times 10^3\text{ m} \times \dfrac{1\text{ km}}{10^3\text{ m}} = \underline{7.8\text{ km}}$ $7.8\text{ km} \times \dfrac{1\text{ mi}}{1.609\text{ km}} = \underline{4.8\text{ mi}}$

$4.8\text{ mi} \times \dfrac{5280\text{ ft}}{\text{mi}} = \underline{2.5 \times 10^4\text{ ft}}$

(b) $0.450\text{ mi} \times \dfrac{5280\text{ ft}}{\text{mi}} = \underline{2380\text{ ft}}$ $0.450\text{ mi} \times \dfrac{1.609\text{ km}}{\text{mi}} = \underline{0.724\text{ km}}$

$0.724\text{ km} \times \dfrac{10^3\text{ m}}{\text{km}} = \underline{724\text{ m}}$

(c) $8.98 \times 10^3\text{ ft} \times \dfrac{1\text{ mi}}{5280\text{ ft}} = \underline{1.70\text{ mi}}$ $1.70\text{ mi} \times \dfrac{1.609\text{ km}}{\text{mi}} = \underline{2.74\text{ km}}$

$2.74\text{ km} \times \dfrac{10^3\text{ m}}{\text{km}} = \underline{2740\text{ m}}$

(d) $6.78\text{ km} \times \dfrac{1\text{ mi}}{1.609\text{ km}} = \underline{4.21\text{ mi}}$ $4.21\text{ mi} \times \dfrac{5280\text{ ft}}{\text{mi}} = \underline{2.22 \times 10^4\text{ ft}}$

$6.78\text{ km} \times \dfrac{10^3\text{ m}}{\text{km}} = \underline{6780\text{ m}}$

1-65 (a) $6.78\text{ gal} \times \dfrac{3.785\text{ L}}{\text{gal}} = \underline{25.7\text{ L}}$ $25.7\text{ L} \times \dfrac{1.057\text{ qt}}{\text{L}} = \underline{27.2\text{ qt}}$

(b) $670\text{ qt} \times \dfrac{1\text{ L}}{1.057\text{ qt}} = \underline{630\text{ L}}$ $670\text{ qt} \times \dfrac{1\text{ gal}}{4\text{ qt}} = \underline{170\text{ gal}}$

(c) $7.68 \times 10^3\text{ L} \times \dfrac{1.057\text{ qt}}{\text{L}} = \underline{8.12 \times 10^3\text{ qt}}$ $8.12 \times 10^3\text{ qt} \times \dfrac{1\text{ gal}}{4\text{ qt}} = \underline{2.03 \times 10^3}\text{ gal}$

1-67 $122\text{ lb} \times \dfrac{453.6\text{ g}}{\text{lb}} \times \dfrac{1\text{ kg}}{10^3\text{ g}} = \underline{55.3\text{ kg}}$

1-69 $28.0\text{ m} \times \dfrac{10^2\text{ cm}}{\text{m}} \times \dfrac{1\text{ in.}}{2.54\text{ cm}} \times \dfrac{1\text{ ft}}{12\text{ in.}} \times \dfrac{1\text{ yd}}{3\text{ ft}} = \underline{30.6\text{ yd}}$ (New punter is needed.)

1-71 $0.375\text{ qt} \times \dfrac{1\text{ L}}{1.057\text{ qt}} = \underline{0.355\text{ L}}$

1-73 6 ft 10 1/2 in. = 82.5 in. $82.5\text{ in.} \times \dfrac{2.54\text{ cm}}{\text{in.}} \times \dfrac{1\text{ m}}{10^2\text{ cm}} = \underline{2.10\text{ m}}$

$212\text{ lb} \times \dfrac{1\text{ kg}}{2.205\text{ lb}} = \underline{96.1\text{ kg}}$

1-74 $55.0 \text{ L} \times \dfrac{1.057 \text{ qt}}{\text{L}} \times \dfrac{1 \text{ gal}}{4 \text{ qt}} = \underline{14.5 \text{ gal}}$

1-76 $0.200 \text{ gal} \times \dfrac{4 \text{ qt}}{\text{gal}} = 0.800 \text{ qt}$ $\qquad$ $0.800 \text{ qt} \times \dfrac{1 \text{ L}}{1.057 \text{ qt}} \times \dfrac{1 \text{ mL}}{10^{-3} \text{ L}} = \underline{757 \text{ mL}}$

There is slightly more in a "fifth" than in 750 mL.

1-78 $\dfrac{65.0 \text{ mi}}{\text{hr}} \times \dfrac{1.609 \text{ km}}{\text{mi}} = \underline{105 \text{ km/hr}}$

1-82 $\dfrac{\$0.899}{\text{gal}} \times \dfrac{1 \text{ gal}}{4 \text{ qt}} \times \dfrac{1.057 \text{ qt}}{\text{L}} = \$0.238/\text{L}$ $\qquad$ $80.0 \text{ L} \times \dfrac{\$0.238}{\text{L}} = \underline{\$19.04}$

$80.0 \text{ L} \times \dfrac{1 \text{ gal}}{3.785 \text{ L}} \times \dfrac{\$2.759}{\text{gal}} = \underline{\$58.31}$

1-83 $551 \text{ mi} \times \dfrac{1 \text{ gal}}{21.0 \text{ mi}} \times \dfrac{\$2.759}{\text{gal}} = \underline{\$72.39}$

$482 \text{ km} \times \dfrac{1 \text{ mi}}{1.609 \text{ km}} \times \dfrac{1 \text{ gal}}{21.0 \text{ mi}} \times \dfrac{\$2.759}{\text{gal}} = \underline{\$39.36}$

1-84 $\$75.00 \times \dfrac{1 \text{ gal}}{\$2.759} \times \dfrac{21.0 \text{ mi}}{\text{gal}} \times \dfrac{1.609 \text{ km}}{\text{mi}} = \underline{919 \text{ km}}$

1-87 $\$2.50 \times \dfrac{1 \text{ lb}}{\$0.95} \times \dfrac{145 \text{ nails}}{\text{lb}} = \underline{382 \text{ nails}}$

1-88 $5670 \text{ nails} \times \dfrac{1 \text{ lb}}{185 \text{ nails}} \times \dfrac{\$0.92}{\text{lb}} = \underline{\$28.20}$

1-89 $858 \text{ km} \times \dfrac{1 \text{ mi}}{1.609 \text{ km}} \times \dfrac{1 \text{ gal}}{38.5 \text{ mi}} \times \dfrac{\$2.759}{\text{gal}} = \underline{\$38.21}$

$858 \text{ km} \times \dfrac{1 \text{ mi}}{1.609 \text{ km}} \times \dfrac{1 \text{ gal}}{18.5 \text{ mi}} \times \dfrac{\$2.759}{\text{gal}} = \underline{\$79.53}$

1-92 $442 \text{ mi} \times \dfrac{1.609 \text{ km}}{\text{mi}} \times \dfrac{1 \text{ hr}}{215 \text{ km}} = \underline{3.31 \text{ hr}}$

1-93 (a) $\$6.50 \times \dfrac{1.12\text{ euro}}{\$} = \underline{7.28\text{ euro}}$ (b) $12.65\text{ euro} \times \dfrac{\$1.12}{\text{euro}} = \underline{\$11.29}$

(c) $\$15.58\text{ euro} \times \dfrac{\$1.27}{\text{euro}} = \underline{\$19.79}$

1-94 $\dfrac{0.649\text{ pd}}{\$} \times \dfrac{\$1.27}{\text{euro}} = \dfrac{0.824\text{ pd}}{\text{euro}}$ $25{,}600\text{ euro} \times \dfrac{0.824\text{ pd}}{\text{euro}} = \underline{21{,}100\text{ pd}}$

1-97 $4.0 \times 10^8\text{ mi} \times \dfrac{5280\text{ ft}}{\text{mi}} \times \dfrac{12\text{ in.}}{\text{ft}} \times \dfrac{2.54\text{ cm}}{\text{in.}} \times \dfrac{1\text{ s}}{3.0 \times 10^{10}\text{ cm}} = \underline{2100\text{ s}}$

$2100\text{ s} \times \dfrac{1\text{ min}}{60\text{ s}} \times \dfrac{1\text{ hr}}{60\text{ min}} = \underline{0.58\text{ hr}}$

1-98 $T(^\circ F) = 1.8(300^\circ C) + 32 = 520 + 32 = 572^\circ F$

1-99 $T(^\circ C) = \dfrac{76 - 32}{1.8} = \dfrac{44}{1.8} = 24^\circ C$

1-101 $T(^\circ F) = [(-39) \times 1.8] + 32 = -38^\circ F$

1-103 $T(^\circ F) = [(35.0 \times 1.8) + 32.0] = 95.0^\circ F$

1-104 (a) $-98^\circ C$ (b) $22^\circ C$ (c) $27^\circ C$ (d) $-48^\circ C$ (e) $600^\circ C$

1-105 (a) 320 K (b) 296 K (c) 200 K (d) 261 K (e) 291 K (f) 244 K

1-107 Since $T(^\circ C) = T(^\circ F)$, substitute $T(^\circ C)$ for $T(^\circ F)$ and set the two equations equal.

$T(^\circ C) \times 1.8] + 32 = \dfrac{T(^\circ C) - 32}{1.8}$ $(1.8)^2 T(^\circ C) - T(^\circ C) = -32 - 32(1.8)$

$T(^\circ C) = -40^\circ$

1-108 (a) 3×10^2 (b) $8.26\text{ g} \cdot \text{cm}$ (c) 5.24 g/mL (d) 19.1

1-109 (a) $\dfrac{1\text{ g}}{10^3\text{ mg}}$, $\dfrac{1\text{ lb}}{453.6\text{ g}}$ (b) $\dfrac{1.057\text{ qt}}{L}$, $\dfrac{2\text{ pt}}{\text{qt}}$

(c) $\dfrac{1\text{ km}}{10\text{ hm}}$, $\dfrac{1\text{ mi}}{1.609\text{ km}}$ (d) $\dfrac{1\text{ in.}}{2.54\text{ cm}}$, $\dfrac{1\text{ ft}}{12\text{ in.}}$

1-110 $5.34 \times 10^{10}\text{ ng} \times \dfrac{10^{-9}\text{ g}}{\text{ng}} \times \dfrac{1\text{ lb}}{453.6\text{ g}} = \underline{0.118\text{ lb}}$

1-112 $1.00\text{ kg} \times \dfrac{10^3\text{ g}}{\text{kg}} \times \dfrac{1\text{ tr lb}}{373\text{ g}} \times \dfrac{12\text{ oz}}{\text{tr lb}} \times \dfrac{\$1249}{\text{oz}} = \underline{\$40{,}182}$

1-113 $\dfrac{247\ \text{lb}}{82.3\ \text{doz}}$ = $\underline{3.00\ \text{lb/doz}}$ $\qquad$ $\dfrac{82.3\ \text{doz}}{247\ \text{lb}}$ = $\underline{0.333\ \text{doz/lb}}$

1-114 $12.0\ \text{fur}$ x $\dfrac{1\ \text{mi}}{8\ \text{fur}}$ x $\dfrac{5280\ \text{ft}}{\text{mi}}$ x $\dfrac{12\ \text{in.}}{\text{ft}}$ x $\dfrac{1\ \text{hand}}{4\ \text{in.}}$ = $\underline{2.38 \times 10^4\ \text{hands}}$

1-116 $0.500\ \text{lb}$ x $\dfrac{453.6\ \text{g}}{\text{lb}}$ x $\dfrac{10^3\ \text{mg}}{\text{g}}$ x $\dfrac{1\ \text{cig.}}{11.0\ \text{mg}}$ x $\dfrac{1\ \text{pkg}}{20\ \text{cig.}}$ = $\underline{1030\ \text{pkgs}}$

$\qquad$ $1030\ \text{pkgs}$ x $\dfrac{1\ \text{day}}{2\ \text{pkg}}$ x $\dfrac{1\ \text{yr}}{365\ \text{day}}$ = $\underline{1.41\ \text{years}}$

1-117 $1\ \text{in.} = 2.54\ \text{cm}$ $\qquad$ $1\ \text{in.}^3 = 16.4\ \text{cm}^3 = 16.4\ \text{mL}$

$\qquad$ $306\ \text{in.}^3$ x $\dfrac{16.4\ \text{mL}}{\text{in.}^3}$ x $\dfrac{1\ \text{L}}{10^3\ \text{mL}}$ = $\underline{5.02\ \text{L}}$

1-119 $5.4 \times 10^7\ ^\circ\text{F}$ $\qquad$ $3.0 \times 10^7\ ^\circ\text{C} + 273 = 3.0 \times 10^7\ \text{K}$

1-121 $1\ \text{ft}$ x $30.0\ \text{ft}$ x $50.0\ \text{ft} = 1500\ \text{ft}^3\ \text{snow}$ x $\dfrac{0.100\ \text{ft}^3\ \text{water}}{\text{ft}^3\ \text{snow}}$ = $150\ \text{ft}^3\ \text{water}$

$\qquad$ $150\ \text{ft}^3$ x $\dfrac{62.0\ \text{lb}}{\text{ft}^3}$ = $\underline{9300\ \text{lb}}$ $\qquad$ $9300\ \text{lb}$ x $\dfrac{1\ \text{ton}}{2000\ \text{lb}}$ = $\underline{4.65\ \text{ton}}$

2

Elements and Compounds

Review of Part A *The Elements and Their Composition*

OBJECTIVES AND DETAILED TABLE OF CONTENTS

2-1   The Elements

OBJECTIVE *List the names and symbols of common elements.*

2-1.1 Free Elements in Nature
2-1.2 The Names of Elements
2-1.3 The Distribution of the Elements
2-1.4 The Symbols of the Elements

2-2   The Composition of Elements: Atomic Theory

OBJECTIVE   *List the postulates of the Atomic Theory.*

   2-2.1 The Atomic Theory
   2-2.2 The Size of an Atom

2-3   Composition of the Atom

OBJECTIVE   *List the components of an atom and their relative masses, charges, and location in the atom.*

   2-3.1 The Electron and Electrostatic Forces
   2-3.2 The Nuclear Model of the Atom
   2-3.3 The Particles in the Nucleus

2-4   Atomic Number, Mass Number, and Atomic Mass

OBJECTIVE   *Using the table of elements, determine the number of protons, neutrons, electrons, or atomic mass of any isotope of an element.*

   2-4.1 Atomic Number, Mass Number, and Isotopes
   2-4.2 Isotopic Mass and Atomic Mass

SUMMARY OF PART A

All **matter** in the universe is made up of less than 90 **elements**. **Compounds** are combinations of the various elements. Over 99% of the crust of the earth is composed of only ten elements. The names of the elements have a variety of sources, but all are designated by a **symbol**. Symbols may have one or two letters. Elements that have not yet been given a formal name have three letters.

   Elements are composed of basic particles called atoms. An **atom** is the smallest particle of an element that has the characteristics of that element and which can enter into chemical reactions. Even before powerful microscopes could create images of atoms, scientists had presumed that they existed. In fact, since the early 1800s, the **atomic theory** of John Dalton has been the accepted way to explain the nature of matter.

   At first it was assumed that atoms were hard spheres, but in the late 1800s, experiments proved that the atom was considerably more complex. The experiments of Thomson and Rutherford, among others, led to the **nuclear model** of the atom. In this model, the atom is composed of three particles: **electrons**, **protons**, and **neutrons**. The neutrons and protons (called **nucleons**) exist in a small, dense core of the atom called the **nucleus**. Protons carry a positive charge and the nucleus contains nearly all of the mass of the atom. The comparatively small, negatively charged electrons exist in the vast, mostly empty space of the atom outside of the nucleus. The electrons in the atom are attracted to the positively charged nucleus by **electrostatic forces**.

   The atoms of a particular element are characterized by the **atomic number**, which is the number of protons (or the total positive charge) in the nucleus. The atoms of the same element may have different **mass numbers**, however; the mass number is the total number of protons and neutrons. Atoms of the same element with different mass numbers are known as **isotopes** of that element. Neutral atoms have the same number of electrons as protons. For example, consider the isotope $^{40}_{18}\text{Ar}$:

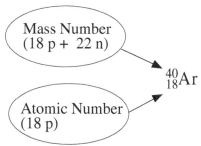

The **isotopic mass** of an isotope is obtained by comparing the mass of the isotope to the mass of $^{12}_{6}C$ the standard, which is defined as having a mass of exactly 12 **atomic mass units** (**amu**). Elements generally occur in nature as mixtures of isotopes. The **atomic mass** of an element is the weighted average of the isotopic masses of the naturally occurring isotopes.

An electrical charge on an atom results when there are more or less electrons present than protons. For example, an ion of sulfur exists that has 18 electrons and 16 protons and thus has a charge of -2.

ASSESSMENT OF OBJECTIVES

A-1 Fill in the Blanks

Write the chemical symbols for:

_____ nitrogen   _____ magnesium   _____ lead

_____ aluminum   _____ sodium   _____ chlorine

Write the name of the element for:

_____ P   _____ Fe   _____ S

_____ Ca   _____ K   _____ B

A-2 Multiple Choice

____ 1. Which of the following describes an electron?

(a) mass = 1 amu, charge = -1   (b) mass = 0 amu, charge = 0
(c) mass = 0 amu, charge = -1   (d) mass = 1 amu, charge = +1

___ 2. The nucleus of an atom contains:

(a) protons and electrons
(b) most of the mass and all of the positive charge
(c) neutrons and electrons
(d) most of the positive charge and all of the mass

___ 3. How many neutrons are in $^{239}_{94}$Pu?

(a) 94   (b) 239   (c) 333   (d) 145

___ 4. One isotope of sulfur is $^{32}_{16}$S. A second isotope of sulfur contains atoms with

(a) 16 protons and 17 neutrons   (b) 32 nucleons with 17 protons
(c) 16 neutrons with 17 protons   (d) 32 protons and 16 neutrons

___ 5. An element has an atomic mass of about one-third of the standard. The element is

(a) Cl   (b) O   (c) He   (d) Li

___ 6. How many electrons are in $^{128}_{52}$Te^{2-}?

(a) 52   (b) 54   (c) 50   (d) 76   (e) 78

___ 7. A certain ion has a +2 charge and contains 23 electrons. This is an ion of what element?

(a) Mn   (b) V   (c) Co   (d) Sc

A-3 Matching

___ Proton (a) The number of neutrons in an atom

___ Neutron (b) Has a -1 charge and a mass of about 1 amu

___ Nucleon (c) The number of nucleons in an atom

___ Atomic number (d) The mass of an atom compared to ^{12}C

___ Mass number (e) Has a +1 charge and a mass of about 1 amu

___ Neutral atom (f) An atom with equal numbers of electrons and protons

___ Atomic mass (g) Has a 0 charge and a mass of about 1 amu

(h) The total number of neutrons and electrons in an atom

(i) Has either a +1 or 0 charge and a mass of about 1 amu

(j) The number of protons in an atom

(k) An atom with equal numbers of protons and neutrons

(l) The mass of a naturally occurring element compared to ^{12}C

A-4 Problems

1. $^{89}_{39}Y$ has _____ protons, _____ neutrons, and _____ electrons.

   The mass number is _____. The atomic number is _____.

2. ^{91}Zr has _____ protons, _____ neutrons, and _____ electrons.

   The mass number is _____. The atomic number is _____.

3. $^{197}_{78}?$ has _____ protons, _____ neutrons, and _____ electrons.

   The symbol of the element is _____.

4. The naturally-occurring isotopic mixture of an element has a mass 2.58 times that of the standard, ^{12}C. What is the atomic mass of this element and what is the element?

5. If the isotopic mass of ^{12}C were defined as exactly five, what are the approximate isotopic masses of the following?

(a) ^{14}N (b) ^{238}U

6. Magnesium occurs in nature as a mixture of three isotopes: ^{24}Mg isotopic mass = 23.99 amu, ^{25}Mg isotopic mass = 24.99, and ^{26}Mg isotopic mass = 25.98 amu. If magnesium is 79.0% ^{24}Mg, 10.0% ^{25}Mg, and 11.0% ^{26}Mg, what is the atomic mass of magnesium?

7. Iridium (Ir) occurs in nature as a mixture of two isotopes: ^{191}Ir (isotopic mass = 191.04 amu) and ^{193}Ir (isotopic mass = 193.04 amu). Using the atomic mass of iridium from the periodic table, determine the percent of each isotope present.

Review of Part B *Compounds and Their Composition*

OBJECTIVES AND DETAILED TABLE OF CONTENTS

2-5   Molecular Compounds

OBJECTIVE *List the characteristics of compounds composed of molecules.*

2-6   Ionic Compounds

OBJECTIVE *Write the formulas of simple ionic compounds given the charges on the ions.*

SUMMARY OF PART B

We now look at compounds, the other form of **pure substances**. **Molecular compounds** are composed of **molecules** that contain atoms of different elements. Molecules are composed of two or more atoms that are bound to each other by **covalent bonds**. The **formula** of a compound expresses the number of atoms of each element comprising the particular molecule. Some elements, such as chlorine and oxygen, are also composed of molecules in their normal state (i.e., Cl_2 and O_2). **Structural formulas** show more detail, such as the sequence of the bonded atoms. The formula of ethyl alcohol is as follows:

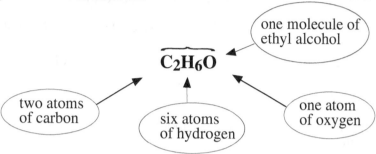

Another important category of compounds exists, known as **ionic compounds**. They are composed of **ions** rather than discrete molecules. Ions are atoms or groups of atoms that have a net electrical charge. The charge on an atom (or group of atoms forming a **polyatomic ion**), is represented in the upper right-hand corner of the symbol or symbols (e.g., S^{2-}, Ba^{2+}, SO_3^{2-}).

The ions in ionic compounds are held together by attractive electrostatic forces. The formula of an ionic compound represents one **formula unit**, which expresses the ratio of **cations** and **anions** present in the compound. For example, the formula of potash indicates the following:

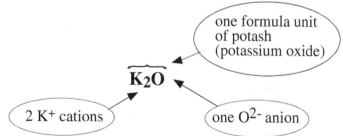

Eventually, we will discuss why ionic compounds form in some cases and why molecular compounds form in others.

ASSESSMENT OF OBJECTIVES

B-1 Multiple Choice

_____ 1. Which of the following formulas represents molecules of an element?

    (a) Ni   (b) N_2 (c) H_2O (d) Ne   (e) NaCl

___ 2. Which of the following formulas represents molecules of a compound?

(a) Mn   (b) P_4 (c) PF_3 (d) Br_2 (e) K

___ 3. Which of the following is a polyatomic anion?

(a) NH_4^+ (b) SO_3 (c) F^- (d) SO_3^{2-}

___ 4. An ionic compound contains one Ni^{2+} ion and two ClO_3^- ions. Which of the following is the proper representation of the formula?

(a) $NiCl_2O_6$ (b) $Ni_2Cl_2O_6$ (c) $Ni(ClO_3)_2$ (d) $(Ni)(ClO_3)_2$

___ 5. How many atoms of carbon are in one formula unit of $Fe_2(C_2O_4)_3$?

(a) 6   (b) 2   (c) 3   (d) 4   (e) 12

___ 6. A certain formula of an ionic compound contains two K^+ cations and one anion. The anion could be:

(a) Br^- (b) S^{2-} (c) N^{3-} (d) Ca^{2+}

___ 7. A certain formula of an ionic compound contains three CrO_4^{2-} ions and two cations. The cation could be:

(a) Na^+ (b) Mg^{2+} (c) Al^{3+} (d) Mn^{4+}

B-2 Matching

___ A monatomic anion

___ A monatomic cation

___ A molecule

___ A polyatomic ion

(a) A sulfur atom with chemical bonds to two oxygen atoms containing a total of 32 electrons

(b) A strontium atom with 36 electrons

(c) An oxygen atom with 8 electrons

(d) An iodine atom with 54 electrons

(e) A chlorine atom bound to an oxygen atom containing a total of 26 electrons

Chapter Summary Assessment

S-1 Matching

____ A proton

____ Molecule of an element

____ An isotope with 13 neutrons

____ An isotope of sodium

____ Formula of a molecular compound

____ An anion

____ An electron

____ Formula of an ionic compound

(a) NH_4^+

(b) F_2

(c) Na^+

(d) $Ca(ClO)_2$

(e) $HClO_4$

(f) $_1^1H^+$

(g) $_{11}^{24}X$

(h) A neutral particle with a mass of about 1 amu

(i) A negatively charged particle in an atom

(j) I^-

(k) $_{12}^{23}X$

(l) Xe

Answers to Assessments of Objectives

A-1 Fill in the Blanks

N - nitrogen, Mg - magnesium, Pb - lead, Al - aluminum, Na - sodium, Cl – chlorine, phosphorus - P, iron - Fe, sulfur - S, calcium - Ca, potassium - K, boron - B

A-2 Multiple Choice

1. **c** mass = 0 amu, charge = -1

2. **b** most of the mass and all of the positive charge

3. **d** 145

5. **c** He

6. **b** 54

7. **a** Mn

4. **a** 16 protons and 17 neutrons

A-3 Matching

e A **proton** has a +1 charge and a mass of about 1 amu.

g A **neutron** has 0 charge and a mass of about 1 amu.

i A **nucleon** is either a proton or a neutron.

j The number of protons in an atom is the **atomic number**.

c The number of nucleons in an atom is the **mass number**.

f A **neutral atom** contains equal numbers of protons and electrons.

l **Atomic mass** is the mass of a naturally occurring element compared to ^{12}C.

A-4 Problems

1. $^{89}_{39}Y$ has 39 protons, 50 neutrons, and 39 electrons. The mass number is 89. The atomic number is 39.

2. ^{91}Zr has 40 protons, 51 neutrons, and 40 electrons. The mass number is 91. The atomic number is 40.

3. $^{197}_{78}?$ has 78 protons, 119 neutrons, and 78 electrons. The symbol of the element is Pt.

4. $^{12}C = 12.0000$ amu
   2.58×12.000 amu = 30.96 amu. The element must be phosphorus (P).

5. (a) ^{14}N has a mass $\frac{14}{12}$ that of ^{12}C. If $^{12}C = 5.00$,

   $$^{14}N = \frac{14}{12} \times 5.00 = 5.83$$

   (b) ^{238}U has a mass $\frac{238}{12}$ that of ^{12}C. If $^{12}C = 5.00$,

   $$^{238}U = \frac{238}{12} \times 5.00 = 99.2$$

6. Calculate the weighted average of all isotopes.

   ^{24}Mg: $0.790 \times 23.99 =$ 18.95 amu
   ^{25}Mg: $0.100 \times 24.99 =$ 2.50 amu
   ^{26}Mg: $0.110 \times 25.98 =$ 2.85 amu
                                   24.30 amu = 24.3 amu

7. At. wt. of Ir = 192.22 amu

Let X = decimal fraction of ^{191}Ir

Then (1 - X) = decimal fraction of ^{193}Ir

191.04 X = mass due to ^{191}Ir, 193.04 (1 - X) = mass due to ^{193}Ir

191.04 X + 193.04 (1 - X) = 192.22

$\qquad$ -2.00 X = -0.82

X = 0.41 $\quad$ 0.41 x 100% = 41% $\quad$ ^{191}Ir 100% - 41% = 59% ^{193}Ir

B-1 Multiple Choice

1. **b** N_2

2. **c** PF_3

3. **d** $SO_3{}^{2-}$

4. **c** $Ni(ClO_3)_2$

5. **a** 6

6. **b** S^{2-} $(2K^+)(S^{2-})$ = K_2S

7. **c** Al^{3+} $(2Al^{3+})(3CrO_4{}^{2-})$ = $Al_2(CrO_4)_3$

B-2 Matching

d An iodine atom with 54 electrons would be a **monatomic anion** with a -1 charge.

b A strontium atom with 36 electrons would be a **monatomic cation** with a +2 charge.

a SO_2 would be a neutral **molecule** with 32 electrons. (Two oxygens and one sulfur would have a total positive charge of +32 in their nuclei.)

e A chlorine atom bound to an oxygen atom with a total of 26 electrons is a **polyatomic ion**. (The positive charge in a neutral oxygen and chlorine is +25.)

S-1 Matching

f **A proton** is the same as $^{1}_{1}H$ $^+$.

b F_2 represents a **molecule of an element**.

g $^{24}_{11}X$ has 24 - 11 = **13 neutrons**.

k $^{23}_{12}X$ is an **isotope of sodium**.

e $HClO_4$ represents the **formula of a molecular compound**.

j I^- is **an anion**.

i An **electron** is a negatively charged particle in an atom.

d $Ca(ClO)_2$ is the **formula of an ionic compound**.

Answers and Solutions to Green Text Problems

2-2   cadmium - Cd, calcium - Ca, californium - Cf, carbon - C, cerium - Ce, cesium - Cs, chlorine - Cl, chromium - Cr, cobalt - Co, copper - Cu, curium - Cm

2-5   (a) barium - Ba   (b) neon - Ne   (c) cesium - Cs   (d) platinum - Pt   (e) manganese - Mn

         (f) tungsten - W

2-7   (a) B - boron   (b) Bi - bismuth   (c) Ge - germanium   (d) U - uranium   (e) Co - cobalt
         (f) Hg - mercury   (g) Be - beryllium   (h) As - arsenic

2-8   (b) and (e)

2-9   (c) zero charge, 1 amu

2-12   (a) 21 p, 21 e, 24 n   (b) 90 p, 90 e, 142 n   (c) 87 p, 87 e, 136 n   (d) 38 p, 38 e, 52 n

2-14

Isotope name	Isotope symbol	At. no.	Mass no.	p	n	e
(a) silver-108	$^{108}_{47}\text{Ag}$	47	108	47	61	47
(b) silicon-28	$^{28}_{14}\text{Si}$	14	28	14	14	14
(c) potassium-39	$^{39}_{19}\text{K}$	19	39	19	20	19
(d) cerium-140	$^{140}_{58}\text{Ce}$	58	140	58	82	58
(e) iron-56	$^{56}_{26}\text{Fe}$	26	56	26	30	26
(f) tin-110	$^{110}_{50}\text{Sn}$	50	110	50	60	50
(g) iodine-118	$^{118}_{53}\text{I}$	53	118	53	65	53
(h) mercury-196	$^{196}_{80}\text{Hg}$	80	196	80	116	80

2-16 ^{59}Co

2-18   (a) The identity of a specific element is determined by its atomic number. An element is a basic form of matter and its atomic number relates to the number of protons in the nuclei of its atoms.

         (b) Both relate to the particles in the nucleus of the atoms of an element. The atomic mass is the mass of an average atom, since elements are usually composed of more than one isotope. The atomic number is the number of protons.

(c) Both relate to the number of particles in a nucleus. The mass number relates to the total number of protons and neutrons in an isotope of an element, while the atomic number is the number of protons in the atoms of a specific element.

(d) All isotopes of the same element have the same atomic number.

(e) Different isotopes of a specific element have the same number of protons but different numbers of neutrons or mass numbers.

2-19 (a) Re: at. no. 75, at. mass 186.2
(b) Co: at. no. 27, at. mass 58.9332
(c) Br: at. no. 35, at. mass 79.904
(d) Si: at. no. 14, at. mass 28.086

2-20 copper (Cu)

2-22 O - at. no. 8, mass no 16   Si - at. no. 14, mass no. 28
N - at. no. 7, mass no. 14   Ca - at. no. 20, mass no. 40

2-24 $5.81 \times 12.00 = 69.7$ amu   The element is Ga.

2-26 $0.505 \times 78.92 = 39.85$
$0.495 \times 80.92 = \underline{40.06}$
$79.91 = \underline{79.9\ amu}$

2-27 ^{28}Si $0.9221 \times 27.98 = 25.80$
^{29}Si $0.0470 \times 28.98 = 1.362$
^{30}Si $0.0309 \times 29.97 = \underline{0.926}$
$\phantom{^{30}Si\ 0.0309 \times 29.97 = }28.088\ = \underline{28.09\ amu}$

2-29 Let X = decimal fraction of ^{35}Cl and Y = decimal fraction of ^{37}Cl. Since there are two isotopes present, X + Y = 1, Y = 1 - X.
$(X \times 35) + (Y \times 37) = 35.5$
$(X \times 35) + [(1 - X) \times 37] = 35.5$
$X = 0.75\ \underline{(75\%\ ^{35}Cl)}$ $Y = 0.25\ \underline{(25\%\ ^{37}Cl)}$

2-31 (a) They are both basic units of matter. Most elements are composed of individual atoms and many compounds are composed of individual molecules. Molecules are composed of atoms chemically bonded together.

(b) A compound is a pure form of matter. It is composed of individual units called molecules.

(c) They are both pure forms of matter. Compounds, however, are composed of two or more
elements chemically combined.

(d) Most elements are composed of individual atoms. Some elements, however, are composed of molecules which, in most cases, contain two atoms.

2-32 (b) Br_2 (b) S_8 (f) P_4

2-33 P - phosphorus   O - oxygen   Br - bromine
F - fluorine   S - sulfur   Mg - magnesium

2-34 Hf is the symbol of the element hafnium. HF is the formula of a compound composed of one atom of hydrogen and one atom of fluorine.

2-36 (b) CO, diatomic compound (e) N_2, diatomic element

2-39 (a) six carbons, four hydrogens, and two chlorines
(b) two carbons, six hydrogens, and one oxygen
(c) one copper, one sulfur, 18 hydrogens, and 13 oxygens
(d) nine carbons, eight hydrogens, four oxygens
(e) two aluminums, three sulfurs, and 12 oxygens
(f) two nitrogens, eight hydrogens, one carbon and three oxygens

2-40 (a) 12 (b) 9 (c) 33 (d) 21 (e) 17 (f) 14

2-41 (a) 8 (b) 7 (c) 4 (d) 3

2-42 (a) SO_2 (b) CO_2 (c) H_2SO_4 (d) C_2H_2

2-44 (a) They are both basic forms of matter. Atoms are neutral but ions are atoms that have acquired an electrical charge. Positive and negative ions always are found together.

(b) They are both basic forms of matter containing more than one atom. Molecules are neutral, but polyatomic ions have acquired an electrical charge.

(c) Both have an electrical charge. Cations have a positive charge and anions have a negative charge.

(d) Both are classified as compounds which are composed of the atoms of two or more elements. The basic unit of a molecular compound is a neutral molecule, but the basic units of ionic compounds are cations and anions.

(e) Both are the basic units of compounds. Molecules are the basic entities of molecular compounds and an ionic formula unit represents the smallest whole number of cations and anions representing a net charge of zero.

2-46 (a) $Ca(ClO_4)_2$ (b) $(NH_4)_3PO_4$ (c) $FeSO_4$

2-47 (a) one calcium, two chlorines, and eight oxygens
(b) three nitrogens, 12 hydrogens, one phosphorus, and four oxygens
(c) one iron, one sulfur, and four oxygens

2-50 (c) S^{2-}

2-52 (d) Li^+

2-54 FeS, Li_2SO_3

2-56 SO_3 represents the formula of a compound. It could be a gas. SO_3^{2-} is an anion and does not exist independently. It is part of an ionic compound with the other part being a cation.

2-57 (a) K^+ 19 p, 18 e
(b) Br^-: 35 p, 36 e
(c) S^{2-}: 16 p, 18e
(d) NO_2^-: 7 + 16 = 23 p, 24 e
(e) Al^{3+}: 13 p, 10 e
(f) NH_4^+: 7 + 4 = 11 p, 10 e

2-59 (a) Ca^{2+} (b) Te^{2-} (c) PO_3^{3-} (d) NO_2^+

2-61 This is the Br^- ion. It is part of an ionic compound.

2-64 (a) $^{90}_{38}Sr^{2+}$ (b) $^{52}_{24}Cr^{3+}$ (c) $^{79}_{34}Se^{2-}$ (d) $^{14}_{7}N^{3-}$ (e) $^{139}_{57}La^{3+}$

2-66 (a) Na - 11 protons, 12 neutrons, and 11 electrons; Na^+ - 10 electrons
(b) Ca - 20 protons, 20 neutrons, and 20 electrons; Ca^{2+} - 18 electrons
(c) F - 9 protons, 10 neutrons, and 9 electrons; F^- - 10 electrons
(d) SS- 21 protons, 24 neutrons, and 21 electrons; Sc^{3+} - 18 electrons

2-68 Let x = mass no. of I and y = mass no. of Tl. Then (1) x + y = 340 or x = 340 - y
Also, (2) $x = \frac{2}{3}y - 10$ Substituting for x from (1) and solving for y,
y = 210 amu (Tl)   and x = 340 - 210 = 130 amu (I)

2-70 121.8 (Sb) Sb^{3+} has 51 - 3 = 48 electrons
^{121}Sb has 121 - 51 = 70 neutrons ^{123}Sb has 123 - 51 = 72 neutrons
^{121}Sb =57.9% due to neutrons; ^{123}Sb 58.5% due to neutrons

2-71 0.602 x 196 amu = 118 neutrons   (196 - 118= 78 protons) Element is platinum.
[platinum (Pt)] 78 - 2 = 76 electrons for Pt^{2+}

2-73 Mass of other atom = 16 (oxygen) NO^+ = (7 + 8) = 15 - 1 = 14 electrons

2-75 (a) H: $\frac{1.008}{12.00}$ x 8.000 = 0.672   (b) N: $\frac{14.01}{12.00}$ x 8.000 = 9.34
(c) Na: $\frac{22.99}{12.00}$ x 8.000 = 15.3   (d) Ca: $\frac{40.08}{12.00}$ x 8.000 = 26.7

2-76 $\frac{43.3}{10.0}$ = 4.33 times as heavy as ^{12}C

4.33 x 12.0 amu = 52.0 amu The element is Cr.

3

The Properties of Matter and Energy

Review of Part A   *The Properties of Matter*

OBJECTIVES AND DETAILED TABLE OF CONTENTS

3-1 The Physical and Chemical Properties of Matter

OBJECTIVE *List several properties of matter and distinguish them as physical or chemical.*

3-2   Density: A Physical Property

OBJECTIVE *Perform calculations involving the density of liquids and solids.*

   3-2.1 Density as a Physical Property
   3-2.2 Density as a Conversion Factor
   3-2.3 Specific Gravity

3-3   The Properties of Mixtures

OBJECTIVE *Perform calculations involving percent of a pure substance in a mixture.*

   3-3.1 Heterogeneous Mixtures
   3-3.2 Homogeneous Mixtures and Solutions
   3-3.3 Percent Composition of Solutions

SUMMARY OF PART A

Elements and compounds can be distinguished by their properties. For example, all matter exists in one of three physical states: **solid**, **liquid**, or **gas**. The physical state is an example of a **physical property**, and the change of a substance from one physical state (**phase**) to another is an example of a **physical change**. This involves **melting**, **freezing**, **boiling**, and **condensation**. The **melting point** of a solid (the **freezing point** of a liquid) and the **boiling point** of a liquid are all distinct physical properties. All pure elements and compounds have distinguishing physical properties.

Substances have **chemical properties** as well as physical properties. Chemical properties relate to the profound changes (called **chemical changes**) that the substance undergoes. Pure substances also have distinguishing chemical properties. In any chemical process the **law of conservation of mass** can be demonstrated.

All properties are either **intensive** or **extensive**. An intensive property is **density** and is an important physical property of homogeneous matter that relates mass to a specified volume. For solids and liquids, density is usually expressed in grams per milliliter (g/mL). For gases, density is expressed in grams per liter (g/L). Density can be used as a conversion factor, which converts mass to an equivalent volume or vice versa. An example follows:

Example A-1 Density as a Conversion Factor

What is the volume in mL of 0.23 lb of table salt? (density = 2.16 g/mL)

PROCEDURE

A two-step conversion is necessary as shown by the following unit map:

$$\text{lb} \longrightarrow \text{g} \longrightarrow \text{mL}$$

SOLUTION

$$0.23\ \cancel{\text{lb}} \quad \times \quad \frac{453.6\ \text{g}}{\cancel{\text{lb}}} \quad \times \quad \frac{1\ \text{mL}}{2.16\ \cancel{\text{g}}} \quad = \quad \underline{48\ \text{mL}}$$

converts
lb to g

density factor
converts g to mL

The density of a liquid can also be expressed as a **specific gravity**. Specific gravity is the density expressed without units. It is used more in the medical field.

Much of what we see around us in nature is a mixture of pure substances, at least to some extent. Intimate mixtures in which the pure substances are mixed at the basic particle level and exist in one **phase** are known as **homogeneous mixtures** or **solutions**. Pure substances themselves are examples of homogeneous matter (when they occur in one phase). **Heterogeneous matter**, on the other hand, exists in two or more phases with identifiable boundaries between phases. The classification of matter from heterogeneous mixtures down to the most basic substances (the elements) can be illustrated as follows:

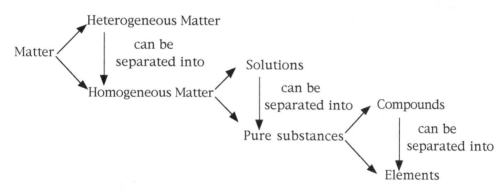

In the laboratory, heterogeneous mixtures of a solid in a liquid can be separated by *filtration*. Solutions may be separated into components by *distillation*. It is not always possible to distinguish a mixture, especially a solution, from a pure substance by casual observation. Often, one must closely examine properties such as melting point and boiling point to tell the difference. Pure substances have distinct and unchanging properties. A mixture, on the other hand, melts and boils at different temperatures than any of its pure components would individually.

Solutions are generally composed of a substance dissolved in a liquid, which is usually water. The water is considered the **solvent** and the substance dissolved is the **solute**. The amount of solute present in a solution is conveniently expressed as **percent by mass**.

ASSESSMENT OF OBJECTIVES

A-1 Multiple Choice

_____ 1. Which of the following is an example of an element?

(a) water
(b) carbon
(c) concrete
(d) carbon dioxide

41

___ 2. Which of the following distinguishes an element from a compound?

      (a) An element can be broken down into compounds.
      (b) A compound has variable properties.
      (c) An element cannot be broken down by chemical means.
      (d) An element has definite properties.

___ 3. Which of the following describes the liquid state?

      (a) It fills the entire container.
      (b) It has definite dimensions.
      (c) It fills the lower part of the container.
      (d) It has a definite shape and volume.

___ 4. Which of the following is not a physical property of sodium?

      (a) It is a solid.
      (b) It is soft.
      (c) It is shiny.
      (d) It forms a compound in the presence of chlorine.
      (e) It melts at 98°C.

___ 5. Which of the following is not a chemical property of oxygen?

      (a) It is a colorless gas.
      (b) It supports combustion (burning).
      (c) It reacts with almost all other elements.
      (d) It is used in animal metabolism.
      (e) It reacts with iron to form rust.

___ 6. Which of the following is an example of a physical change?

      (a) burning
      (b) decaying
      (c) boiling
      (d) tarnishing
      (e) decomposing

___ 7. Which of the following is a unit of density?

      (a) g/m^2 (b) lb/ft (c) qt/lb (d) lb/gal (e) m/L

___ 8. Solid X floats in liquid A but sinks in liquid B. The order of densities of the three substances is therefore:

      (a) A > B > X   (b) A > X > B   (c) B > X > A

      (d) X > B > A   (e) X > A > B

___ 9. A 55-g quantity of a substance has a volume of 10.0 mL. Its specific gravity is:

(a) 5.5 g/mL   (b) 0.55 g/Ml   (c) 0.18

(d) 1.8   (e) 5.5

A-2 Matching

_____ Density	(a) Cannot be decomposed into simpler substances
_____ Solid	(b) Same as freezing point
_____ Compound	(c) Has a definite volume but not a definite shape
_____ Melting point	(d) Observation does not require a change
_____ Pure substance	(e) Temperature at which condensation occurs
_____ Physical property	(f) The ratio of the volume to the mass
	(g) Has definite dimensions
	(h) Made up of two or more elements
	(i) Has variable properties
	(j) Requires the observation of a change
	(k) The ratio of the mass to the volume
	(l) Element or compound

A-3 Problems

1. The volume of a 65.5-g sample of a pure substance is 70.8 mL. What is its density?

2. A sample of metal has rectangular sides which measure 20.0 cm, 15.0 cm, and 35.0 cm. The sample has a mass of 84.6 kg. What is its density?

3. If the density of a substance is 2.70 g/mL, what is the mass of 1.12 L of the substance?

4. If the density of a substance is 0.954 g/mL, what is the volume of a 1.00-kg sample?

A-4 Multiple Choice

____ 1. Which of the following is an example of a homogeneous mixture?

(a) vinegar   (b) oxygen gas
(c) table salt   (d) oil and vinegar

____ 2. Which of the following is an example of heterogeneous matter?

(a) air   (b) aluminum
(c) foamy shaving cream   (d) sugar

____ 3. Which of the following is NOT true about a solution?

(a) It is homogeneous.
(b) It has definite properties.
(c) It exists in one phase.
(d) It can often be separated into its components by distillation.

____ 4. Which of the following is a solution?

(a) oil and water   (b) vinegar
(c) chromium   (d) sodium sulfide

____ 5. A certain solution contains 3.5% by mass sugar. What mass of solution can be formed from 2.34 kg of sugar?

(a) 82 kg   (b) 8.2 kg
(c) 2600 kg   (d) 67 kg

A-5 Matching

_____ One solid phase, definite melting point, can be decomposed further

_____ A liquid and a solid phase, definite melting point, can be decomposed further

_____ One liquid phase, melting point range, can be separated into other substances

_____ Two solid phases, melting point range

_____ One liquid phase, sharp boiling point, cannot be decomposed further

(a) Ice water (heterogeneous - pure)

(b) Sodium chloride (homogeneous - pure)

(c) Bromine (homogeneous - pure)

(d) Sugar mixed with salt (heterogeneous - mixture)

(e) Sugar water solution (homogeneous - mixture)

Review of Part B *The Properties of Energy*

OBJECTIVES AND DETAILED TABLE OF CONTENTS

3-4 The Forms and Types of Energy

OBJECTIVE *Define terms associated with energy exchanges in physical and chemical processes.*

3-5 Energy Measurement and Specific Heat

OBJECTIVE *Perform calculations involving the specific heat of a substance.*

SUMMARY OF PART B

Chemistry is also concerned with the energy that accompanies chemical changes. **Energy** (the ability to do work), however, comes in various forms, each of which can be converted into the others. In these changes, energy is neither created nor destroyed, which is known as the law of **conservation of energy**. When chemical energy is transformed into heat energy in a chemical reaction, the reaction is said to be **exothermic**. When heat energy is changed into chemical energy, the reaction is said to be **endothermic**. There are also two types of energy: **potential energy** (energy of position or composition) and **kinetic energy** (energy of motion).

Heat may interact with matter in three ways. It may be involved in a chemical reaction, it may cause a phase change, or it may just change the temperature of a specific phase of a substance. The **specific heat** of a homogeneous substance relates how much heat (in **calories** or **joules**) it takes to change one gram of the substance by one degree Celsius. Water is the standard which has a specific heat of $1°$ cal/g $\cdot$ $°C$ or 4.184 J/g $\cdot$ $°C$. The nutritional **Calorie** is actually one kilocalorie. The "c" in calorie is capitalized to indicate the nutritional calorie.

Example B-1 Specific Heat and Temperature Change

How many joules does it take to change 10.0 g of water a total of 5.2 Celsius degrees?

> **PROCEDURE**
>
> Sp ht = $\:$ J/g $\cdot$ $°C$ $\qquad$ Solving algebraically for *joule* we obtain
>
> joule = Sp ht $\:$ x $\:$ g $\:$ x $\:$ $°C$(T)
>
> **SOLUTION**
>
> 4.18 J/g $\cdot$ $°C$ x 10.0 g x $5.2°C$ $=$ <u>220 J</u>

In the final section, we observed two very important laws relating to chemical reactions. Mass and energy are neither created nor destroyed in a chemical reaction. Much of the quantitative relationships that we will study later are based on the law of **conservation of mass**. However, according to Einstein's law, there is a relationship between mass and energy. Fortunately, though, the conversion that occurs in chemical reactions is so small that it can be ignored entirely.

ASSESSMENT OF OBJECTIVES

B-1 Multiple Choice

_____ 1. Which of the following is an example of potential energy?

    (a) a running halfback
    (b) a bowling ball at the top of the stairs
    (c) water flowing over a falls
    (d) a moving train

____ 2. When wood burns, which two forms of energy are released?

      (a) heat, chemical   (b) electrical, heat   (c) heat, light
      (d) light, chemical   (e) light, mechanical

____ 3. The _____ energy of gasoline is converted in an automobile engine into
      the _____ energy of the moving tires.

      (a) heat, electrical   (b) chemical, mechanical   (c) light, heat
      (d) chemical, light   (e) light, mechanical

____ 4. An exothermic chemical reaction results when:

      (a) electrical energy is changed into heat energy.
      (b) heat energy is changed into chemical energy.
      (c) heat energy is changed into mechanical energy.
      (d) chemical energy is changed into heat energy.

____ 5. Which of the following are possible units of specific heat?

      (a) $J/°C$ (b) $g/cal \cdot °C$
      (c) $J/g \cdot K$ (d) $K/g \cdot cal$

____ 6. How many grams of water can be heated by 50 calories a total of 2.0
      Celsius degrees?

      (a) 25 g   (b) 50 g   (c) 100 g   (d) 200 g   (e) 75 g

B-2 Problems

1. It takes 36.0 calories to heat a 10.0-g sample of a metal from 0.0°C to 31.0°C.
   What is the specific heat of the metal?

2. The specific heat of a substance is 0.523 J/g · °C. How many joules are liberated when
   65.0 g of the substance cools a total of 12.0 Celsius degrees?

47

Chapter Summary Assessment

S-1 Matching

_____ Potential energy

_____ A form of energy

_____ Liquid

_____ Element

_____ Property

_____ Exothermic reaction

_____ Phase

_____ Distillation

_____ Solution

(a) Has a definite volume and a definite shape

(b) A chemical reaction that gives off heat

(c) Electricity

(d) A pure substance

(e) Contains two or more phases

(f) A unique, observable characteristic or trait

(g) Energy of position

(h) A chemical reaction that absorbs heat

(i) A laboratory process used to separate the components of a solution

(j) Has a definite volume and can be poured

(k) A homogeneous mixture

(l) A laboratory procedure used to separate a liquid from a solid phase

(m) Energy of motion

(n) Any physical state with uniform properties and identifiable boundaries

S-2 Problems

1. The density of a metal is 2.73 g/mL. What is the volume in liters occupied by 225 kg of the metal?

2. The specific heat of a substance is 1.15 J/g · °C. What mass in grams is heated from 25.0°C to 35.0°C by 209 joules?

Answers to Assessments of Objectives

A-1 Multiple Choice

1. **b**	2. **c**	3. **c**	4. **d**	5. **a**	6. **c**
7. **d**	8. **b**	9. **c**			

A-2 Matching

k   **Density** is the ratio of the mass to the volume.

g   A **solid** has definite dimensions.

h   A **compound** is made up of two or more elements.

b   The **melting point** is the same as the freezing point.

l   A **pure substance** is an element or a compound.

d   The observation of a **physical property** does not require a change.

A-3 Problems

1. 65.6 g ~ 70.8 mL

    ? g ~ 1.00 mL $\dfrac{65.5 \text{ g}}{70.8 \text{ mL}}$ = <u>0.925 g/mL</u>

2. Volume = 20.0 cm x 15.0 cm x 35.0 cm = 10,500 cm^3 = 10,500 mL
    Mass = 84.6 kg = 84,600 g

    $$\text{Density} = \frac{84600 \text{ g}}{10500 \text{ mL}} = \underline{8.06 \text{ g/mL}}$$

3.   L $\longrightarrow$ mL $\longrightarrow$ g

    1.12 L̶ x $\dfrac{10^3 \text{ m̶L̶}}{\text{L̶}}$ x $\dfrac{2.70 \text{ g}}{\text{m̶L̶}}$ = 3.02 x 10^3 g

4.   kg $\longrightarrow$ g $\longrightarrow$ mL

    1.00 k̶g̶ x $\dfrac{1 \text{ g̶}}{10^{-3} \text{ k̶g̶}}$ x $\dfrac{1 \text{ mL}}{0.954 \text{ g̶}}$ = 1.05 x 10^3 mL

A-4 Multiple Choice

1. **a**	2. **c**	3. **b**	4. **b**	5. **d**

A-5 Matching

b   **Sodium chloride** is one solid phase that can be decomposed into elements. It has a definite melting point.

a   **Ice water** is composed of a solid and liquid phase (heterogeneous) but both phases are the same pure compounds.

e   A **sugar water solution** is one homogeneous liquid phase but is a mixture that can be separated into other substances.

f   A mixture of sugar and salt is a **heterogeneous mixture** with two identifiable solid phases.

c   **Bromine** is an element that is composed of a single liquid phase with a sharp boiling point. It cannot be separated into more basic components.

B-1 Multiple Choice

1. **b**   2. **c**   3. **b**   4. **d**   5. **c**   6. **a**

B-2 Problems

1. $\dfrac{36.0 \text{ cal}}{10.0 \text{ g} \times 31.0 \ ^{\circ}\text{C}} = 0.116 \text{ cal/g} \cdot {}^{\circ}\text{C}$

2. $\dfrac{0.523 \text{ J}}{\text{g} \cdot {}^{\circ}\text{C}} \times 65.0 \text{ g} \times 12.0 \ ^{\circ}\text{C} = \underline{408 \text{ J}}$

S-1 Matching

g   **Potential energy** is energy of position.

c   Electricity is a **form of energy**.

j   A **liquid** has a definite volume and can be poured.

d   An **element** is a pure substance.

f   A **property** is a unique, observable characteristic or trait.

b   An **exothermic reaction** is a reaction that gives off heat.

n   A **phase** is a physical state with uniform properties and identifiable boundaries.

i   **Distillation** is a laboratory process used to separate the components of a solution.

k   A **solution** is a homogeneous mixture.

Problems

1. $225 \, \cancel{kg} \, \times \, \dfrac{1 \, \cancel{g}}{10^{-3} \, \cancel{kg}} \, \times \, \dfrac{1 \, \cancel{mL}}{2.73 \, \cancel{g}} \, \times \, \dfrac{10^{-3} \, L}{\cancel{mL}} \, = \underline{82.4 \, L}$

2. $T = (35.0 - 25.0) \, °C = 10.0°C$

$$ \text{mass} = \dfrac{J}{\text{sp heat} \times °C} \ = \ \dfrac{209 \, \cancel{J}}{1.15 \, \dfrac{\cancel{J}}{\cancel{g} \cdot \cancel{°C}} \times 10.0 \, \cancel{°C}} = \underline{18.2 \, g} $$

Answers and Solutions to Green Text Problems

3-1   (c) It has a definite volume but not a definite shape.

3-2   The gaseous state is compressible because the basic particles are very far apart and thus the volume of a gas is mostly empty space.

3-4   (a) physical   (b) chemical   (c) physical   (d) chemical   (e) chemical   (f) physical
        (g) physical   (h) chemical   (i) physical

3-6   (a) chemical   (b) physical   (c) physical   (d) chemical   (e) physical

3-9   Physical Property: melts at 660°C   Physical Change: melting
        Chemical Property: burns in oxygen
        Chemical Change: formation of aluminum oxide

3-10   Original substance: green, solid (physical); can be decomposed (chemical). Substance is a compound. Gas: gas, colorless (physical); can be decomposed (chemical). Substance is a compound since it can be decomposed. Solid: shiny, solid (physical); cannot be decomposed (chemical). Substance is an element since it cannot be decomposed.

3-12 $208 \, g / 80.0 \, mL = \underline{2.60 \, g/mL}$

3-14 $1064 \, g / 657 \, mL = \underline{1.62 \, g/mL}$ (carbon tetrachloride)

3-17 $671 \, \cancel{mL} \, \times \, \dfrac{2.16 \, g}{\cancel{mL}} \, = \underline{1450 \, g}$

3-18 $1.00 \, \cancel{L} \, \times \, \dfrac{1 \, \cancel{mL}}{10^{-3} \, \cancel{L}} \, \times \, \dfrac{0.67 \, g}{\cancel{mL}} \, = \underline{670 \, g}$

3-19 $1.00 \, \cancel{gal} \, \times \, \dfrac{3.785 \, \cancel{L}}{\cancel{gal}} \, \times \, \dfrac{\cancel{mL}}{10^{-3} \, \cancel{L}} \, \times \, \dfrac{0.67 \, \cancel{g}}{\cancel{mL}} \, \times \, \dfrac{1 \, lb}{453.6 \, \cancel{g}} \, = \underline{5.6 \, lb}$

3-21 $1.00 \, \cancel{kg} \, \times \, \dfrac{10^{3} \, \cancel{g}}{\cancel{kg}} \, \times \, \dfrac{1.00 \, mL}{1.60 \, \cancel{g}} \, = \underline{625 \, mL}$

3-22   Vol. = 92.45 - 14.00 = 78.45 mL

density = 136.5 g/78.45 mL = <u>1.74 g/mL</u> (magnesium)

3-23 $1.05 \, \cancel{lb} \times \dfrac{453.6 \, g}{\cancel{lb}} = 476 \, g$ $476 \, g/10^3 \, mL = \underline{0.476 \, g/mL}$

Yes, it floats.

3-24 $155 \, g/163 \, mL = 0.951 \, g/mL$ $4.56 \, \cancel{kg} \times \dfrac{10^3 \cancel{g}}{\cancel{kg}} \times \dfrac{1.00 \, mL}{0.951 \, \cancel{g}} = \underline{4790 \, mL}$

Pumice floats in water but sinks in alcohol.

3-27 $3.00 \, cm \times 8.50 \, cm \times 6.00 \, cm = 153 \, cm^3 = 153 \, mL$

$153 \, \cancel{mL} \times \dfrac{13.6 \, g}{\cancel{mL}} = \underline{2080 \, g}$

3-28   mass of liquid = 143.5 - 32.5 = 111.0 g $111.0 \, g/125 \, mL = \underline{0.888 \, g/mL}$

3-30   Water: $1.00 \, \cancel{L} \times \dfrac{1 \, \cancel{mL}}{10^{-3} \cancel{L}} \times \dfrac{1.00 \, g}{\cancel{mL}} = 1000 \, g$

Gasoline: $1.00 \, \cancel{L} \times \dfrac{1 \, \cancel{mL}}{10^{-3} \cancel{L}} \times \dfrac{0.67 \, g}{\cancel{mL}} = 670 \, g$
One liter of water has a greater mass.

3-31 $5.65 \, \cancel{oz} \times \dfrac{1 \, \cancel{lb}}{16 \, \cancel{oz}} \times \dfrac{453.6 \, g}{\cancel{lb}} = 160 \, g$ $33.3 - 25.0 = 8.3 \, mL$

$160 \, g/8.3 \, mL = \underline{19 \, g/mL}$ It's gold.

3-34   One needs a conversion factor between mL (cm^3) and ft^3.

$(\dfrac{2.54 \, cm}{in.})^3 = \dfrac{16.4 \, cm^3}{in.^3} = \dfrac{16.4 \, mL}{in.^3}$ $(\dfrac{12 \, in.}{ft})^3 = \dfrac{1728 \, in.^3}{ft^3}$

$\dfrac{1.00 \, \cancel{g}}{\cancel{mL}} \times \dfrac{1 \, lb}{453.6 \, \cancel{g}} \times \dfrac{16.4 \, \cancel{mL}}{\cancel{in.^3}} \times \dfrac{1728 \, \cancel{in.^3}}{ft^3} = \underline{62.5 \, lb/ft^3}$

3-35 $4.5 \, \cancel{mL} \times \dfrac{2.0 \times 10^7 \cancel{g}}{\cancel{mL}} \times \dfrac{1 \, lb}{453.6 \, \cancel{g}} = \underline{2.0 \times 10^5 \, lb \, (100 \, tons)}$

3-36   Carbon dioxide is a compound composed of carbon and oxygen. It can be prepared from a mixture of carbon and oxygen but the compound is no longer a mixture of the two elements.

3-38   Ocean water is the least pure because it contains a large amount of dissolved compounds. That is why it is not drinkable and cannot be used for crop irrigation. Drinking water also contains chlorine and some dissolved compounds but not nearly as much as ocean water. Rain water is the most pure but still contains some dissolved gases from the air.

3-40   (a) liquid only

3-41   (a) gasoline - homogeneous   (b) dirt - heterogeneous   (c) smog - heterogeneous   (d) alcohol - homogeneous   (e) a new nail - homogeneous   (f) vinegar - homogeneous solution   (g) aerosol spray - heterogeneous   (h) air - homogeneous

3-43   (a) liquid   (b) various solid phases   (c) gas and liquid   (d) liquid   (e) solid   (f) liquid   (g) liquid and gas   (h) gas

3-45   A mixture of all three would have carbon tetrachloride on the bottom, water in the middle and kerosene on top. Water and kerosene float on carbon tetrachloride; kerosene floats on water.

3-47   Ice is less dense than water. An ice-water mixture is pure but heterogeneous.

3-48   (a) solution (a solid dissolved in a liquid) (b) heterogeneous mixture (probably a solid suspended in a liquid such as dirty water) (c) element (d) compound (e) solution (two liquids)

3-50   Mass of mixture = 22.6 + 855 = 878 g
22.6/878 x 100% = $\underline{2.57\%\ NaCl}$ (100% - 2.57) = $\underline{97.4\%\ water}$

3-52   255 ~~kg solution~~ x $\dfrac{25\ kg\ solute}{100\ kg\ solution}$ = $\underline{64\ kg\ solute}$

3-54   122 ~~lb iron~~ x $\dfrac{100\ lb\ duriorn}{86\ lb\ iron}$ = $\underline{140\ lb\ duriorn}$

3-56   (a) exothermic   (b) endothermic   (c) endothermic   (d) exothermic   (e) exothermic

3-57   Gasoline is converted into heat energy when it burns. The heat energy moves the pistons which is mechanical energy. The mechanical energy turns the alternator, which generates electrical energy. The electrical energy is converted into chemical energy in the battery.

3-59   (a) potential   (b) kinetic   (c) potential (It is stored because of its composition.)   (d) kinetic   (e) kinetic

3-62   Kinetic energy is at a maximum nearest the ground when the swing is moving the fastest. Potential energy is at a maximum when the swing has momentarily stopped at the highest point. Assuming no gain or loss of energy, the total of the two energies is constant.

3-63 $\dfrac{73.2\ J}{10.0\ g \cdot 8.58\ ^oC}$ = 0.853 J/g · °C

3-64 $\dfrac{56.6 \text{ cal}}{365 \text{ g} \cdot 5.0 \,^{\circ}\text{C}} = 0.031 \text{ cal/g} \cdot {}^{\circ}\text{C}$ (gold)

3-66 $^{\circ}\text{C} = \dfrac{\text{cal}}{\text{sp. heat x g}} = \dfrac{150 \text{ cal}}{0.092 \dfrac{\text{cal}}{\text{g} \cdot {}^{\circ}\text{C}} \times 50.0 \text{ g}} = 33 \,^{\circ}\text{C rise}$

$T(^{\circ}\text{C}) = 25 + 33 = 58 \,^{\circ}\text{C}$

This compares to a 3.0 °C rise in temperature for 50.0 g of water.

3-68 heat (J) = sp heat x g x (°C)

$\dfrac{0.895 \text{ J}}{\text{g} \cdot {}^{\circ}\text{C}} \times 43.5 \text{ g} \times 13 \,^{\circ}\text{C} = \underline{506 \text{ J}}$

3-69 58 - 25 = 33 °C rise in temperature

$g = \dfrac{\text{cal}}{\text{sp heat} \cdot {}^{\circ}\text{C}} = \dfrac{16.0 \text{ cal}}{0.106 \dfrac{\text{cal}}{\text{g} \cdot {}^{\circ}\text{C}} \cdot 33 \,^{\circ}\text{C}} = \underline{4.6 \text{ g}}$

3-70 The copper skillet because it has a lower specific heat. The same amount of applied heat will
heat the copper skillet more than the iron.

3-72 $^{\circ}\text{C} = \dfrac{\text{J}}{\text{g x sp heat}}$ $\qquad \dfrac{50.0 \text{ J}}{12.0 \text{ g} \times 0.444 \dfrac{\text{J}}{\text{g} \cdot {}^{\circ}\text{C}}} = 9.38 \,^{\circ}\text{C rise (iron)}$

$\dfrac{50.0 \text{ J}}{12.0 \text{ g} \times 0.13 \dfrac{\text{J}}{\text{g} \cdot {}^{\circ}\text{C}}} = 32 \,^{\circ}\text{C rise (gold)}$

$\dfrac{50.0 \text{ J}}{12.0 \text{ g} \times 4.18 \dfrac{\text{J}}{\text{g} \cdot {}^{\circ}\text{C}}} = 0.997 \,^{\circ}\text{C rise (water)}$

3-74 $^{\circ}\text{C} = \dfrac{\text{cal}}{\text{g x sp heat}}$ $\qquad \dfrac{1000 \text{ cal}}{50.0 \text{ g} \times 1.00 \dfrac{\text{cal}}{\text{g} \cdot {}^{\circ}\text{C}}} = 20 \,^{\circ}\text{C rise}$

$T(^{\circ}\text{C}) = 20 + 25 = \underline{45 \,^{\circ}\text{C}}$

3-76 $\dfrac{0.895 \text{ J}}{\text{g} \cdot {}^{\circ}\text{C}} \times 50.0 \text{ g} \times 65 \,^{\circ}\text{C} = \underline{2910 \text{ J}}$

3-80 Heat gained by water $100 \text{ mL} \times \dfrac{1.00 \text{ g}}{\text{mL}} \times (28.7 - 25.0 \text{ °C}) \times \dfrac{4.184 \text{ J}}{\text{g} \cdot \text{°C}} = 1550 \text{ J}$

Heat gained by lead $\text{g lead} \times (42.8 - 28.7 \text{ °C}) \times \dfrac{0.128 \text{ J}}{\text{g} \cdot \text{°C}} = 1550 \text{ J}$

g lead = <u>860 g</u>

3-81 Heat lost by metal = heat gained by water
$100.0 \text{ g} \times 68.7 \text{ °C} \times \text{sp ht.} = 100.0 \text{ g} \times 6.3 \text{ °C} \times 4.184 \text{ j/g} \cdot \text{°C}$
specific heat = <u>0.38 J/g · °C</u> The metal is <u>copper</u>.

3-82 Density of A = 0.86 g/mL; density of B = 0.89 g/mL
Liquid A floats on liquid B.

3-84 $100 \text{ mL} \times \dfrac{1.06 \text{ g}}{\text{mL}} = 106 \text{ g solution}$

$106 \text{ g solution} \times \dfrac{14 \text{ g sugar}}{100 \text{ g solution}} = \underline{15 \text{ g of sugar}}$

3-85 volume of metal = 24.96 - 18.22 = 6.74 mL

62.485 g/6.74 mL = <u>9.27 g/mL</u>

3-87 $305 \text{ mL} \times \dfrac{1.00 \text{ g}}{\text{mL}} = 305 \text{ g water}$

$\dfrac{10.0 \text{ g}}{(305 + 10.0) \text{ g}} \times 100\% = \underline{3.17\% \text{ salt}}$

3-89 $50.0 \text{ mL gold} \times \dfrac{19.3 \text{ g}}{\text{mL gold}} = 965 \text{ g gold}$

$50.0 \text{ mL alum.} \times \dfrac{2.70 \text{ g}}{\text{mL alum.}} = 135 \text{ g alum.}$ $\dfrac{965 \text{ g gold}}{(965 + 135) \text{ g alloy}} \times 100\% = \underline{87.7\% \text{ gold}}$

3-91 $\dfrac{215 \text{ J}}{25.0 \text{ g} \times 66\text{°C}} = 0.13 \dfrac{\text{J}}{\text{g} \cdot \text{°C}}$ (gold) $25.0 \text{ g gold} \times \dfrac{1 \text{ mL}}{19.3 \text{ g gold}} = \underline{1.30 \text{ mL}}$

3-94 When a log burns, most of the compounds formed in the combustion are gases and dissipate into the atmosphere. Only some solid residue (ashes) is left. When zinc and sulfur (both solids) combine, the only product is a solid so there is no weight change. When iron burns, however, its only product is a solid. It weighs more than the original iron because the iron has combined with the oxygen gas from the air.

4

The Periodic Table and
Chemical Nomenclature

Review of Part A *Relationships Among the Elements and the Periodic Table*

OBJECTIVES AND DETAILED TABLE OF CONTENTS

4-1   The Origin of the Periodic Table

OBJECTIVE *Describe the origins of the periodic table.*

4-1.1 Construction of the Periodic Table
4-1.2 Metals and Nonmetals

4-2   Using the Periodic Table

OBJECTIVE *Using the periodic table, identify a specific element with its group number and name, period, and physical state.*

4-2.1 Periods
4-2.2 Groups
4-2.3 Physical States and the Periodic Table

SUMMARY OF PART A

One of the most important classifications of the elements is that of **metals** and **nonmetals**. A third class of elements (**metalloids**) is sometimes employed to describe elements with intermediate behavior. Metals may be subdivided further as either chemically reactive, such as sodium, or chemically inert, such as gold. In fact, there are many other metals and nonmetals that are chemically similar. The **periodic table** shows these relationships in the form of a chart of the elements. This chart displays the elements by increasing atomic number with families of elements falling into vertical columns. This is known as the **periodic law**. This table was first displayed in this fashion by Mendeleev and Meyer around 1870. In any given **period** (a horizontal row of elements ending with a noble gas), nonmetals are found toward the end of the period. If a **group** (a vertical column of elements) contains nonmetals, they are found at the top. Groups are traditionally numbered using Roman numerals (i.e., VIA) or consecutive numbers (i.e., 1 through 18).

The groups of elements can be classified into four main categories.

1. **Representative elements (Groups IA - VIIA) (1, 2, 13-17)**
    The elements in several of the groups of representative elements have enough common properties that they are entitled to specific names. These include the **alkali metals** (Group IA), the **alkaline earth metals** (Group IIA), and the **halogens** (Group VIIA).

2. **Noble gases (Group VIIIA) (18)**
    These are the nonmetallic gases at the end of each period. The heavier of these (Kr, Xe, and Rn) form a few compounds, but generally these elements are chemically unreactive.

3. **Transition metals (Groups IIIB - IIB) (3-12)**
    These elements are all metals and contain many of the familiar structural and coinage metals such as iron, chromium, nickel, silver, and gold.

4. **Inner transition metals**
    There are two categories of these elements containing 14 elements each: the **lanthanides** (beginning with cerium) and the **actinides** (beginning with thorium).

At **room temperature** (25°C) all three physical states are found among the elements, with the solid state being the most common by far. The rest are gases, except for two liquids—one is a metal (mercury) and the other is a nonmetal (bromine). The gases are to the right and top in the periodic table. Some of the nonmetals exist in nature as molecules. All elements of Group VIIA, as well as nitrogen, oxygen, and hydrogen, exist as diatomic molecules. The most common form of phosphorus is P_4, while the most common form of sulfur is S_8.

ASSESSMENT OF OBJECTIVES

A-1 Multiple Choice

___ 1. Which of the following elements exists (at room temperature) as a liquid?

   (a) Ce   (b) Hf   (c) Br   (d) N   (e) Ne

___ 2. Which of the following is a halogen?

   (a) H   (b) Li   (c) Ca   (d) I   (e) Ar

___ 3. Which of the following elements appears at the end of the third period?

   (a) Cl   (b) K   (c) Zn   (d) Ne   (e) Ar

___ 4. Which of the following elements is a metalloid?

   (a) Sn (#50)   (b) Sb (#51)   (c) C (#6)

   (d) Mg (#24)   (e) Zn (#30)

___ 5. How many elements are in the third period?

   (a) 8   (b) 18   (c) 2   (d) 12   (e) 32

___ 6. Which of the following elements is a representative element?

   (a) Zn   (b) Se   (c) Sc   (d) Kr   (e) Ce

___ 7. Which of the following groups does not contain a metal?

   (a) IIIB   (b) IA   (c) VIA   (d) VIIA   (e) VIIB

___ 8. Which of the following is a representative element metal?

   (a) Si   (b) Fe   (c) Ga   (d) Cu   (e) U

A-2 Matching

___ Lanthanide	(a) Ru	(g) H
___ Representative element gas at room temperature	(b) K	(h) Br
___ Noble gas	(c) Sr	(i) Kr
___ An alkali metal	(d) Sm	(j) C
___ Transition metal	(e) As	(k) W
___ An actinide	(f) Sn	(l) U

Review of Part B *The Formulas and Names of Compounds*

OBJECTIVES AND DETAILED TABLE OF CONTENTS

4-3 Naming and Writing Formulas of Metal–Nonmetal Binary Compounds

OBJECTIVE *Write the names and formulas of ionic compounds between metals and nonmetals using the IUPAC conventions.*

4-4 Naming and Writing Formulas of Compounds with Polyatomic Ions

OBJECTIVE *Write the names and formulas of compounds containing polyatomic ions.*

4-5 Naming Nonmetal–Nonmetal Binary Compounds

OBJECTIVE *Name binary molecular (nonmetal-nonmetal) compounds using proper Greek prefixes or common names.*

4-6   Naming Acids

OBJECTIVE *Write the names and formulas of acids.*

4-6.1 Binary Acids
4-6.2 Oxyacids

SUMMARY OF PART B

Before proceeding into other aspects of chemistry, we take some time in this chapter to systematize the naming of compounds (**chemical nomenclature**). In the first three sections, we focus on the nomenclature of metal compounds that are either **binary** (two element) or contain a polyatomic ion.

A potential problem we encounter is that some metals form a variety of compounds with the same nonmetal. For example, the following oxides of manganese are known: MnO, Mn_2O_3, MnO_2, and Mn_2O_7. Obviously, we need some system to distinguish among these compounds. The system used (the **Stock method**) is to indicate the positive charge on the metal (or the positive charge that it would have if it were an ion).

The naming of **salts** is summarized as follows:

1. Metal-nonmetal binary compounds

(a) metals that have only one charge (IA, IIA, and Al)

Example: $CaBr_2$ calcium bromide
            (Calcium forms only a +2 charge.)

(b) metals that can have more than one charge (most other metals)

Example: Cr_2O_3 chromium(III) oxide
            (The Roman numeral III indicates that Cr has a +3 charge.)

            CrO chromium(II) oxide
            (In this case, the Cr has a +2 charge.)

2. Metal-polyatomic ions (**oxyanions**)

(a) metals that have one charge

Example: $Sr_3(PO_4)_2$ strontium phosphate
            (To arrive at the formula, we recall that the total positive charge on the metal must cancel the total negative charge on the anions.)
            Thus, $(Sr^{2+} \times 3) + (PO_4^{3-} \times 2) = 0$

(b) metals that form more than one charge

Example: $Ti_2(SO_4)_3$ <u>titanium(III) sulfate</u>
(By solving the following algebra equation, we can arrive at the charge on the metal.) $(Tl^x \times 2) + (SO_4^{2-} \times 3) = 0$ $x = \underline{+3}$.

The formula can also be determined by the "cross charge" method. That is, the number of each respective charge becomes the subscript of the other ion as follows.

barium chloride Ba^{2+} ⤬ Cl^{1-} $BaCl_2$

iron(III) sulfate Fe^{3+} ⤬ (SO_4^{2-}) $Fe_2(SO_4)_3$

Some unique problems are encountered with the naming of nonmetal-nonmetal compounds. For example, which element is written and named first? With a few exceptions for several common compounds, we write and name the element that is closest to being a metal first. The Stock system is not generally used because of certain ambiguities that can result. Instead, we make use of Greek prefixes.

Most anions combined with hydrogen form a class of compounds called **acids**. We will see in a later chapter why these compounds are so special that they have their own nomenclature.

The naming of nonmetal-nonmetal compounds is summarized as follows:

1. Nonmetal-nonmetal binary compounds

Example: N_2O_3 <u>dinitrogen trioxide</u>
[Name the most metallic element first, then the other element in the same manner as an anion. Use Greek prefixes to indicate number of atoms present.]

2. Acids

(a) **binary acids**

Example: H_2Te <u>hydrotelluric acid</u>
[If parent anion ends in *ide* (e.g., telluride), add *hydro* to root plus *ic* ending.]

(b) **oxyacids**

Example: HNO_3 <u>nitric acid</u>
        HNO_2 <u>nitrous acid</u>
[If parent anion ends in *ate* (i.e, nitrate), add *ic* ending. If parent anion ends in *ite* (i.e., nitrite) add *ous* ending.]

ASSESSMENT OF OBJECTIVES

B-1 Writing Formulas and Names

1. Write the formulas of the following ions including the charges.

   (a) carbonate _____   (f) chlorate _____

   (b) bromide _____   (g) nitrite _____

   (c) bisulfate _____   (h) nitride _____

   (d) sulfide _____   (i) nitrate _____

   (e) acetate _____   (j) ammonium _____

2. Write the correct formulas in the boxes.

	HSO_3^-	CrO_4^{2-}	PO_4^{3-}
Ba^{2+}	$Ba(HSO_3)_2$		
Fe(III)			
NH_4^+			

3. Write the correct names for the compounds in problem 2.

	HSO_3^-	CrO_4^{2-}	PO_4^{3-}
Ba^{2+}	_____	_____	_____
Fe(III)	_____	_____	_____
NH_4^+	_____	_____	_____

4. Of the following sets of two elements, choose which would be written and named first in a binary compound formed by the two elements:

(a) Si and Cl _____   (b) Xe and O _____

(c) O and S _____   (d) I and Te _____

(e) F and H _____   (f) N and P _____

(g) Cl and Br _____   (h) Cl and N _____

5. Write the number that corresponds to each prefix:

(a) di _____   (b) penta _____

(c) mono _____   (d) hepta _____

(e) tetra _____   (f) hexa _____

6. Write the formulas and names of the acids formed by the following anions:

	Formula	Name
(a) $Cr_2O_7^{2-}$	_____	_____
(b) HSO_3^-	_____	_____
(c) PO_4^{3-}	_____	_____
(d) Br^-	_____	_____

Chapter Summary Assessment

S-1 Problems

1. Write the correct names for:

(a) CaO _____

(b) BaI_2 _____

(c) Al_2S_3 _____

(d) VO_2 _____

(e) $FeCl_3$ _____

(f) Au_2S_3 _____

(g) $LiClO_4$ _____

(h) Zr(OH)$_4$ _____

(i) (NH$_4$)$_2$Se   _____

(j) Ni(HCO$_3$)$_2$ _____

(k) ICl$_3$ _____

(l) AsF$_3$ _____

(m) I$_2$O$_7$ _____

(n) HI   _____

(o) HClO   _____

(p) H$_2$CrO$_4$ _____

2. Write formulas for the following:

   (a) rubidium oxide   _____

   (b) aluminum iodide   _____

   (c) beryllium selenide   _____

   (d) manganese(II) oxide   _____

   (e) manganese(VII) oxide   _____

   (f) platinum(IV) chloride   _____

   (g) potassium oxalate   _____

   (h) cesium bisulfate   _____

   (i) radium cyanide   _____

(j) cadmium(II) phosphate _____

(k) cobalt(II) nitrate _____

(l) tetraphosphorus hexoxide _____

(m) dibromine trioxide _____

(n) disulfur dichloride _____

(o) hydrofluoric acid _____

(p) sulfuric acid _____

3. The following compounds have names that are not usually accepted. Write the correct name and explain the error.

Formula	Wrong Name	Correct Name
K_2CrO_4	dipotassium chromate	_____

Reason: _____

| $Ca(HSO_4)_2$ | calcium(II) bisulfate | |
| | | _____ |

Reason: _____

| H_2SO_4 | hydrogen sulfate | _____ |

Reason: _____

| PbO | lead oxide | _____ |

Reason: _____

| SO_3 | sulfur(VI) oxide | _____ |

Reason: _____

| $Sr(NO_3)_2$ | strontium nitrite | _____ |

Reason: _____

Answers to Assessments of Objectives

A-1 Multiple Choice

1. **c** Br (exists in nature as Br_2)   5. **a** 8

2. **d** I   6. **b** Se

3. **e** Ar   7. **d** VIIA

4. **b** Sb (#51)   8. **c** Ga

A-2 Matching

d   Samarium (Sm) is a **lanthanide**.

g   Hydrogen (H) is a **representative element gas**.

i   Krypton (Kr) is a **noble gas**.

b   Potassium (K) is a **representative element metal**.

a   Ruthenium (Ru) is a **transition metal**.

l   Uranium (U) is an **actinide**.

B-1 Writing Formulas and Names

1.   (a) CO_3^{2-} (f) ClO_3^-

(b) Br^- (g) NO_2^-

(c) HSO_4^- (h) N^{3-}

(d) S^{2-} (i) NO_3^-

(e) $C_2H_3O_2^-$ (j) NH_4^+

2.

	$BaCrO_4$	$Ba_3(PO_4)_2$
$Fe(HSO_3)_3$	$Fe_2(CrO_4)_3$	$FePO_4$
NH_4HSO_3	$(NH_4)_2CrO_4$	$(NH_4)_3PO_4$

3.   barium bisulfate   barium chromate   barium phosphate
iron(III) bisulfite   iron(III) chromate   iron(III) phosphate
ammonium bisulfite   ammonium chromate   ammonium phosphate

4.   (a) Si   (b) Xe
          (c) S   (d) Te
          (e) H   (f) P
          (g) Br   (h) N

5.   (a) di - 2   (b) penta - 5
          (c) mono - 1   (d) hepta - 7
          (e) tetra - 4   (f) hexa - 6

6.   (a) $H_2Cr_2O_7$ dichromic acid   (b) H_2SO_3 sulfurous acid

          (c) H_3PO_4 phosphoric acid   (d) HBr   hydrobromic acid

S-1 Problems

1.   (a) calcium oxide   (b) barium iodide
          (c) aluminum sulfide   (d) vanadium(IV) oxide
          (e) iron(III) chloride   (f) gold(III) sulfide
          (g) lithium perchlorate   (h) zirconium(IV) hydroxide
          (i) ammonium selenide   (j) nickel(II) bicarbonate
          (k) iodine trichloride   (l) arsenic trifluoride
          (m) diiodine heptoxide   (n) hydroiodic acid
          (o) hypochlorous acid   (p) chromic acid

2.   (a) Rb_2O (b) AlI_3
          (c) BeSe   (d) MnO
          (e) Mn_2O_7 (f) $PtCl_4$
          (g) $K_2C_2O_4$ (h) $CsHSO_4$
          (i) $Ra(CN)_2$ (j) $Cd_3(PO_4)_2$
          (k) $Co(NO_3)_2$ (l) P_4O_6
          (m) Br_2O_3 (n) S_2Cl_2
          (o) HF   (p) H_2SO_4

3.

(a) K_2CrO_4 potassium chromate   Since chromate has a -2 charge and potassium
                                                  is always +1, the Greek prefix is unnecessary.

(b) $Ca(HSO_4)_2$ calcium bisulfate   Since calcium has only a +2 charge,
                                                  the (II) is unnecessary.

(c) H_2SO_4 sulfuric acid   Oxyacids have only an acid name.

(d) PbO	lead(II) oxide	Lead forms ions with more than one charge, so the Stock method is generally used to indicate charge.
(e) SO_3	sulfur trioxide	Greek prefixes are used in nonmetal–nonmetal binary compounds rather than the Stock method.
(f) $Sr(NO_3)_2$	strontium nitrate	The anion is a nitrate not a nitrite.

Answers and Solutions to Green Text Problems

4-1 An active metal reacts with water and air. A noble metal is not affected by air, water, or most acids.

4-2 32

4-3 (c) I_2 and (g) Br_2

4-5 (a) Fe and (d) La - transition elements
(b) Te, (f) H, and (g) In - representative elements
(c) Pm - inner transition element
(e) Xe - noble gas

4-7 (b) Ti, (e) Pd, and (g) Ag

4-8 The most common physical state is a solid and metals are more common than nonmetals.

4-10 (a) Ne   (d) Cl   (f) N

4-12 (a) Ru   (b) Sn   (c) Hf   (h) W

4-13 (d) Te   (f) B

4-14 (a) Ar   (b) Hg   (c) N_2 (d) Be

4-16 Element 118 would be a nonmetal noble gas

4-18 (a) lithium fluoride   (b) barium telluride   (c) strontium nitride
(d) barium hydride   (e) aluminum chloride

4-20 (a) Rb_2Se (b) SrH_2 (c) RaO   (d) Al_2Te_3 (e) BeF_2

4-22 (a) bismuth(V) oxide   (b) tin(II) sulfide   (c) tin(IV) sulfide
(d) copper(I) telluride   (e) titanium(IV) oxide

4-24 (a) Cu_2S (b) V_2O_3 (c) AuBr (d) Ni_3P_2 (e) CrO_3

4-26 In 4-22 (a) Bi_2O_5, (c) SnS_2, and (e) TiO_2 In 4-24 (e) CrO_3

4-28 (c) ClO_3^-

4-30 ammonium, NH_4^+

4-31 (b) permanganate (MnO_4^-)   (c) perchlorate (ClO_4^-)

(e) phosphate (PO_4^{3-})   (f) oxalate ($C_2O_4^{2-}$)

4-32 (a) chromium(II) sulfate   (b) aluminum sulfite   (c) iron(II) cyanide
(d) rubidium hydrogen carbonate   (e) ammonium carbonate   (f) ammonium nitrate
(g) bismuth(III) hydroxide

4-34 (a) $Mg(MnO_4)_2$ (b) $Co(CN)_2$ (c) $Sr(OH)_2$
(d) Tl_2SO_3 (e) $Fe_2(C_2O_4)_3$
(f) $(NH_4)_2Cr_2O_7$ (g) $Hg_2(C_2H_3O_2)_2$

4-36

Cation/Anion	HSO_3^-	Te^{2-}	PO_4^{3-}
NH_4^+	NH_4HSO_3 ammonium bisulfite	$(NH_4)_2Te$ ammonium telluride	$(NH_4)_3PO_4$ ammonium phosphate
Co^{2+}	$Co(HSO_3)_2$ cobalt(II) bisulfite	CoTe cobalt(II) telluride	$Co_3(PO_4)_2$ cobalt(II) phosphate
Al^{3+}	$Al(HSO_3)_3$ aluminum bisulfite	Al_2Te_3 aluminum telluride	$AlPO_4$ aluminum phosphate

4-38 (a) sodium chloride   (b) sodium hydrogen carbonate
(c) calcium carbonate   (d) sodium hydroxide
(e) sodium nitrate   (f) ammonium chloride
(g) aluminum oxide   (h) calcium hydroxide
(i) potassium hydroxide

4-39 (a) Ca_2XeO_6 (b) K_4XeO_6 (c) $Al_4(XeO_6)_3$

4-40 (a) phosphonium fluoride   (b) potassium hypobromite
(c) cobalt(III) iodate   (d) calcium silicate (actual name is calcium metasilicate)
(e) aluminum phosphite   (f) chromium(II) molybdate

4-41 (a) Si (b) I (c) H (d) Kr (e) H (f) As

4-43 (a) carbon disulfide   (b) boron trifluoride
(c) tetraphosphorus decoxide   (d) dibromine trioxide
(e) methane   (f) dichlorine oxide or dichlorine monoxide
(g) phosphorus pentachloride   (h) sulfur hexafluoride

4-45 (a) P_4O_6 (b) CCl_4 (c) IF_3 (d) C_6H_{14} (e) SF_6 (f) XeO_2

4-47 (a) hydrochloric acid   (b) nitric acid
(c) hypochlorous acid   (d) permanganic acid
(e) periodic acid   (f) hydrobromic acid

4-48 (a) HCN (b) H_2Se (c) $HClO_2$ (d) H_2CO_3 (e) HI (f) $HC_2H_3O_2$

4-50 (a) hypobromous acid   (b) iodic acid   (c) phosphorous acid
(d) molybdic acid   (e) perxenic acid

4-52 ClO_2 chlorine dioxide

4-53 A = F   X = Br BrF_5 (Br is more metallic.) bromine pentafluoride

4-55 Gas = N_2 Al forms only +3 aluminum nitride AlN (N^{3-})
Ti_3N_2 titanium (II) nitride

4-57 Co^{2+} and Br^- $CoBr_2$ cobalt (II) bromide

4-58 Metal = Mg, Nonmetal = S MgH_2 magnesium hydride H_2S hydrogen sulfide or hydrosulfuric acid

4-60 NiI_2 - nickel (II) iodide H_3PO_4 - phosphoric acid
$Sr(ClO_3)_2$ - strontium chlorate H_2Te - hydrogen telluride or hydrotelluric acid
As_2O_3 - diarsenic trioxide SnC_2O_4 - tin(II) oxalate
Sb_2O_3 - antimony (III) oxide

4-62 tin(II) hypochlorite - $Sn(ClO)_2$ chromic acid - H_2CrO_4
xenon hexafluoride - XeF_6 barium nitride - Ba_3N_2
hydrofluoric acid - HF iron(III) telluride - Fe_2Te_3 lithium phosphate - Li_3P

4-63 (e) $Rb_2C_2O_4$ (Rb^+ and $C_2O_4^{2-}$)

4-65 Rb_2O_2, MgO_2, $Al_2(O_2)_3$, $Ti(O_2)_2$, H_2O_2, hydrogen peroxide, hydroperoxic acid

4-67 (a) Na_2CO_3 (b) $CaCl_2$ (c) $KClO_4$ (d) $Al(NO_3)_3$ (e) $Ca(OH)_2$ (f) $AlCl_3$

4-69 N^{3-} - nitride, NO_2^- - nitrite, NO_3^- - nitrate, CN^- - cyanide, NH_4^+ - ammonium

4-71   (e) chromium(III) carbonate

4-73   (c) barium chlorite

5

Quantities in Chemistry

Review of Part A *The Measurement of Masses of Elements and Compounds*

OBJECTIVES AND DETAILED TABLE OF CONTENTS

5-1 Relative Masses of Elements

OBJECTIVE *Calculate the masses of equivalent numbers of atoms of different elements.*

    5-1.1 Counting by Weighing
    5-1.2 The Mass of an Average Atom

5-2   The Mole and the Molar Mass of Elements

OBJECTIVE *Define the mole in terms of numbers of atoms and atomic masses of elements.*

     5-2.1 The Definition of the Mole
     5-2.2 The Meaning of a Mole
     5-2.3 Relationships Among Dozens (the Unit), Number, and Mass
     5-2.4 Relationships Among Moles (the Unit), Number, and Mass

5-3   The Molar Mass of Compounds

OBJECTIVE *Perform calculations involving masses, moles and number of molecules or formula units for a compound.*

     5-3.1 The Formula Weight of a Compound
     5-3.2 The Formula Weight of Hydrates
     5-3.3 The Molar Mass of a Compound

SUMMARY OF PART A

The atom is such a small thing that we must deal with an almost incomprehensible number of them to manipulate in the laboratory. Still, although we do not need to know the actual number of atoms present in a sample, we often need to know the relative number present. In this endeavor, the chemist is not alone. The grocer (vegetables and fruit), the builder (nails), and the farmer (seeds) must often sell, purchase, or plant large numbers of items. Since counting is out of the question, we rely on a mass scale to provide a count or a comparative number.

In Chapter 2, we described how the mass of one atom compares to that of another. We found that the atomic masses of the elements are found by comparison to the standard (^{12}C = 12.000··· amu). The atomic mass of an element can be thought of as the mass of an "average" atom of that element, although an "average" atom does not exist. When the atomic masses of the elements are expressed in mass units such as grams rather than amu, there are a large number of atoms present. However, the number of atoms is the *same* for each element. In chemistry, this unit is known as "the **mole**" and the numerical quantity that it represents is referred to as **Avogadro's number**.

$$1.000 \text{ mole} = 6.022 \times 10^{23} \text{ objects or particles}$$

The mass of one mole of the atoms of an element is the atomic mass expressed in grams and is known as the **molar mass of the element**. Thus, one mole of an element stands for a certain number of atoms (always 6.022×10^{23}) and a mass in grams (which depends on the element).

	Number of atoms	Mass
1.000 mole of atoms	6.022×10^{23}	32.07 g of S
	6.022×10^{23}	238.0 g of U
	6.022×10^{23}	55.85 g of Fe
	6.022×10^{23}	4.003 g of He

The mathematical conversions among moles, mass, and number turn out to be surprisingly straightforward operations. The confusion often centers on the scientific notation necessary in the definition of the mole. Students using dimensional analysis (and a little neatness and organization) have little trouble with these problems. This section is extremely important, however, since many of the following chapters require a familiarity with mole relationships.

The mole unit is easily applied to molecules as well as atoms. Again, one mole of molecules implies both a number and a mass. For molecules, the mass of one mole (known as the **molar mass of the compound**) is the **formula weight** expressed in grams. The formula weight of a specific compound is determined from the atomic mass of each element and the number of atoms of that element in one molecule of the compound. For example, consider the compound, N_2O_3:

Element	Number of Atoms		Atomic Mass		Total Mass of Element
N	2	x	14.01 amu	=	28.02 amu
O	3	x	16.00 amu	=	48.00 amu
			Formula Weight	=	76.02 amu

	Number of Molecules or Formula Units	Mass
	6.022×10^{23}	76.02 g of N_2O_3
1.000 mole of molecules or formula units	6.022×10^{23}	28.01 g of CO
	6.022×10^{23}	138.2 g of K_2CO_3
	6.022×10^{23}	352.0 g of UF_6

Conversions among moles, mass, and number of molecules are carried out in the same manner as with moles of atoms. Two examples of the calculations follow.

Example A-1 Moles to Mass

What is the mass of 0.175 moles of N_2F_4?

PROCEDURE

The molar mass of the compound is used as a conversion factor between moles and mass. The molar mass of N_2F_4 is

N	2 x 14.01 amu =	28.02 amu
F	4 x 19.00 amu =	76.00 amu
	Formula Weight =	104.02 amu
	Molar Mass = 104.0 g/mol	

In this case, the conversion factor has grams in the numerator (requested) and moles in the denominator (given).

SOLUTION

$$0.175 \text{ mol } N_2F_4 \text{ } \times \text{ } \frac{104.0 \text{ g } N_2F_4}{\text{mol } N_2F_4} \text{ } = \text{ } \underline{18.2 \text{ g } N_2F_4}$$

Given

This conversion factor converts moles to grams.

Requested

Example A-2 Mass to Number of Molecules

How many individual molecules are in 651 grams of PCl_3?

PROCEDURE

In this case, we must convert from grams of PCl_3 to moles of PCl_3 using the molar mass of PCl_3 as a conversion factor. In a second step, we convert moles of PCl_3 to a number of molecules using Avogadro's number as a conversion factor.

SOLUTION

The molar mass of PCl_3 is

P	1 x 30.97 amu =	30.97 amu
Cl	3 x 35.45 amu =	106.35 amu
	Formula weight =	137.32 amu
	Molar mass = 137.3 g/mol	

$$651 \text{ g } PCl_3 \text{ } \times \text{ } \frac{1 \text{ mol } PCl_3}{137.3 \text{ g } PCl_3} \text{ } \times \text{ } \frac{6.022 \times 10^{23} \text{ molecules}}{\text{mol } PCl_3} = 2.86 \times 10^{24} \text{ molecule}$$

Given

This factor converts grams to moles.

This factor converts moles to molecules.

Requested

SELF-ASSESSMENT

A-1 Multiple Choice

___ 1. The molar mass of oxygen atoms is

    (a) 32.00 g/mol   (d) 16.00 amu
    (b) 16.00 g/mol   (e) 32.00 amu
    (c) 8.00 g/mol

___ 2. A 0.50-mole quantity of Ne atoms has a mass of about

    (a) 20 g   (b) 5.0 g   (c) 10 g   (d) 10 amu   (e) 40 g

___ 3. The number 6.022×10^{22} is

    (a) 1.000 mole   (d) 5.000 mole
    (b) 0.1000 mole   (e) no such number exists
    (c) 10.00 mole

___ 4. The mass of one atom of carbon is about

    (a) 2×10^{-23} g   (d) 12.0 g
    (b) 1×10^{-23} g   (e) 6.0 g
    (c) 2×10^{23} g

___ 5. How many moles of O atoms are in two moles of $Fe_2(C_2O_4)_3$?

    (a) 4   (b) 3   (c) 24   (d) 12   (e) 6

___ 6. The molar mass of SO_2 is

    (a) 64.07 g   (d) 44.07 g
    (b) 32.00 g   (e) 48.00 amu
    (c) 64.07 amu

___ 7. How many moles of P atoms are in 62 g of P_4?

    (a) 0.50   (b) 2.0   (c) 4.0   (d) 1.0   (e) 8.0

___ 8. Which of the following statements is false about 1.00 mole of O_2?

    (a) It contains 16.0 g of O atoms.

    (b) It contains 1.20×10^{24} atoms of O.

    (c) It contains 6.02×10^{23} molecules of O_2.

    (d) It contains two moles of O atoms.

    (e) It has a mass of 32.0 g.

A-2 Problems

1. Relative numbers of items

    (a) A farmer wishes to plant a certain field in corn rather than soybeans. An average soybean seed weighs 25.00 mg and an average corn seed weighs 47.50 mg. In the previous year, the farmer used 148 lb of soybeans. How many pounds of corn are needed to provide the same number of seeds as was present in the soybeans?

    (b) What mass of calcium contains the same number of atoms as 54.8 kg of carbon?

    (c) The formula of a compound is K_2O. What mass of potassium is present for each 10.0 lb of oxygen?

2. Find the mass of the following.

    (a) 0.50 mole of K

    (b) 125 moles of Ca

    (c) 3.22×10^{21} atoms of Br

3. Find the number of moles in the following.

    (a) 765 g of Al

    (b) 4.42×10^{18} atoms of B

4. Find the formula weight of the following.

    (a) $HMnO_4$

    (b) $Al_2(SO_3)_3$

    (c) $C_6H_{12}O_6$

5. Calculate the following.

    (a) the number of moles in 45.5 g of CO

    (b) the mass of 15.2 moles of N_2O

    (c) the number of moles in 4.75×10^{23} molecules of SO_2

    (d) the mass of 7.20×10^{24} formula units of K_2O

    (e) the number of molecules in 324 kg of UO_2

Review of Part B *The Component Elements of Compounds*

DETAILED TABLE OF CONTENTS AND OBJECTIVES

5-4 The Composition of Compounds

OBJECTIVE *Given the formula of a compound, determine the mole, mass, and percent composition of its elements.*

5-5   Empirical and Molecular Formulas

OBJECTIVE *Given the percent or mass composition, determine the empirical formula and also the molecular formula if the molar mass is provided.*

   5-5.1 Empirical Formulas
   5-5.2 Calculating Empirical Formulas
   5-5.3 Determining the Molecular Formula of a Compound

SUMMARY OF PART B

To complete the discussion of the quantitative relationships implied by a chemical formula, we now examine the composition of compounds. The moles and mass of a certain element present in a given amount of compound can be calculated using the formula as a conversion factor between moles of a component atom and moles of compound as illustrated in the following example.

Example B-1 Mole and Mass Composition of a Compound

(a) How many moles of aluminum and (b) what mass of chromium are present in 382 g of $Al_2(Cr_2O_7)_3$?

PROCEDURE

Calculate the molar mass, and from this, the number of moles of $Al_2(Cr_2O_7)_3$ present in 382 g of $Al_2(Cr_2O_7)_3$. For part (a), calculate the number of moles of Al present using the conversion factor $\dfrac{2\ \text{mol Al}}{1\ \text{mol } Al_2(Cr_2O_7)_3}$. For part (b), calculate the mass of Cr present using $\dfrac{6\ \text{mol Cr}}{1\ \text{mol } Al_2(Cr_2O_7)_3}$ and $\dfrac{52.00\ \text{g Cr}}{\text{mol Cr}}$.

SOLUTION

$(2 \times 26.98\ \text{amu}) + (6 \times 52.00\ \text{amu}) + (21 \times 16.00\ \text{amu}) = 702.0\ \text{amu}$ (702.0 g/mol)

$$382\ \text{g } Al_2(Cr_2O_7)_3 \times \frac{1\ \text{mol } Al_2(Cr_2O_7)_3}{702.0\ \text{g } Al_2(Cr_2O_7)_3} = 0.544\ \text{mol } Al_2(Cr_2O_7)_3$$

(a) $0.544\ \text{mol } Al_2(Cr_2O_7)_3 \times \dfrac{2\ \text{mol Al}}{1\ \text{mol } Al_2(Cr_2O_7)_3} \cdot = \underline{1.09\ \text{mol Al}}$

(b) $0.544\ \text{mol } Al_2(Cr_2O_7)_3 \times \dfrac{6\ \text{mol Cr}}{1\ \text{mol } Al_2(Cr_2O_7)_3} \times \dfrac{52.00\ \text{g Cr}}{\text{mol Cr}} = \underline{170\ \text{g Cr}}$

The **percent composition** can be calculated directly from the molecular formula. Given the percent composition, however, only the **empirical formula** can be determined. The empirical formula and the molar mass are both needed to determine the **molecular formula**.

The percent composition of a compound is easily found from the formula of a compound.

$$\% \text{ composition} = \frac{\text{total mass of component element}}{\text{molar mass of compound}} \times 100\%$$

The percent composition of the elements in a compound is the same for all compounds with the same empirical formula. The empirical formula of a compound is the simplest whole number ratio of atoms.

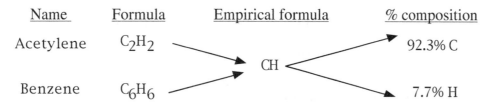

Name	Formula	Empirical formula	% composition
Acetylene	C_2H_2		92.3% C
		CH	
Benzene	C_6H_6		7.7% H

Conversely, if percent composition is given (or the masses of the elements comprising a compound), the empirical formula can be determined. A sample calculation following the procedure outlined in Section 5-4 in the text follows.

Example B-2 Empirical and Molecular Formulas

A component is composed of 56.4% P and 43.6% O. (a) What is the empirical formula of the compound? (b) If the molar mass is 220 g/mol, what is the molecular formula of the compound?

PROCEDURE

In a 100-g sample of the compound, there are 56.4 g of P and 43.6 g of O. The element ratio is the same as the mole ratio, so the next step is to convert the masses of each element to moles.

SOLUTION

(a)   P $56.4 \text{ g P} \times \dfrac{1 \text{ mol P}}{30.97 \text{ g P}} = 1.82 \text{ mol P}$

O $43.6 \text{ g O} \times \dfrac{1 \text{ mol O}}{16.00 \text{ g O}} = 2.72 \text{ mol O}$

The ratios of the moles of the elements are:

P: $\dfrac{1.82}{1.82} = 1.0$ O: $\dfrac{2.72}{2.72} = 1.5$ $PO_{1.5} = P_2O_3$

(b) The mass of one empirical unit is $(2 \times 30.97) + (3 \times 16.00) = 109.9$ g/emp. unit

$$\frac{220 \text{ g/mol}}{109.9 \text{ g/emp. unit}} = 2 \text{ emp. unit/mol}$$

The molecular formula is therefore P_4O_6, which contains two empirical units.

ASSESSMENT OF OBJECTIVES

B-1 Multiple Choice

____ 1. The percent composition of carbon in carbon monoxide is

(a) 50.0%   (b) 27.3%   (c) 42.9%   (d) 60.0%

____ 2. There is 140 g of boron in a 450-g sample of a compound. What is the percent composition of boron in the compound?

(a) 23.7%   (b) 31.1%   (c) 45.2%   (d) 68.9%

____ 3. Which of the following is an empirical formula?

(a) $C_4H_8O_2$ (d) N_2H_4
(b) C_9H_{15} (e) $C_3H_6O_2$
(c) $B_2(OH)_4$

____ 4. The molar mass of a compound with the empirical formula CH_2 is 56 g. The molecular formula is

(a) $C_2N_2H_4$ (d) C_3H_4O
(b) C_4H_8 (e) C_7H_{14}
(c) C_3H_6

B-2 Problems

1. How many moles of each atom are present in 38.0 g of $Ba(ClO_3)_2$?

2. What mass of sulfur is present in 782 g of $K_2S_2O_3$?
   (Calculate by converting to moles of sulfur, then to mass of sulfur.)

3. What is the percent composition of the elements in:

   (a) $NaClO_2$?

   (b) Ammonium carbonate?

4. What is the mass of nitrogen in 195 lb of ammonium carbonate?
   (Calculate using percent composition.)

5. A 5.67-g quantity of chlorine reacts with 8.94 g of oxygen to form a compound.
   What is its empirical formula?

6. A compound is 53.3% C, 11.1% H, and 35.6% O. Its molar mass is 90 g/mol.
   What is its molecular formula?

Chapter Summary Assessment

S-1 Problems

1. Fill in the blanks.

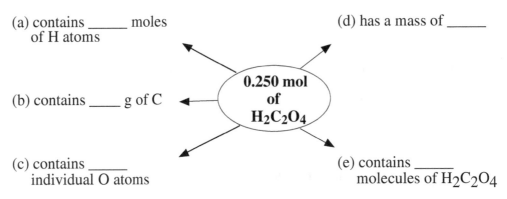

(a) contains _____ moles of H atoms

(b) contains _____ g of C

(c) contains _____ individual O atoms

(d) has a mass of _____

(e) contains _____ molecules of $H_2C_2O_4$

(f) The empirical formula for $H_2C_2O_4$ is _____.

(g) The percent composition of oxygen in $H_2C_2O_4$ _____.

2.   (a) contains _____ moles of O atoms

(b) contains _____ g of C

(c) contains _____ individual Na ions

(d) has a mass of _____

(e) contains _____ formula units of Na_2CO_3

3.   Ethylenediaminetetraacetic acid (EDTA) is a compound that is 41.1% C, 5.5% H, 9.6% N, and 43.8% O. There are two empirical units per molecular unit.

(a) Its empirical formula is _____.

(b) Its molecular formula is _____.

(c) 1.22 mole has a mass of _____.

(d) 5.00×10^{24} molecules have a mass of _____.

Answers to Assessments of Objectives

A-1 Multiple Choice

1. **b**   16.00 g/mol

2. **c**   10 g [0.50 $\cancel{\text{mol}}$ x $\dfrac{20\ \text{g}}{\cancel{\text{mol}}}$ = 10 g]

3. **b**   0.100 mol [6.02 x 10^{22} $\cancel{\text{atoms}}$ x $\dfrac{1\ \text{mol}}{6.02\ \text{x}\ 10^{23}\ \cancel{\text{atoms}}}$ = 0.100 mol]

4. **a**   2 x 10^{-23} g [$\dfrac{12\ \text{g}}{\cancel{\text{mol}}}$ x $\dfrac{1\ \cancel{\text{mol}}}{6.0\ \text{x}\ 10^{23}\ \text{atoms}}$ = 2 x 10^{-23} g/atom]

5. **c**   24 [2 $\cancel{\text{mol Fe}_2(\text{C}_2\text{O}_4)_3}$ x $\dfrac{12\ \text{mol O}}{\cancel{\text{mol Fe}_2(\text{C}_2\text{O}_4)_3}}$ = 24 mol O]

6. **a**   64.07 g [(S) 32.07 g + (O) (2 x 16.00 g)] = 64.07 g]

7. **b**   2.0 [62 $\cancel{\text{g P}_4}$ x $\dfrac{1\ \cancel{\text{mol P}_4}}{123.9\ \cancel{\text{g P}_4}}$ x $\dfrac{4\ \text{mol P atoms}}{1\ \cancel{\text{mol P}_4}}$ = 2.0 mol P atoms]

8. **a**   It contains 16 g of O atoms. (It contains 32.0 g of O_2 molecules and 32.0 g of O atoms.)

A-2 Problems

1.   (a) 148 $\cancel{\text{lb soybean}}$ x $\dfrac{47.50\ \text{lb corn}}{25.00\ \cancel{\text{lb soybean}}}$ = $\underline{281\ \text{lb corn}}$

   (b) 54.8 $\cancel{\text{kg carbon}}$ x $\dfrac{40.08\ \text{kg calcium}}{12.01\ \cancel{\text{kg carbon}}}$ = $\underline{183\ \text{lb calcium}}$

   (c) 10.0 $\cancel{\text{lb oxygen}}$ x $\dfrac{39.10\ \text{lb potassium x 2}}{16.00\ \cancel{\text{lb oxygen}}\ \text{x 1}}$ = $\underline{48.9\ \text{lb oxygen}}$

2.   (a) 0.50 $\cancel{\text{mol K}}$ x $\dfrac{39.10\ \text{g K}}{\cancel{\text{mol K}}}$ = $\underline{20\ \text{g K}}$

   (b) 125 $\cancel{\text{mol Ca}}$ x $\dfrac{40.08\ \text{g}}{\cancel{\text{mol Ca}}}$ = $\underline{5010\ \text{g Ca}}$

   (c) 3.22 x 10^{21} $\cancel{\text{atoms}}$ x $\dfrac{1\ \cancel{\text{mol Br}}}{6.022\ \text{x}\ 10^{23}\ \cancel{\text{atoms}}}$ x $\dfrac{79.90\ \text{g}}{\cancel{\text{mol Br}}}$ = $\underline{0.427\ \text{g}}$

3.   (a) $765 \text{ g Al} \times \dfrac{1 \text{ mol Al}}{26.98 \text{ g Al}} = \underline{28.4 \text{ mol Al}}$

        (b) $4.42 \times 10^{18} \text{ atoms B} \times \dfrac{1 \text{ mol B}}{6.022 \times 10^{23} \text{ atoms B}} = 7.34 \times 10^{-6} \text{ mol B}$

4.   (a) $HMnO_4$ [1.008 (H) + 54.94 (Mn) + (4 $\times$ 16.00) (O)] = $\underline{119.9 \text{ amu}}$

        (b) $Al_2(SO_3)_3$ [(2 $\times$ 26.98)(Al) + (3 $\times$ 32.07)(S) + (9 $\times$ 16.00) (O)] = $\underline{294.2 \text{ amu}}$

        (c) $C_6H_{12}O_6$ [(6 $\times$ 12.01)(C) + (12 $\times$ 1.008)(H) + (6 $\times$ 16.00)(O)] = $\underline{180.2 \text{ amu}}$

5.   (a) $45.5 \text{ g CO} \times \dfrac{1 \text{ mol}}{28.01 \text{ g CO}} = \underline{1.62 \text{ mol}}$

        (b) $15.2 \text{ mol N}_2\text{O} \times \dfrac{44.02 \text{ g}}{\text{mol N}_2\text{O}} = \underline{669 \text{ g}}$

        (c) $4.75 \times 10^{23} \text{ molecules} \times \dfrac{1 \text{ mol}}{6.022 \times 10^{23} \text{ molecules}} = \underline{0.789 \text{ mol}}$

        (d) $7.20 \times 10^{24} \text{ form. units} \times \dfrac{1 \text{ mol K}_2\text{O}}{6.022 \times 10^{23} \text{ form. units}}$

            $\times \dfrac{94.20 \text{ g K}_2\text{O}}{\text{mol K}_2\text{O}} = 1.13 \times 10^3 \text{ g K}_2\text{O}$

        (e) $324 \text{ kg UO}_2 \times \dfrac{1 \text{ g UO}_2}{10^{-3} \text{ kg UO}_2} \times \dfrac{1 \text{ mol UO}_2}{270.0 \text{ g UO}_2}$

            $\times \dfrac{6.022 \times 10^{23} \text{ molecules}}{\text{mol UO}_2} = 7.23 \times 10^{26} \text{ molecules}$

B-1 Multiple Choice

1. **c**   12.0 g/28.0 g $\times$ 100% = 42.9%

2. **b**   140 g/450 g $\times$ 100% = 31.1%

3. **e** $C_3H_6O_2$

4. **b** C_4H_8 [CH_2 = 14 g/emp. unit; $\dfrac{56 \text{ g/mol}}{14 \text{ g/emp. unit}} = 4$ (CH_2) $\times$ 4 = C_4H_8]

B-2 Problems

1. $Ba(ClO_3)_2$ formula weight = $[137.3 + (2 \times 35.45) + (6 \times 16.00)] = 304.2$ g/mol

$$38.0 \text{ g } Ba(ClO_3)_2 \times \frac{1 \text{ mol } Ba(ClO_3)_2}{304.2 \text{ g } Ba(ClO_3)_2} = 0.125 \text{ mol } Ba(ClO_3)_2$$

$$0.125 \text{ mol } Ba(ClO_3)_2 \times \frac{1 \text{ mol } Ba}{\text{mol } Ba(ClO_3)_2} = \underline{0.125 \text{ mol } Ba}$$

$$0.125 \text{ mol } Ba(ClO_3)_2 \times \frac{2 \text{ mol } Cl}{\text{mol } Ba(ClO_3)_2} = \underline{0.250 \text{ mol } Cl}$$

$$0.125 \text{ mol } Ba(ClO_3)_2 \times \frac{6 \text{ mol } O}{\text{mol } Ba(ClO_3)_2} = \underline{0.750 \text{ mol } O}$$

2. $K_2S_2O_3$: formula weight = $[(2 \times 39.10) + (2 \times 32.07) + (3 \times 16.00)]$
$$= 190.3 \text{ g/mol}$$

$$782 \text{ g } K_2S_2O_3 \times \frac{1 \text{ mol } K_2S_2O_3}{190.3 \text{ g } K_2S_2O_3} \times \frac{2 \text{ mol } S}{\text{mol } K_2S_2O_3} \times \frac{32.07 \text{ g } S}{\text{mol } S} = \underline{264 \text{ g } S}$$

3. (a) Na - 25.4%, Cl - 39.2%, O - 35.4%

   (b) $(NH_4)_2CO_3$ N - 29.2%, H - 8.3%, C - 12.5%, O - 50.0%

4. $195 \text{ lb } (NH_4)_2CO_3 \times \dfrac{29.2 \text{ lb } N}{100 \text{ lb } (NH_4)_2CO_3} = \underline{56.9 \text{ lb } N}$

5. Convert the mass of each element to moles.

   Cl $5.67 \text{ g } Cl \times \dfrac{1 \text{ mol } Cl}{35.45 \text{ g } Cl} = 0.160 \text{ mol } Cl$

   O $8.94 \text{ g } O \times \dfrac{1 \text{ mol } O}{16.00 \text{ g } O} = 0.559 \text{ mol } O$

   Cl: $\dfrac{0.160}{0.160} = 1.0$ O: $\dfrac{0.559}{0.160} = 3.5$ $ClO_{3.5} = \underline{Cl_2O_7}$

6. Convert percent to moles. In 100 g of compound there are 53.3 g C, 11.1 g of H, and 35.6 g of O.

$$C \quad 53.3 \; g\text{-}C \times \frac{1 \; mol \; C}{12.01 \; g\text{-}C} = 4.44 \; mol \; C$$

$$H \quad 11.1 \; g\text{-}H \times \frac{1 \; mol \; H}{1.008 \; g\text{-}H} = 11.0 \; mol \; H$$

$$O \quad 35.6 \; g\text{-}O \times \frac{1 \; mol \; O}{16.00 \; g\text{-}O} = 2.22 \; mol \; O$$

$$C: \; \frac{4.44}{2.22} = 2.0; \quad H: \; \frac{11.0}{2.22} = 5.0; \quad O: \; \frac{2.22}{2.22} = 1.0$$

Empirical formula = C_2H_5O Emp. wt. = 45.06 g/emp. unit

$$\frac{90 \; g/mol}{45.06 \; g/emp. \; unit} = 2.0 \; emp. \; unit/mole$$

Molecular formula = $\underline{C_4H_{10}O_2}$

S-1 Problems

1. 0.250 mole $H_2C_2O_4$

(a) $0.250 \; mol \; H_2C_2O_4 \times \dfrac{2 \; mol \; H \; atoms}{mol \; H_2C_2O_4} = \underline{0.500 \; mol \; H \; atom}$

(b) $0.250 \; mol \; H_2C_2O_4 \times \dfrac{2 \; mol \; C \; atoms}{mol \; H_2C_2O_4} \times \dfrac{12.01 \; g \; C}{mol \; C \; atoms} = \underline{6.01 \; g \; C}$

(c) $0.250 \; mol \; H_2C_2O_4 \times \dfrac{4 \; mol \; O \; atoms}{mol \; H_2C_2O_4} \times$

$\dfrac{6.022 \times 10^{23} \; atoms \; O}{mol \; O \; atoms} = \underline{6.02 \times 10^{23} \; atoms \; O}$

(d) $0.250 \; mol \; H_2C_2O_4 \times \dfrac{90.04 \; g}{mol \; H_2C_2O_4} = \underline{22.5 \; g}$

(e) $0.250 \; mol \; H_2C_2O_4 \times \dfrac{6.022 \times 10^{23} \; molecules}{mol \; H_2C_2O_4} = \underline{1.51 \times 10^{23} \; molecules}$

(f) The empirical formula is HCO_2.

(g) 64.0 g/90.0 g x 100% = $\underline{71.1\% \; oxygen}$

2. 8.75×10^{-5} mole Na_2CO_3

(a) $8.75 \times 10^{-5} \; mol \; Na_2CO_3 \times \dfrac{3 \; mol \; O \; atoms}{mol \; Na_2CO_3} = \underline{2.63 \times 10^{-4} \; mol \; O}$

(b) $8.75 \times 10^{-5} \; mol \; Na_2CO_3 \times \dfrac{1 \; mol \; C \; atoms}{1 \; mol \; Na_2CO_3} \times$

$\dfrac{12.01 \; g \; C}{mol \; C \; atoms} = \underline{1.05 \times 10^{-3} \; g}$

88

(c) 8.75×10^{-5} ~~mol Na₂CO₃~~ x $\dfrac{2 \text{ mol Na ions}}{1 \text{ mol Na₂CO₃}}$

x $\dfrac{6.022 \times 10^{23} \text{ Na ions}}{\text{mol Na ions}}$ = 1.05×10^{20} Na ions

(d) 8.75×10^{-5} ~~mol Na₂CO₃~~ x $\dfrac{106.0 \text{ g Na₂CO₃}}{\text{mol Na₂CO₃}}$ = 9.28×10^{-3} g Na₂CO₃

(e) 8.75×10^{-5} ~~mol Na₂CO₃~~ x $\dfrac{6.022 \times 10^{23} \text{ form. units}}{\text{mol Na₂CO₃}}$ =

5.27×10^{19} form. units

3. (a) C: 41.1 ~~g~~ x $\dfrac{1 \text{ mol}}{12.01 \text{ g}}$ = 3.42 mol   H: 5.5 ~~g~~ x $\dfrac{1 \text{ mol}}{1.008 \text{ g}}$ = 5.5 mol

N: 9.6 ~~g~~ x $\dfrac{1 \text{ mol}}{14.01 \text{ g}}$ = 0.69 mol   O: 43.8 ~~g~~ x $\dfrac{1 \text{ mol}}{16.00 \text{ g}}$ = 2.74 mol

C: $\dfrac{3.42}{0.69}$ = 5.0   H: $\dfrac{5.5}{0.69}$ = 8.0   N: $\dfrac{0.69}{0.69}$ = 1.0   O: $\dfrac{2.74}{0.69}$ = 4.0

The empirical formula is $C_5H_8NO_4$.

(b) The molecular formula is $C_{10}H_{16}N_2O_8$.

(c) The molar mass of EDTA is $(10 \times 12.01) + (16 \times 1.008) + (2 \times 14.01)$
$+ (8 \times 16.00) = 292.2$ g

1.22 ~~mol~~ x $\dfrac{292.2 \text{ g}}{\text{mol}}$ = 356 g

(d) 5.00×10^{24} ~~molecules~~ x $\dfrac{1 \text{ mol}}{6.022 \times 10^{23} \text{ molecules}}$

x $\dfrac{292.2 \text{ g}}{\text{mol}}$ = 2.43×10^3 g

Answers and Solutions to Green Text Problems

5-1 12.4 ~~lb nickels~~ x $\dfrac{2.47 \text{ lb pennies}}{5.03 \text{ lb nickels}}$ = <u>6.09 lb of pennies</u>

5-3 145 ~~g Au~~ x $\dfrac{107.9 \text{ g Ag}}{197.0 \text{ g Au}}$ = <u>79.4 g Ag</u>

5-4 212 ~~lb Al~~ x $\dfrac{12.01 \text{ lb C}}{26.98 \text{ lb Al}}$ = <u>94.4 lb C</u>

5-6 $18.0 \text{ g } \cancel{O} \times \dfrac{63.55 \text{ g Cu}}{16.00 \text{ g } \cancel{O}} = \underline{71.5 \text{ g Cu}}$

5-8 $25.0 \text{ g } \cancel{C} \times \dfrac{x \text{ g}}{12.01 \text{ g } \cancel{C}} = 33.3 \text{ g} \qquad x = 16.0 \text{ g (O)} \qquad$ The compound is CO.

5-10 $60.0 \text{ } \cancel{\text{lb O}} \times \dfrac{32.07 \text{ lb S}}{16.00 \text{ } \cancel{\text{lb O}}} \times \dfrac{1}{3} = \underline{40.1 \text{ lb S}}$

5-11 $6.022 \times 10^{23} \text{ } \cancel{\text{units}} \times \dfrac{1 \text{ } \cancel{\text{sec}}}{2 \text{ } \cancel{\text{units}}} \times \dfrac{1 \text{ } \cancel{\text{min}}}{60 \text{ } \cancel{\text{sec}}} \times \dfrac{1 \text{ } \cancel{\text{hr}}}{60 \text{ } \cancel{\text{min}}} \times \dfrac{1 \text{ } \cancel{\text{day}}}{24 \text{ } \cancel{\text{hr}}}$

$\times \dfrac{1 \text{ year}}{365 \text{ } \cancel{\text{day}}} = 9.548 \times 10^{15} \text{ years (9.548 quadrillion)}$

$\dfrac{9.548 \times 10^{15} \text{ years}}{6.8 \times 10^{9}} = 1.4 \times 10^{6} \text{ years (1.4 million)}$

5-13 6.022×10^{26} (if mass in kg); 6.022×10^{20} (if mass in mg)

5-14 (a) P: $14.5 \text{ g } \cancel{P} \times \dfrac{1 \text{ mol P}}{30.97 \text{ g } \cancel{P}} = \underline{0.468 \text{ mol P}}$

$0.468 \text{ } \cancel{\text{mol P}} \times \dfrac{6.022 \times 10^{23} \text{ atoms}}{\cancel{\text{mol P}}} = 2.82 \times 10^{23} \text{ atoms P}$

(b) Rb: $1.75 \text{ } \cancel{\text{mol Rb}} \times \dfrac{85.47 \text{ g Rb}}{\cancel{\text{mol Rb}}} = \underline{150 \text{ g Rb}}$

$1.75 \text{ } \cancel{\text{mol Rb}} \times \dfrac{6.022 \times 10^{23} \text{ atoms}}{\cancel{\text{mol Rb}}} = 1.05 \times 10^{24} \text{ atoms}$

(c) Al: 27.0 g, 1.00 mol, 6.02×10^{23} atoms

(d) $3.01 \times 10^{24} \text{ } \cancel{\text{atoms}} \times \dfrac{1 \text{ mol X}}{6.022 \times 10^{23} \text{ } \cancel{\text{atoms}}} = 5.00 \text{ mol X}$

$\dfrac{363 \text{ g X}}{5.00 \text{ mol X}} = 72.6 \text{ g/mol};$ From the periodic table, $\underline{X = Ge}$

(e) Ti: $1 \text{ } \cancel{\text{atom}} \times \dfrac{1 \text{ mol Ti}}{6.022 \times 10^{23} \text{ } \cancel{\text{atoms}}} = 1.66 \times 10^{-24} \text{ mol}$

$1.66 \times 10^{-24} \text{ } \cancel{\text{mol Ti}} \times \dfrac{47.88 \text{ g Ti}}{\cancel{\text{mol Ti}}} = 7.95 \times 10^{-23} \text{ g Ti}$

5-16 (a) 63.5 g Cu   (b) 16 g S   (c) 40.1 g Ca

5-18 (a) $32.1 \text{ } \cancel{\text{mol S}} \times \dfrac{6.022 \times 10^{23} \text{ atoms}}{\cancel{\text{mol S}}} = 1.93 \times 10^{25} \text{ atoms}$

(b) $32.1 \text{ g S} \times \dfrac{1 \text{ mol S}}{32.07 \text{ g S}} \times \dfrac{6.022 \times 10^{23} \text{ atoms}}{\text{mol S}} = 6.03 \times 10^{23} \text{ atoms}$

(c) $32.0 \text{ g O} \times \dfrac{1 \text{ mol O}}{16.00 \text{ g O}} \times \dfrac{6.022 \times 10^{23} \text{ atoms}}{\text{mol O}} = 1.20 \times 10^{24} \text{ atoms}$

5-20 $50.0 \text{ g Al} \times \dfrac{1 \text{ mol Al}}{26.98 \text{ g Al}} = 1.85 \text{ mol Al}$

$50.0 \text{ g Fe} \times \dfrac{1 \text{ mol Fe}}{55.85 \text{ g Fe}} = 0.895 \text{ mol Fe}$ There are more moles of atoms in 50.0 g of Al.

5-21 $20.0 \text{ g Ni} \times \dfrac{1 \text{ mol Ni}}{58.69 \text{ g Ni}} = 0.341 \text{ mol Ni}$

$2.85 \times 10^{23} \text{ atoms} \times \dfrac{1 \text{ mol Ni}}{6.022 \times 10^{23} \text{ atoms}} = 0.473 \text{ mol Ni}$

The 2.85×10^{23} atoms of Ni contain more atoms than 20.0 g.

5-23 $1.40 \times 10^{21} \text{ atoms} \times \dfrac{1 \text{ mol}}{6.022 \times 10^{23} \text{ atoms}} = 2.32 \times 10^{-3} \text{ mol}$

$0.251 \text{ g}/2.32 \times 10^{-3} \text{ mol} = \underline{108 \text{ g/mol (silver)}}$

5-25 (a) $KClO_2$ $39.10 + 35.45 + (2 \times 16.00) = \underline{106.6 \text{ amu}}$
(b) SO_3 $32.07 + (3 \times 16.00) = \underline{80.07 \text{ amu}}$
(c) N_2O_5 $(2 \times 14.01) + (5 \times 16.00) = \underline{108.0 \text{ amu}}$
(d) H_2SO_4 $(2 \times 1.008) + 32.07 + (4 \times 16.00) = \underline{98.09 \text{ amu}}$
(e) Na_2CO_3 $(2 \times 22.99) + 12.01 + (3 \times 16.00) = \underline{106.0 \text{ amu}}$
(f) CH_3COOH $(C_2H_4O_2)$ $(2 \times 12.01) + (4 \times 1.008) + (2 \times 16.00) = \underline{60.05 \text{ amu}}$
(g) $Fe_2(CrO_4)_3$ $(2 \times 55.85) + (3 \times 52.00) + (12 \times 16.00) = \underline{459.7 \text{ amu}}$

5-27 $Cr_2(SO_4)_3$ $(2 \times 52.00) + (3 \times 32.07) + (12 \times 16.00) = \underline{392.2 \text{ amu}}$

5-29 (a) H_2O: $10.5 \text{ mol } H_2O \times \dfrac{18.01 \text{ g } H_2O}{\text{mol } H_2O} = \underline{189 \text{ g } H_2O}$

$10.5 \text{ mol } H_2O \times \dfrac{6.022 \times 10^{23} \text{ molecules}}{\text{mol } H_2O} = \underline{6.32 \times 10^{24} \text{ molecules}}$

(b) BF_3: $3.01 \times 10^{21} \text{ molecules} \times \dfrac{1 \text{ mol}}{6.022 \times 10^{23} \text{ molecules}} = \underline{5.00 \times 10^{-3} \text{ mol } BF_3}$

$5.00 \times 10^{-3} \text{ mol } BF_3 \times \dfrac{67.81 \text{ g } BF_3}{\text{mol } BF_3} = \underline{0.339 \text{ g } BF_3}$

(c) SO_2: $14.0 \text{ g } SO_2 \times \dfrac{1 \text{ mol } SO_2}{64.07 \text{ g } SO_2} = \underline{0.0219 \text{ mol } SO_2}$

$0.219 \text{ mol } SO_2 \times \dfrac{6.022 \times 10^{23} \text{ molecules}}{\text{mol } SO_2} = \underline{1.32 \times 10^{23} \text{ molecules}}$

(d) K_2SO_4: $1.20 \times 10^{-4} \text{ mol } K_2SO_4 \times \dfrac{174.3 \text{ g } K_2SO_4}{\text{mol } K_2SO_4} = \underline{0.0209 \text{ g } K_2SO_4}$

$$1.20 \times 10^{-4} \text{ mol } K_2SO_4 \times \frac{6.022 \times 10^{23} \text{ formula units}}{\text{mol } K_2SO_4} = \underline{7.23 \times 10^{19} \text{ formula units}}$$

(e) SO_3:

$$4.50 \times 10^{24} \text{ molecules} \times \frac{1 \text{ mol } SO_3}{6.022 \times 10^{23} \text{ molecules}} = \underline{7.47 \text{ mol } SO_3}$$

$$7.47 \text{ mol } SO_3 \times \frac{80.07 \text{ g } SO_3}{\text{mol } SO_3} = \underline{598 \text{ g } SO_3}$$

(f) $N(CH_3)_3$: $0.450 \text{ g } N(CH_3)_3 \times \frac{1 \text{ mol } N(CH_3)_3}{59.11 \text{ g } N(CH_3)_3} = \underline{7.61 \times 10^{-3} \text{ mol}}$

$$7.61 \times 10^{-3} \text{ mol } N(CH_3)_3 \times \frac{6.022 \times 10^{23} \text{ molecules}}{\text{mol } N(CH_3)_3} = \underline{4.58 \times 10^{21} \text{ molecules}}$$

5-31 $21.5 \text{ g}/0.0684 \text{ mol} = \underline{314 \text{ g/mol}}$

5-33 $1.07 \times 10^{24} \text{ molecules} \times \frac{1 \text{ mol}}{6.022 \times 10^{23} \text{ molecules}} = 1.78 \text{ mol} \qquad \frac{287 \text{ g}}{1.78 \text{ mol}} = \underline{161 \text{ g/mol}}$

5-35 C: $2.55 \text{ mol } C_2H_6O \times \frac{2 \text{ mol C}}{\text{mol } C_2H_6O} = 5.10 \text{ mol C}$

H: 15.3 mol; O: 2.55 mol   Total = 23.0 mol of atoms

C: $5.10 \text{ mol C} \times \frac{12.01 \text{ g C}}{\text{mol C}} = 61.3 \text{ g C}$

H: 15.4 g; O: 40.8 g   Total mass = $\underline{117.5 \text{ g}}$

5-36 $28.0 \text{ g } Ca(ClO_3)_2 \times \frac{1 \text{ mol}}{207.0 \text{ g } Ca(ClO_3)_2} = 0.135 \text{ mol } Ca(ClO_3)_2$

Ca: $0.135 \text{ mol } Ca(ClO_3)_2 \times \frac{1 \text{ mol Ca}}{\text{mol } Ca(ClO_3)_2} = 0.135 \text{ mol Ca}$

Cl: 0.270 mol Cl   O: 0.810 mol O   Total = $\underline{1.215 \text{ mol of atoms}}$

5-38 $1.50 \text{ mol } H_2SO_3 \times \frac{2 \text{ mol H}}{\text{mol } H_2SO_3} \times \frac{1.008 \text{ g H}}{\text{mol H}} = \underline{3.02 \text{ g H}}$

$1.50 \text{ mol } H_2SO_3 \times \frac{1 \text{ mol S}}{\text{mol } H_2SO_3} \times \frac{32.07 \text{ g S}}{\text{mol S}} = \underline{48.1 \text{ g S}}$

$1.50 \text{ mol } H_2SO_3 \times \frac{3 \text{ mol O}}{\text{mol } H_2SO_3} \times \frac{16.00 \text{ g O}}{\text{mol O}} = \underline{72.0 \text{ g O}}$

5-40 $1.20 \times 10^{22} \text{ molecules} \times \frac{1 \text{ mol } O_2}{6.022 \times 10^{23} \text{ molecules}} = \underline{0.0199 \text{ mol } O_2}$

$0.0199 \text{ mol } O_2 \times \frac{2 \text{ mol O atoms}}{\text{mol } O_2} = 0.0398 \text{ mol O atoms}$

$0.0199 \text{ mol } O_2 \times \frac{32.00 \text{ g } O_2}{\text{mol } O_2} = 0.637 \text{ g } O_2$ $\underline{\text{The mass is the same.}}$

5-42 Total mass of compound = 1.375 + 3.935 = 5.310 g

N: $\frac{1.375 \text{ g}}{5.310 \text{ g}}$ x 100% = <u>25.89% N</u>　　　　O: $\frac{3.935 \text{ g}}{5.310 \text{ g}}$ x 100% = <u>74.11% O</u>

5-43 Si: $\frac{2.27 \text{ g}}{4.86 \text{ g}}$ x 100% = <u>46.7% Si</u>　　　　O: 100% - 46.7% = <u>53.3%O</u>

5-45 (a) C_2H_6O Formula weight = (2 x 12.01) + (6 x 1.008) + 16.00 = 46.07 amu

C: $\frac{24.02 \text{ amu}}{46.07 \text{ amu}}$ x 100% = <u>52.14% C</u>　　　　H: $\frac{6.048 \text{ amu}}{46.07 \text{ amu}}$ x 100% = <u>13.13% H</u>

O: 100.00% - (52.14 + 13.13)% = <u>34.73% O</u>

(b) C_3H_6 Formula weight = (3 x 12.01) + (6 x 1.008) = 42.08 amu

C: $\frac{36.03 \text{ amu}}{42.08 \text{ amu}}$ x 100% = <u>85.62% C</u>　　　　H: 100.00% - 85.62% = <u>14.38% H</u>

(c) C_9H_{18} Formula weight =(9 x 12.01) + (18 x 1.008)= 126.2 amu

C: $\frac{108.1 \text{ amu}}{126.2 \text{ amu}}$ x 100% = <u>85.66% C</u>　　　　H = 100% - 85.66% = <u>14.34% H</u>

(b) and (c) are actually the same. The difference comes from rounding off.

(d) Na_2SO_4 Formula weight = (2 x 22.99) + 32.07 + (4 x 16.00) = 142.1 amu

Na: $\frac{45.98 \text{ amu}}{142.1 \text{ amu}}$ x 100% = <u>32.36% Na</u>　　　　S: $\frac{32.07 \text{ amu}}{142.1 \text{ amu}}$ x 100% = <u>22.57% S</u>

O: 100.00% - (32.36 + 22.57)% = <u>45.07% O</u>

(e) $(NH_4)_2CO_3$ Formula weight = (2 x 14.01) + (8 x 1.008) + 12.01 + (3 x 16.00)= 96.09 amu

N: $\frac{28.02 \text{ amu}}{96.09 \text{ amu}}$ x 100% = <u>29.16% N</u>　　　　H: $\frac{8.064 \text{ amu}}{96.09 \text{ amu}}$ x 100% = <u>8.392% H</u>

C: $\frac{12.01 \text{ amu}}{96.09 \text{ amu}}$ x 100% = <u>12.50%C</u>　　　　O: $\frac{48.00 \text{ amu}}{96.09 \text{ amu}}$ x 100% = <u>49.95% O</u>

5-47 $Na_2B_4O_7$ $10H_2O$ (or $Na_2B_4O_{17}H_{20}$) Formula weight = (2 x 22.99) + (4 x 10.81) + (17 x 16.00) + (20 x 1.008) = 381.4 amu

Na: $\frac{45.98 \text{ amu}}{381.4 \text{ amu}}$ x 100% = <u>12.06% Na</u>　　　　B: $\frac{43.24 \text{ amu}}{381.4 \text{ amu}}$ x 100% = <u>11.34% B</u>

O: $\frac{272.0 \text{ amu}}{381.4 \text{ amu}}$ x 100% = <u>71.31% O</u>　　　　H: $\frac{20.16 \text{ amu}}{381.4 \text{ amu}}$ x 100% = <u>5.286% H</u>

5-49 $C_7H_5SNO_3$ Formula weight = (7 x 12.01) + (5 x 1.008)+ 32.07+ 14.01 + (3 x 16.00) = 183.2 amu

C: $\frac{84.07 \text{ amu}}{183.2 \text{ amu}}$ x 100% = <u>45.89% C</u>　　　　H: $\frac{5.040 \text{ amu}}{183.2 \text{ amu}}$ x 100% = <u>2.751% H</u>

S: $\frac{32.07 \text{ amu}}{183.2 \text{ amu}}$ x 100% = <u>17.51% S</u>　　　　N: $\frac{14.01 \text{ amu}}{183.2 \text{ amu}}$ x 100% = <u>7.647% N</u>

O: 100% - (45.89 + 2.751 + 17.51 + 7.647) = <u>26.20% O</u>

5-51 $Na_2C_2O_4$ Formula weight = (2 x 22.99) + (2 x 12.01) + (4 x 16.00) = 134.0 amu

$$125 \text{ g } Na_2C_2O_4 \times \frac{1 \text{ mol } Na_2C_2O_4}{134.0 \text{ g } Na_2C_2O_4} \times \frac{2 \text{ mol C}}{\text{mol } Na_2C_2O_4} \times \frac{12.01 \text{ g C}}{\text{mol C}} = \underline{22.4 \text{ g C}}$$

5-52 Na_3PO_4 Formula weight = (3 x 22.99) + 30.97 + (4 x 16.00) = 163.9 amu
There is 30.97 lb of P in 163.9 lb of compound.

$$25.0 \text{ lb } Na_3PO_4 \times \frac{30.97 \text{ lb P}}{163.9 \text{ lb } Na_3PO_4} = \underline{4.72 \text{ lb P}}$$

5-54 Fe_2O_3 Formula weight = (2 x 55.85) + (3 x 16.00) = 159.7 amu
There is 111.7 lb of Fe (2 x 55.85) in 159.7 lb of Fe_2O_3.

$$2000 \text{ lb } Fe_2O_3 \times \frac{111.7 \text{ lb Fe}}{159.7 \text{ lb } Fe_2O_3} = \underline{1.40 \times 10^3 \text{ lb Fe}}$$

5-55   (a) N_2O_4 and (d) $H_2C_2O_4$

5-56   (a) FeS   (b) SrI_2 (c) $KClO_3$ (d) I_2O_5

(e) $Fe_2O_{2.66} = Fe_{6/3}O_{8/3} = Fe_6O_8 = Fe_3O_4$

(f) C: $\frac{4.22}{4.22} = 1.0$ H: $\frac{7.03}{4.22} = 1.66$ Cl: $\frac{4.22}{4.22} = 1.0$ $CH_{1.66}Cl = CH_{5/3}Cl = \underline{C_3H_5Cl_3}$

5-58   Assume 100 g of compound. There are then 63.2 g of O and 36.8 g of N per 100 g.

$$63.2 \text{ g O} \times \frac{1 \text{ mol O}}{16.00 \text{ g O}} = 3.95 \text{ mol O}$$

$$36.8 \text{ g N} \times \frac{1 \text{ mol N}}{14.01 \text{ g N}} = 2.63 \text{ mol N}$$

O: 3.95/2.63 = 1.5   N: 2.63/2.63 = 1.0 $NO_{1.5} = \underline{N_2O_3}$

5-60 $8.25 \text{ g K} \times \frac{1 \text{ mol K}}{39.10 \text{ g K}} = 0.211 \text{ mol K}$

$6.75 \text{ g O} \times \frac{1 \text{ mol O}}{16.00 \text{ g O}} = 0.422 \text{ mol O}$ $\frac{0.211}{0.211} = 1.0$ $\frac{0.422}{0.211} = 2.0$ $\underline{KO_2}$

5-62   In 100 g of compound there are 21.6 g of Mg, 21.4 g of C, and 57.0 g of O.

$21.6 \text{ g Mg} \times \frac{1 \text{ mol Mg}}{24.31 \text{ g Mg}} = 0.889 \text{ mol Mg}$ $21.4 \text{ g C} \times \frac{1 \text{ mol C}}{12.01 \text{ g C}} = 1.78 \text{ mol C}$

$57.0 \text{ g O} \times \frac{1 \text{ mol O}}{16.00 \text{ g O}} = 3.56 \text{ mol O}$

Mg: $\frac{0.889}{0.889} = 1.0$ C: $\frac{1.78}{0.889} = 2.0$ O: $\frac{3.56}{0.889} = 4.0$ $\underline{MgC_2O_4}$

5-63   C: $9.90 \text{ g C} \times \frac{1 \text{ mol C}}{12.01 \text{ g C}} = 0.824 \text{ mol C}$ H: $1.65 \text{ g H} \times \frac{1 \text{ mol H}}{1.008 \text{ g H}} = 1.64 \text{ mol H}$

Cl: $29.3 \text{ g Cl} \times \frac{1 \text{ mol Cl}}{35.45 \text{ g Cl}} = 0.827 \text{ mol Cl}$

C: $\frac{0.824}{0.824} = 1.0$ H: $\frac{1.64}{0.824} = 2.0$ Cl: $\frac{0.827}{0.824} = 1.0$ $\underline{CH_2Cl}$

5-65 $\% \text{ O} = 100\% - (24.1 + 6.90 + 27.6)\% = 41.4\% \text{ O}$

N: $24.1 \text{ g N} \times \dfrac{1 \text{ mol N}}{14.01 \text{ g N}} = 1.72 \text{ mol N}$ H: $6.90 \text{ g H} \times \dfrac{1 \text{ mol H}}{1.008 \text{ g H}} = 6.85 \text{ mol H}$

S: $27.6 \text{ g S} \times \dfrac{1 \text{ mol S}}{32.07 \text{ g S}} = 0.861 \text{ mol S}$ O: $41.4 \text{ g O} \times \dfrac{1 \text{ mol O}}{16.00 \text{ g O}} = 2.59 \text{ mol O}$

N: $\dfrac{1.72}{0.861} = 2.0$ H: $\dfrac{6.85}{0.861} = 8.0$ S: $\dfrac{0.861}{0.861} = 1.0$ O: $\dfrac{2.59}{0.861} = 3.0$

$\underline{N_2H_8SO_3}$

5-66   C: $63.2 \text{ g C} \times \dfrac{1 \text{ mol C}}{12.01 \text{ g C}} = 5.26 \text{ mol C}$ O: $31.6 \text{ g O} \times \dfrac{1 \text{ mol O}}{16.00 \text{ g O}} = 1.98 \text{ mol O}$

H: $5.26 \text{ g H} \times \dfrac{1 \text{ mol H}}{1.008 \text{ g H}} = 5.22 \text{ mol H}$

C: $\dfrac{5.26}{1.98} = 2.66 \left(\dfrac{8}{3}\right)$ O: $\dfrac{1.98}{1.98} = 1.0$ H: $\dfrac{5.22}{1.98} = 2.63 \left(\dfrac{8}{3}\right)$

$C_{8/3}H_{8/3}O = \underline{C_8H_8O_3}$

5-68   In 100 g of compound: 20.0 g of C, 2.2 g of H, and 77.8 g of Cl.

$20.0 \text{ g C} \times \dfrac{1 \text{ mol C}}{12.01 \text{ g C}} = 1.66 \text{ mol C}$ $2.2 \text{ g H} \times \dfrac{1 \text{ mol H}}{1.008 \text{ g H}} = 2.18 \text{ mol H}$

$77.8 \text{ g Cl} \times \dfrac{1 \text{ mol Cl}}{35.45 \text{ g Cl}} = 2.19 \text{ mol Cl}$

C: $\dfrac{1.66}{1.66} = 1.0$ H: $\dfrac{2.18}{1.66} = 1.31 \left(\dfrac{4}{3}\right)$ Cl: $\dfrac{2.19}{1.66} = 1.32 \left(\dfrac{4}{3}\right)$

$CH_{4/3}Cl_{4/3} = C_3H_4Cl_4$ (empirical formula)

empirical mass $= (3 \times 12.01) + (4 \times 1.0) + (4 \times 35.5) = 182 \text{ amu} = 182 \text{ g/emp. unit}$

$\dfrac{545 \text{ g/mol}}{182 \text{ g/emp. unit}} = 3 \text{ emp. unit/mol}$ $C_{(3 \times 3)}H_{(3 \times 4)}Cl_{(3 \times 4)} = \underline{C_9H_{12}Cl_{12}}$

5-70   B: $18.7 \text{ g B} \times \dfrac{1 \text{ mol B}}{10.81 \text{ g B}} = 1.73 \text{ mol B}$ C: $20.7 \text{ g C} \times \dfrac{1 \text{ mol C}}{12.01 \text{ g C}} = 1.72 \text{ mol C}$

H: $5.15 \text{ g H} \times \dfrac{1 \text{ mol H}}{1.008 \text{ g H}} = 5.11 \text{ mol H}$ O: $55.4 \text{ g O} \times \dfrac{1 \text{ mol O}}{16.00 \text{ g O}} = 3.46 \text{ mol O}$

B: $\dfrac{1.73}{1.72} = 1.0$ C: $\dfrac{1.72}{1.72} = 1.0$ H: $\dfrac{5.11}{1.72} = 3.0$ O: $\dfrac{3.46}{1.72} = 2.0$

Emp. formula $= BCH_3O_2$ Emp. mass $= 57.84 \text{ g/emp. unit}$

$\dfrac{115 \text{ g/mol}}{57.84 \text{ g/emp. unit}} = 2 \text{ emp. units/mol}$ $\underline{B_2C_2H_6O_4}$ (molecular formula)

5-71 $34.9 \text{ g K} \times \dfrac{1 \text{ mol K}}{39.10 \text{ g K}} = 0.893 \text{ mol K}$ $21.4 \text{ g C} \times \dfrac{1 \text{ mol C}}{12.01 \text{ g C}} = 1.78 \text{ mol C}$

$12.5 \text{ g N} \times \dfrac{1 \text{ mol N}}{14.01 \text{ g N}} = 0.892 \text{ mol N}$ $2.68 \text{ g H} \times \dfrac{1 \text{ mol H}}{1.008 \text{ g H}} = 2.66 \text{ mol H}$

$28.6 \text{ g O} \times \dfrac{1 \text{ mol O}}{16.00 \text{ g O}} = 1.79 \text{ mol O}$

K: $\dfrac{0.893}{0.892} = 1.0$ C: $\dfrac{1.78}{0.892} = 2.0$ N: $\dfrac{0.892}{0.892} = 1.0$ H: $\dfrac{2.66}{0.892} = 3.0$ O: $\dfrac{1.79}{0.892} = 2.0$

Empirical formula = $KC_2NH_3O_2$ Empirical mass = 112.2 g/ emp. unit

$\dfrac{224 \text{ g/mol}}{112.2 \text{ g/emp. unit}} = 2 \text{ emp. units/mol}$ $K_2C_4N_2H_6O_4$ (molecular formula)

5-73 In a 20.0-g sample, there are 18.3 g of I and 1.7 g of C.

$18.3 \text{ g I} \times \dfrac{1 \text{ mol I}}{126.9 \text{ g I}} = 0.144 \text{ mol I}$ $1.7 \text{ g C} \times \dfrac{1 \text{ mol C}}{12.01 \text{ g C}} = 0.14 \text{ mol C}$

Emp. formula = IC   Emp. mass= 138.9 g/emp. unit

$\dfrac{834 \text{ g/mol}}{138.9 \text{ g/emp. unit}} = 6 \text{ emp. units/mol}$ $\underline{I_6C_6}$ (molecular formula)

5-75 $\$4.5 \times 10^{12} \times \dfrac{100 \text{ pennies}}{\$} \times \dfrac{1 \text{ mol pennies}}{6.022 \times 10^{23} \text{ pennies}} = \underline{7.5 \times 10^{-10} \text{ mol pennies}}$

5-76 $0.443 \text{ g N} \times \dfrac{1 \text{ mol N}}{14.01 \text{ g N}} = 0.0316 \text{ mol N}$

1.420 g of M also equals 0.0316 mol M since M and N are present in equimolar amounts.

1.420 g /0.0316 mol = $\underline{44.9 \text{ g/mol [scandium (Sc)]}}$

5-79 $0.344 \text{ g P}_4 \times \dfrac{1 \text{ mol P}_4}{123.9 \text{ g P}_4} = 2.78 \times 10^{-3} \text{ mol P}_4;$

$2.78 \times 10^{-3} \text{ mol P}_4 \times \dfrac{4 \text{ mol P}}{\text{mol P}_4} \times \dfrac{6.022 \times 10^{23} \text{ atoms P}}{\text{mol P}} = \underline{6.70 \times 10^{21} \text{ atoms P}}$

5-80 100 mol H_2 = 202 g H_2, therefore

100 H atoms < 100 H_2 molecules < 100 g H_2 < 100 mol H_2

5-82 $2.84 \times 10^{23} \text{ form units} \times \dfrac{1 \text{ mol}}{6.022 \times 10^{23} \text{ form units}} = 0.472 \text{ mol}$

$\dfrac{56.6 \text{ g}}{0.472 \text{ mol}} = 120 \text{ g/mol}$ 120 - 55.8 = 64 g of S $\dfrac{64 \text{ g S}}{32.07 \text{ g S/mol}} = 2 \text{ mol S}$

Formula = $\underline{FeS_2}$

5-83 (a) $2Na^+$ and $S_4O_6^{2-}$

(b) $10.0 \text{ g Na} \times \dfrac{1 \text{ mol Na}}{22.99 \text{ g Na}} \times \dfrac{4 \text{ mol S}}{2 \text{ mol Na}} \times \dfrac{32.07 \text{ g S}}{\text{mol S}} = \underline{27.9 \text{ g S}}$

(c) NaS_2O_3 (d) 270.3 g/mol

(e) $25.0 \text{ g Na}_2S_4O_6 \times \dfrac{1 \text{ mol Na}_2S_4O_6}{270.3 \text{ g Na}_2S_4O_6} = 0.0925 \text{ mol Na}_2S_4O_6$

$0.0925 \text{ mol Na}_2S_4O_6 \times \dfrac{6.022 \times 10^{23} \text{ form units}}{\text{mol Na}_2S_4O_6} = \underline{5.57 \times 10^{22} \text{ formula units}}$

(f) $\dfrac{(6 \times 16.00) \text{ g O}}{270.3 \text{ g compound}} \times 100\% = 35.5 \text{ \% oxygen}$

5-85 $\dfrac{2N}{2N + xO} = 0.368$ $\dfrac{28.02}{28.02 + 16.00 x} = 0.368$ $x = 3$ $\underline{N_2O_3}$ dinitrogen trioxide

5-87 $4.55 \times 10^{22} \text{ molecules} \times \dfrac{1 \text{ mol dioxin}}{6.022 \times 10^{23} \text{ molecules}} = 0.0756 \text{ mol dioxin}$

$\dfrac{24.3 \text{ g}}{0.0756 \text{ mol}} = 321 \text{ g/mol}$ C: $\dfrac{0.456}{0.076} = 6.0$ H & Cl: $\dfrac{0.152}{0.076} = 2.0$ O: $\dfrac{0.076}{0.076} = 1.0$

empirical formula = $C_6H_2Cl_2O$

emp. mass= 161 g/emp unit $\dfrac{321 \text{ g/mol}}{161.0 \text{ g/emp unit}} = 2$ $\underline{C_{12}H_4Cl_4O_2}$ (molecular formula)

5-89 Cr: $14.9 \text{ g Cr} \times \dfrac{1 \text{ mol Cr}}{52.00 \text{ g Cr}} = 0.287 \text{ mol Cr}$

Cl: $30.4 \text{ g Cl} \times \dfrac{1 \text{ mol Cl}}{35.45 \text{ g Cl}} = 0.858 \text{ mol Cl}$ O: $54.7 \text{ g O} \times \dfrac{1 \text{ mol}}{16.00 \text{ g O}} = 3.42 \text{ mol O}$

Cr: $\dfrac{0.287}{0.287} = 1.0$ Cl: $\dfrac{0.858}{0.287} = 3.0$ O: $\dfrac{3.42}{0.287} = 12.0$

Empirical formula $CrCl_3O_{12}$ Actual formula = $\underline{Cr(ClO_4)_3}$ chromium(III) perchlorate

5-90 Assume exactly 100 g of compound. There are then 51.1 g H_2O and 48.9 g $MgSO_4$.

H_2O: $51.1 \text{ g H}_2O \times \dfrac{1 \text{ mol H}_2O}{18.02 \text{ g H}_2O} = 2.84 \text{ mol H}_2O$

$MgSO_4$: $48.9 \text{ g MgSO}_4 \times \dfrac{1 \text{ mol}}{120.4 \text{ g MgSO}_4} = 0.406 \text{ mol MgSO}_4$

2.94 mol H_2O/0.406 mol $MgSO_4$ = 7.0 mol H_2O/mol $MgSO_4$

The formula is $\underline{MgSO_4 \cdot 7H_2O}$.

5-93 $1.20 \text{ g CO}_2 \times \dfrac{1 \text{ mol CO}_2}{44.01 \text{ g CO}_2} \times \dfrac{1 \text{ mol C}}{\text{mol CO}_2} = 0.0273 \text{ mol C}$

$0.489 \text{ g H}_2\text{O} \times \dfrac{1 \text{ mol H}_2\text{O}}{18.02 \text{ g H}_2\text{O}} \times \dfrac{2 \text{ mol H}}{\text{mol H}_2\text{O}} = 0.0543 \text{ mol H}$

C: $\dfrac{0.0273}{0.0273} = 1.0$ H: $\dfrac{0.0543}{0.0273} = 2.0$ $\underline{\text{CH}_2}$

6

Chemical Reactions

Review of Part A *The Representation of Chemical Changes and Three Types of Changes*

OBJECTIVES AND DETAILED TABLE OF CONTENTS

6-1 Chemical Equations

OBJECTIVE *Write balanced chemical equations for simple reactions from inspection.*

6-1.1 Constructing an Equation
6-1.2 Rules for Balancing Equations

6-2   Combustion, Combination, and Decomposition Reactions

OBJECTIVE *Classify certain chemical reactions as being combustion, combination, or decomposition reactions.*

> 6-2.1 Combustion Reactions
> 6-2.2 Combination Reactions
> 6-2.3 Decomposition Reactions

SUMMARY OF PART A

Learning chemistry is much like learning a foreign language. The letters of the language are the symbols of the elements, and the letters combine into words that are the formulas of compounds. Finally, we see in this chapter, that the words combine into sentences, which are the chemical equations. **Chemical equations** represent not only what chemical changes occur, but *how much*. Before a chemical equation can tell how much, it must follow the law of conservation of mass. This means that the equation must be balanced with the same number of atoms on both sides of the equation. This is accomplished by appropriate **coefficients** placed before specific compounds in the equation. The implications of a typical **balanced equation** are illustrated as follows:

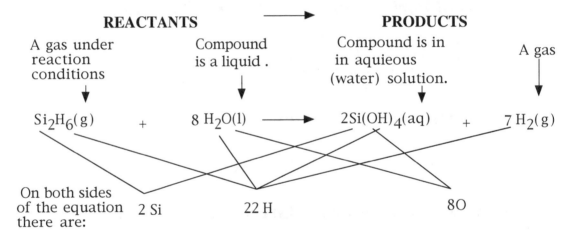

Consider how the above equation was balanced. Elements other than hydrogen or oxygen should be balanced first. Thus silicon is the first element balanced in the above equation by placing a coefficient of "2" before the $Si(OH)_4$. Hydrogen or oxygen is balanced next. Since oxygen appears in the fewer number of compounds and hydrogen is present as an element, we next place an "8" in front of the H_2O to balance the eight oxygens in $2Si(OH)_4$. We now have 22 hydrogens on the left and eight hydrogens on the right in the $2Si(OH)_4$. By placing a coefficent of "7" in front of the H_2, we now have 22 hydrogens on the right. The equation is balanced.

There are several ways to categorize or classify chemical reactions. One simple way that includes a great many chemical reactions is described in this chapter. The five types of reactions described in this chapter do not include all chemical reactions; in later chapters, we will find there are other convenient classifications of chemical reactions that suit certain purposes. In any case, three of the five types and an example of each are as follows:

1. **Combustion reactions**

   Ex. $2C_2H_2(g) + 5O_2(g) \longrightarrow 4CO_2(g) + 2H_2O(l)$

2. **Combination reactions** $(A + B \longrightarrow C)$

   Ex. $LiOH(aq) + CO_2(g) \longrightarrow LiHCO_3(aq)$

3. **Decomposition reactions** $(C \longrightarrow A + B)$

   Ex. $2Au_2O_3(s) \longrightarrow 4Au(s) + 3O_2(g)$

We can recognize these three types quite easily. Oxygen is involved as a reactant in any combustion reaction. A combination reaction results in one product and a decomposition reaction is just the opposite. That is, it has only one reactant.

ASSESSMENT OF OBJECTIVES

A-1 Balancing Equations

1. Balance the following:

   (a) $UO_2 + HF \longrightarrow UF_4 + H_2O$

   (b) $HCl + MgCO_3 \longrightarrow MgCl_2 + CO_2 + H_2O$

   (c) $NaNO_3 \longrightarrow NaNO_2 + O_2$

   (d) $Cu + HNO_3 \longrightarrow Cu(NO_3)_2 + NO_2 + H_2O$

   (e) $Si_2H_6 + O_2 \longrightarrow SiO_2 + H_2O$

   (f) $P_4O_{10} + H_2O \longrightarrow H_3PO_4$

   (g) $UO_2 + CaH_2 \longrightarrow U + CaO + H_2$

2. Change the following into formulas, including all information and balance. Use the following symbols: solid (s), liquid (l), gas (g), aqueous solution (aq), yields or produces

   $(\longrightarrow)$, and heat $(\overset{\triangle}{\longrightarrow})$.

   (a) Liquid alcohol (C_2H_6O) undergoes combustion to produce gaseous carbon dioxide and water.

   (b) Solid iron(III) hydroxide decomposes to iron(III) oxide and water when heated.

(c) Sulfur (S_8) combines with fluorine to produce sulfur hexafluoride gas.

Review of Part B *Ions in Water and How They React*

OBJECTIVES AND DETAILED TABLE OF CONTENTS

6-3   The Formation of Ions in Water

OBJECTIVE *Write the ions formed when ionic compounds or acids dissolve in water.*

6-4   Single-Replacement Reactions

OBJECTIVE *Given the activity series, complete several single-replacement reactions as balanced molecular, total ionic, and net ionic equations.*

6-5   Double-Replacement Reactions-Precipitation

OBJECTIVE *Write balanced molecular, total ionic, and net ionic equations for precipitation reactions.*

6-6   Double-Replacement Reactions-Neutralization

OBJECTIVE *Write balanced molecular, total ionic, and net ionic equations for neutralization reactions.*

SUMMARY OF PART B

When ionic compounds dissolve in water, the water molecules disperse the ions into a homogeneous mixture known as a **solution**. The solution of a typical ionic compound that is **soluble** in water is represented as follows:

$$Na_2CrO_4(s) \xrightarrow{H_2O} 2Na^+(aq) + CrO_4{}^{2-}(aq)$$

Strong acids, although molecular compounds in the pure state, are also ionized by water to produce ions. The solution of a typical strong acid in water is represented as follows:

$$HClO_4(aq) \xrightarrow{H_2O} H^+(aq) + ClO_4{}^-(aq)$$

Compounds with low solubilities in water are labeled as **insoluble**. Regardless of whether a compound is considered soluble or insoluble, there is a limit to how much of any compound dissolves in water. At a certain temperature the limit is known as the compound's **solubility**.

Ions are involved in many chemical reactions that occur in aqueous solution. In this chapter we focused on two types of reactions: single-replacement and double-replacement reactions.

Single replacement reactions $(A + BC \longrightarrow AC + B)$

In ancient times, it was suggested that lead could be changed into gold by immersing a lead bar in an aqueous solution containing gold ions. Actually, what was happening was a single-replacement reaction involving the replacement of Au^{3+} ions for Pb^{2+} ions and lead atoms for gold atoms. The reaction, however, only produced a coating of gold and did not change the whole bar into gold. This reaction is illustrated by the following **molecular equation**:

$$3Pb(s) + 2Au(NO_3)_3(aq) \longrightarrow 3Pb(NO_3)_2(aq) + 2Au(s)$$

Reactions can also be illustrated by equations showing the neutral ionic compounds as separate cations and anions in solution. This is known as the **total ionic equation** and is illustrated as follows:

$$3Pb(s) + 2Au^{3+}(aq) + 6NO_3{}^-(aq) \longrightarrow 3Pb^{2+}(aq) + 6NO_3{}^-(aq) + 2Au(s)$$

The previous equation can be simplified by subtracting out of the equation the ions that are identical on both sides of the equation. These are known as the **spectator ions** and the resulting equation is known as the **net ionic equation**. The net ionic equation allows us to focus on the species that actually changed in the reaction. By subtracting $6NO_3{}^-$ ions from both sides of the equation we have:

$$3Pb(s) + 2Au^{3+}(aq) \longrightarrow 3Pb^{2+}(aq) + 2Au(s)$$

Metals have different abilities to replace the ions of other metals from solutions. With simple experiments, an order can be established that is known as the **activity series**. (See Table 6-2 in the text.) A specific metal will spontaneously react with all of the metal ions underneath that metal in the table. (Notice that Pb is higher than Au^{3+}, so the reaction shown above is spontaneous.) In this table, H_2 is treated as a metal and H^+ (from strong acids) as its metal ion.

103

Double replacement reactions $(AB + CD \longrightarrow AD + CB)$

Another type of reaction involves two ionic compounds in water. If solutions of two soluble compounds are mixed, a chemical reaction may occur if the cation from one solution forms an insoluble compound with the anion from the other solution. Table 6-3 in the text can be used to determine whether some familiar ionic compounds are insoluble. When the solutions are mixed, the ions join together to form the solid state known as a **precipitate**. Such reactions are thus known as **precipitation reactions** and are one type of double-replacement reaction. An example of a precipitation reaction follows.

Example B-1 A Precipitation Reaction Illustrated With Equations

Write the balanced molecular, total ionic, and net ionic equations when solutions of sodium chloride and silver nitrate are mixed.

PROCEDURE

If the reactants are NaCl and $AgNO_3$, an exchange of ions produces possible products AgCl and $NaNO_3$. From Table 6-3 in the text, we note that AgCl is insoluble.

Solution #1   Solution #2   Solution #3 (#1 + #2)

     +   *produces*

[NaCl(aq)]   [$AgNO_3$(aq)]   [AgCl(s) + $NaNO_3$(aq)]

This chemical reaction can be illustrated by equations written in three different forms.

SOLUTION

Molecular equation: NaCl(aq) + $AgNO_3$(aq) $\longrightarrow$ AgCl(s) + $NaNO_3$(aq)

Total ionic equation: Na^+(aq) + Cl^-(aq) + Ag^+(aq) + NO_3^-(aq) $\longrightarrow$ AgCl(s)
                                     + Na^+(aq) + NO_3^-(aq)

Net ionic equation: Ag^+(aq) + Cl^-(aq) $\longrightarrow$ AgCl(s)

In another example, consider what happens when solutions of $(NH_4)_2S$ and $Pb(NO_3)_2$ are mixed. The compounds formed by an exchange of ions are NH_4NO_3 and PbS. Table 6-3 tells us that all nitrates are soluble but most sulfides (including Pb^{2+}) are insoluble. Therefore, the following reaction occurs:

(soluble)	(soluble)	(insoluble)	(soluble)

$$(NH_4)_2S(aq) \ + \ Pb(NO_3)_2(aq) \ \longrightarrow \ PbS(s) \ + \ 2NH_4NO_3(aq)$$

A second type of double-replacement reaction is possible where one of the products is a molecular compound rather than an insoluble ionic compound. **Neutralization reactions** are examples of this latter type of reaction. As mentioned earlier, strong acids are completely ionized by water to form $H^+(aq)$ ions. Another unique type of compound in water is known as a base. Bases obtain their unique character from the production of hydroxide (OH^-) ions in aqueous solution. The common **strong bases** (e.g., NaOH), which are ionic solids when pure, dissolve to form the same ions in solution.

$$NaOH(s) \ \xrightarrow{\ H_2O\ } \ Na^+(aq) \ + \ OH^-(aq)$$

When strong acids and strong bases are mixed, a neutralization reaction occurs. In a neutralization, the cation of the acid and the anion of the base combine to form the molecular compound of water. The spectator ions usually remain in solution forming what is known as a **salt**. A few neutralization reactions occur between a strong acid and specific anions to produce gases. These are known as **gas forming neutralization reactions**. When a strong acid reacts with ionic compounds containing the HCO_3^- or the CO_3^{2-} ion, CO_2 gas is formed. When the anions are HSO_3^- or SO_3^{2-}, SO_2 is formed, and when the anions are HS^- or S^{2-}, H_2S gas is formed.

An example of a neutralization reaction between a strong acid and a strong base follows.

Example B-2 A Neutralization Reaction Illustrated With Equations

Write the balanced molecular, total ionic, and net ionic equations when solutions of nitric acid and sodium hydroxide are mixed.

PROCEDURE

Solution #1	Solution #2		Solution #3 (#1 + #2)
		produces	
[HNO$_3$(aq)] +	[NaOH(aq)]		[H$_2$O + NaNO$_3$(aq)]

NO_3^- (aq) OH^- (aq) $\longrightarrow$ H_2O

H^+ (aq) Na^+ (aq) Na^+ (aq) + NO_3^- (aq)

SOLUTION

Molecular equation: $HNO_3(aq) + NaOH(aq) \longrightarrow H_2O + NaNO_3(aq)$

Total ionic equation: $H^+(aq) + NO_3^-(aq) + Na^+(aq) + OH^-(aq) \longrightarrow H_2O + Na^+(aq)$
$$+ NO_3^-(aq)$$

Net ionic equation: $H^+(aq) + OH^-(aq) \longrightarrow H_2O$
(The net ionic equations for all neutralizations between strong acids and bases are the same.)

The molecular, total ionic, and net ionic equations for another example of a neutralization follow.

$$Acid \ + \ \ \ \ \ Base \longrightarrow Salt \ + \ \ \ \ \ Water$$

Molecular: $H_2SO_4(aq) + 2KOH(aq) \longrightarrow K_2SO_4(aq) + 2H_2O$

Total ionic: $2H^+(aq) + SO_4^{2-}(aq) + 2K^+(aq) + 2OH^-(aq) \longrightarrow 2K^+(aq) +$
$$SO_4^{2-}(aq) + 2H_2O$$

Net ionic: $H^+(aq) + OH^-(aq) \longrightarrow H_2O$

ASSESSMENT OF OBJECTIVES

B-1 Problems

1. Write equations illustrating the solution of the following compounds in water.

    (a) $Sr(ClO_3)_2 \xrightarrow{H_2O}$

    (b) $KNO_2 \xrightarrow{H_2O}$

    (c) $(NH_4)_2SO_3 \xrightarrow{H_2O}$

    (d) $HI \xrightarrow{H_2O}$

2. Write the total ionic and net ionic equations for:

    (a) $Sn(s) \ + \ 2AgNO_3(aq) \longrightarrow Sn(NO_3)_2(aq) \ + \ 2Ag(s)$

(b) $K_2SO_3(aq) + Sr(ClO_3)_2(aq) \longrightarrow SrSO_3(s) + 2KClO_3(aq)$

(c) $2(NH_4)_3PO_4(aq) + 3CoCl_2(aq) \longrightarrow Co_3(PO_4)_2(s) + 6NH_4Cl(aq)$

(d) $HClO_4(aq) + LiOH(aq) \longrightarrow LiClO_4(aq) + H_2O(l)$

3. Write the balanced molecular, total ionic, and net ionic equations illustrating the following:

(a) When cobalt metal is placed in a hydrobromic acid solution, hydrogen gas is evolved and a solution of $CoBr_3$ is formed.

(b) When an aqueous solution of ammonium sulfate is mixed with an aqueous solution of lead(II) acetate, a precipitate of lead(II) sulfate forms.

(c) When a solution of iron(III) chlorate is mixed with a solution of potassium hydroxide, a precipitate of iron(III) hydroxide forms.

(d) When solutions of strontium hydroxide and trifluoroacetic acid ($HC_2F_3O_2$) are mixed, a neutralization occurs. (Trifluoroacetic acid is a strong acid.)

(e) When solutions of sodium sulfide and nitric acid are mixed, a gas forms.

Chapter Summary Assessment

S-1 Types of Reactions

Balance the following equations. Classify the reactions as Combination (CO), Decomposition (D), Combustion (CB), Single-Replacement (SR), Double-Replacement Precipitation (DRP), or Double-Replacement Neutralization (DRN).

1. $Na(s) + O_2(g) \longrightarrow Na_2O_2(s)$

2. $H_2SO_4(aq) + Ca(OH)_2(aq) \longrightarrow CaSO_4(s) + H_2O(l)$

3. $AlI_3(aq) + AgNO_3(aq) \longrightarrow AgI(s) + Al(NO_3)_3(aq)$

4. $Pt(SO_4)_2(aq) + Sn(s) \longrightarrow Pt(s) + SnSO_4(aq)$

5. $B_4H_{10}(g) + O_2(g) \longrightarrow B_2O_3(s) + H_2O(l)$

6. $H_3PO_3(aq) \longrightarrow H_2O(l) + P_4O_6(s)$

7. $C_6H_{12}O_6(s) \longrightarrow C(s) + H_2O\ (l)$

8. $Mg(s) + N_2(g) \longrightarrow Mg_3N_2(s)$

9. $CaCO_3(s) + H_2O(l) + CO_2(g) \longrightarrow Ca(HCO_3)_2(s)$

10. $Cl_2O_3(g) + H_2O(l) \longrightarrow HClO_2(aq)$

11. $HCl(aq) + Mg(OH)_2\ (s) \longrightarrow MgCl_2(aq) + H_2O(l)$

12. $Sc(s) + AgNO_3(aq) \longrightarrow Sc(NO_3)_3(aq) + Ag(s)$

13. $Ag_2O(s) \longrightarrow Ag(s) + O_2(g)$

14. $NH_3(g) + O_2(g) \longrightarrow NO(g) + H_2O(l)$

S-2 Net Ionic Equations

Write the balanced net ionic equations for any single-replacement or double-replacement reactions in Problem S-1.

Answers to Assessments of Objectives

A-1 Balancing Equations

1.
   (a) $UO_2 + 4HF \longrightarrow UF_4 + 2H_2O$

   (b) $2HCl + MgCO_3 \longrightarrow MgCl_2 + CO_2 + H_2O$

   (c) $2NaNO_3 \longrightarrow 2NaNO_2 + O_2$

   (d) $Cu + 4HNO_3 \longrightarrow Cu(NO_3)_2 + 2NO_2 + 2H_2O$

   (e) $2Si_2H_6 + 7O_2 \longrightarrow 4SiO_2 + 6H_2O$

   (f) $P_4O_{10} + 6H_2O \longrightarrow 4H_3PO_4$

   (g) $UO_2 + 2CaH_2 \longrightarrow U + 2CaO + H_2$

2.
   (a) $C_2H_6O(l) + 3O_2(g) \longrightarrow 2CO_2(g) + 3H_2O(l)$

   (b) $2Fe(OH)_3(s) \overset{\Delta}{\longrightarrow} Fe_2O_3(s) + 3H_2O(g)$

   (c) $S_8(s) + 24F_2(g) \longrightarrow 8SF_6(g)$

B-1 Problems

1.   (a) $Sr(ClO_3)_2 \xrightarrow{H_2O} Sr^{2+}(aq) + 2ClO_3^-(aq)$

     (b) $KNO_2 \xrightarrow{H_2O} K^+(aq) + NO_2^-(aq)$

     (c) $(NH_4)_2SO_3 \xrightarrow{H_2O} 2NH_4^+(aq) + SO_3^{2-}(aq)$

     (d) $HI(aq) \xrightarrow{H_2O} H^+(aq) + I^-(aq)$

2.   (a) $Sn(s) + 2Ag^+(aq) + 2NO_3^-(aq) \rightarrow Sn^{2+}(aq) + 2NO_3^-(aq) + 2Ag(s)$

               $Sn(s) + 2Ag^+(aq) \rightarrow Sn^{2+}(aq) + 2Ag(s)$

  (b) $2K^+(aq) + SO_3^{2-}(aq) + Sr^{2+}(aq) + 2ClO_3^-(aq) \rightarrow SrSO_3(s) + 2K^+(aq) + 2ClO_3^-(aq)$

               $Sr^{2+}(aq) + SO_3^{2-}(aq) \rightarrow SrSO_3(s)$

  (c) $6NH_4^+(aq) + 2PO_4^{3-}(aq) + 3Co^{2+}(aq) + 6Cl^-(aq) \rightarrow Co_3(PO_4)_2(s) + 6NH_4^+(aq) + 6Cl^-(aq)$

               $3Co^{2+}(aq) + 2PO_4^{3-}(aq) \rightarrow Co_3(PO_4)_2(s)$

  (d) $H^+(aq) + ClO_4^-(aq) + Li^+(aq) + OH^-(aq) \rightarrow Li^+(aq) + ClO_4^-(aq) + H_2O(l)$

               $H^+(aq) + OH^-(aq) \rightarrow H_2O(l)$

3.   (a) $2Co(s) + 6HBr(aq) \rightarrow 2CoBr_3(aq) + 3H_2(g)$

      $2Co(s) + 6H^+(aq) + 6Br^-(aq) \rightarrow Co^{3+}(aq) + 6Br^-(aq) + 3H_2(g)$

               $2Co(s) + 6H^+(aq) \rightarrow Co^{3+}(aq) + 3H_2(g)$

     (b) $(NH_4)_2SO_4(aq) + Pb(C_2H_3O_2)_2(aq) \rightarrow PbSO_4(s) + 2NH_4C_2H_3O_2(aq)$

      $2NH_4^+(aq) + SO_4^{2-}(aq) + Pb^{2+}(aq) + 2C_2H_3O_2^-(aq) \rightarrow PbSO_4(s) + 2NH_4^+(aq) + 2C_2H_3O_2^-(aq)$

               $Pb^{2+}(aq) + SO_4^{2-}(aq) \rightarrow PbSO_4(s)$

     (c) $Fe(ClO_3)_3(aq) + 3KOH(aq) \rightarrow Fe(OH)_3(s) + 3KClO_3(aq)$

$Fe^{3+}(aq) + 3ClO_3^-(aq) + 3K^+(aq) + 3OH^-(aq) \rightarrow Fe(OH)_3(s) + 3K^+(aq) + 3ClO_3^-(aq)$

               $Fe^{3+}(aq) + 3OH^-(aq) \rightarrow Fe(OH)_3(s)$

(d) $2HC_2F_3O_2(aq) + Sr(OH)_2(aq) \longrightarrow Sr(C_2F_3O_2)_2(aq) + 2H_2O(l)$

$2H^+(aq) + 2C_2F_3O_2^-(aq) + Sr^{2+}(aq) + 2OH^-(aq) \longrightarrow Sr^{2+}(aq) + 2C_2F_3O_2^-(aq) + 2H_2O(l)$

$H^+(aq) + OH^-(aq) \longrightarrow H_2O(l)$

(e) $Na_2S(aq) + 2HNO_3(aq) \longrightarrow 2NaNO_3(aq) + H_2S(g)$

$2Na^+(aq) + S^{2-}(aq) + 2H^+(aq) + 2NO_3^-(aq) \longrightarrow 2Na^+(aq) + 2NO_3^-(aq) + H_2S(g)$

$S^{2-}(aq) + 2H^+(aq) \longrightarrow H_2S(g)$

S-1 Types of Reactions

1. (CO and CB)
2. (DRP and DRN)
3. (DRP)
4. (SR)
5. (CB)

6. (D)
7. (D)
8. (CO)
9. (CO)
10. (CO)

11. (DRN)
12. (SR)
13. (D)
14. (CB)

S-2 Net Ionic Equations

2. $2H^+(aq) + SO_4^{2-}(aq) + Ca^{2+}(aq) + 2OH^-(aq) \longrightarrow CaSO_4(s) + H_2O(l)$

3. $I^-(aq) + Ag^+(aq) \longrightarrow AgI(s)$

4. $Pt^{4+}(aq) + 2Sn(s) \longrightarrow Pt(s) + 2Sn^{2+}(aq)$

11. $2H^+(aq) + Mg(OH)_2(s) \longrightarrow Mg^{2+}(aq) + H_2O(l)$

12. $Sc(s) + 3Ag^+(aq) \longrightarrow Sc^{3+}(aq) + 3Ag(s)$

Answers and Solutions to Green Text Problems

6-1 Refer to Figure 4-5. (a) $Cl_2(g)$ (b) $C(s)$ (c) $K_2SO_4(s)$ (d) $H_2O(l)$ (e) $P_4(s)$ (f) $H_2(g)$ (g) $Br_2(l)$ (h) $NaBr(s)$ (i) $S_8(s)$ (j) $Na(s)$ (k) $Hg(l)$ (l) $CO_2(g)$

6-2 (a) $CaCO_3 \longrightarrow CaO + CO_2$

(b) $4Na + O_2 \longrightarrow 2Na_2O$

(c) $H_2SO_4 + 2NaOH \longrightarrow Na_2SO_4 + 2H_2O$

(d) $2H_2O_2 \longrightarrow 2H_2O + O_2$

6-4 (a) $2Al + 2H_3PO_4 \longrightarrow 2AlPO_4 + 3H_2$

(b) $Ca(OH)_2 + 2HCl \longrightarrow CaCl_2 + 2H_2O$

(c) $3Mg + N_2 \longrightarrow Mg_3N_2$

(d) $2C_2H_6 + 7O_2 \longrightarrow 4CO_2 + 6H_2O$

6-6 (a) $Mg_3N_2 + 6H_2O \longrightarrow 3Mg(OH)_2 + 2NH_3$

(b) $2H_2S + O_2 \longrightarrow 2S + 2H_2O$

(c) $Si_2H_6 + 8H_2O \longrightarrow 2Si(OH)_4 + 7H$

(d) $C_2H_6 + 5Cl_2 \longrightarrow C_2HCl_5 + 5HCl$

6-8 (a) $2B_4H_{10} + 11O_2 \longrightarrow 4B_2O_3 + 10H_2O$

(b) $SF_6 + 2SO_3 \longrightarrow 3O_2SF_2$

(c) $CS_2 + 3O_2 \longrightarrow CO_2 + 2SO_2$

(d) $2BF_3 + 6NaH \longrightarrow B_2H_6 + 6NaF$

6-10 Refer to Chap 4 for review of writing formulas from names.
(a) $2Na(s) + 2H_2O(l) \longrightarrow H_2(g) + 2NaOH(aq)$
(b) $2KClO_3(s) \longrightarrow 2KCl(s) + 3O_2(g)$
(c) $NaCl(aq) + AgNO_3(aq) \longrightarrow AgCl(s) + NaNO_3(aq)$
(d) $2H_3PO_4(aq) + 3Ca(OH)_2(aq) \longrightarrow Ca_3(PO_4)_2(s) + 6H_2O(l)$

6-12 $Ni(s) + 2N_2O_4(l) \longrightarrow Ni(NO_3)_2(s) + 2NO(g)$

6-14 In Problem 6-2, (a) and (d) are decomposition reactions and (b) is a combination and combustion reaction.

In Problem 6-4, (c) is a combination reaction and (d) is a combustion reaction.

6-16 (a) $2C_7H_{14}(l) + 21 O_2(g) \longrightarrow 14CO_2(g) + 14H_2O(l)$

(b) $2LiCH_3(s) + 4O_2(g) \longrightarrow Li_2O(s) + 2CO_2(g) + 3H_2O(l)$

(c) $C_4H_{10}O(l) + 6O_2(g) \longrightarrow 4CO_2(g) + 5H_2O(l)$

(d) $2C_2H_5SH(g) + 9O_2 \longrightarrow 2SO_2(g) + 4CO_2(g) + 6H_2O(l)$

6-18 Recall the formulas and physical states of the nonmetals.

(a) $Ba(s) + H_2(g) \longrightarrow BaH_2(s)$ (b) $8Ba(s) + S_8(s) \longrightarrow 8BaS(s)$

(c) $Ba(s) + Br_2(l) \longrightarrow BaBr_2(s)$ (d) $3Ba(s) + N_2(g) \longrightarrow Ba_3N_2(s)$

6-20 (a) $Ca(HCO_3)_2(s) \longrightarrow CaO(s) + 2CO_2(g) + H_2O(l)$

(b) $2Ag_2O(s) \longrightarrow 4Ag(s) + O_2(g)$

(c) $N_2O_3(g) \longrightarrow NO_2(g) + NO(g)$

6-22 (a) $2K(s) + Cl_2(g) \longrightarrow KCl(s)$

(b) $2C_6H_6(l) + 15O_2(g) \longrightarrow 12CO_2(g) + 6H_2O(l)$

(c) $2Au_2O_3(s) \longrightarrow 4Au(s) + 3O_2(g)$

(d) $2C_3H_8O(l) + 9O_2 \longrightarrow 6CO_2 + 8H_2O(l)$

(e) $P_4(s) + 10F_2(g) \longrightarrow 4PF_5(s)$

6-24 Review Table 4-2, the table of ions.

(a) $Na_2S(s) \longrightarrow 2Na^+(aq) + S^{2-}(aq)$

(b) $Li_2SO_4(s) \longrightarrow 2Li^+(aq) + SO_4^{2-}(aq)$

(c) $K_2Cr_2O_7(s) \longrightarrow 2K^+(aq) + Cr_2O_7^{2-}(aq)$

(d) $CaS(s) \longrightarrow Ca^{2+}(aq) + S^{2-}(aq)$

(e) $(NH_4)_2S(s) \longrightarrow 2NH_4^+(aq) + S^{2-}(aq)$

(f) $Ba(OH)_2(s) \longrightarrow Ba^{2+}(aq) + 2OH^-(aq)$

6-26 (a) $HNO_3(aq) \longrightarrow H^+(aq) + NO_3^-(aq)$

(b) $Sr(OH)_2(s \longrightarrow Sr^{2+}(aq) + 2OH^-(aq)$

6-28 (a) no reaction (b) $Fe + 2H^+ \longrightarrow Fe^{2+} + H_2$

(c) $Cu + 2Ag^+ \longrightarrow Cu^{2+} + 2Ag$ (d) no reaction

6-30 (a) $CuCl_2(aq) + Fe(s) \longrightarrow FeCl_2(aq) + Cu(s)$

$Cu^{2+}(aq) + 2Cl^-(aq) + Fe(s) \longrightarrow Fe^{2+}(aq) + 2Cl^-(aq) + Cu(s)$

$Cu^{2+}(aq) + Fe(s) \longrightarrow Fe^{2+}(aq) + Cu(s)$

(b) no reaction

(c) no reaction

(d) $3Zn(s) + 2Cr(NO_3)_3(aq) \longrightarrow 3Zn(NO_3)_2(aq) + 2Cr(s)$

$3Zn(s) + 2Cr^{3+}(aq) + 6NO_3^-(aq) \longrightarrow 3Zn^{2+}(aq) + 6NO_3^-(aq) + 2Cr(s)$

$3Zn(s) + 2Cr^{3+}(aq) \longrightarrow 3Zn^{2+}(aq) + 2Cr(s)$

6-32 $6Na(l) + Cr_2O_3(s) \longrightarrow 2Cr(s) + 3Na_2O(s)$

$6Na + 2Cr^{3+} \longrightarrow 2Cr + 6Na^+$

6-34 Insoluble compounds are (b) $PbSO_4$, and (d) Ag_2S

6-36 Write the net ionic equations and balance the cation and anion charges.
(a) $Ag^+(aq) + Br^-(aq) \longrightarrow AgBr(s)$
(b) $2Ag^+(aq) + CO_3^{2-}(aq) \longrightarrow Ag_2CO_3(s)$
(c) $3Ag^+(aq) + PO_4^{3-}(aq) \longrightarrow Ag_3PO_4(s)$

6-38 (a) $2Cu^+(aq) + CO_3^{2-}(aq) \longrightarrow Cu_2CO_3(s)$
(b) $Cd^{2+}(aq) + CO_3^{2-}(aq) \longrightarrow CdCO_3(s)$
(c) $2Cr^{3+}(aq) + 3CO_3^{2-}(aq) \longrightarrow Cr_2(CO_3)_3(s)$

6-40 Hg_2Cl_2

6-42 (b) $Ca_3(PO_4)_2$

6-44 First write in all of the reactants and products and then balance.
(a) $2KI(aq) + Pb(C_2H_3O_2)_2(aq) \longrightarrow PbI_2(s) + 2KC_2H_3O_2(aq)$
(b) and (c) no reaction occurs
(d) $BaS(aq) + Hg_2(NO_3)_2(aq) \longrightarrow Hg_2S(s) + Ba(NO_3)_2(aq)$
(e) $FeCl_3(aq) + 3KOH(aq) \longrightarrow Fe(OH)_3(s) + 3KCl(aq)$

6-46 Recall writing formulas from names in Chap 4.

(a) $2K^+(aq) + 2I^-(aq) + Pb^{2+}(aq) + 2C_2H_3O_2^-(aq) \longrightarrow PbI_2(s) + 2K^+(aq) + 2C_2H_3O_2^-(aq)$

$$Pb^{2+}(aq) + 2I^-(aq) \longrightarrow PbI_2(s)$$

(d) $Ba^{2+}(aq) + S^{2-}(aq) + Hg_2^{2+}(aq) + 2NO_3^-(aq) \longrightarrow Hg_2S(s) + Ba^{2+}(aq) + 2NO_3^-(aq)$

$$Hg_2^{2+}(aq) + S^{2-}(aq) \longrightarrow Hg_2S(s)$$

(e) $Fe^{3+}(aq) + 3Cl^-(aq) + 3K^+(aq) + 3OH^-(aq) \longrightarrow Fe(OH)_3(s) + 3K^+(aq) + 3Cl^-(aq)$

$$Fe^{3+}(aq) + 3OH^-(aq) \longrightarrow Fe(OH)_3(s)$$

6-48 (a) $2K^+(aq) + S^{2-}(aq) + Pb^{2+}(aq) + 2NO_3^-(aq) \longrightarrow PbS(s) + 2K^+(aq) + 2NO_3^-(aq)$

$$S^{2-}(aq) + Pb^{2+}(aq) \longrightarrow PbS(s)$$

(b) $2NH_4^+(aq) + CO_3^{2-}(aq) + Ca^{2+}(aq) + 2Cl^-(aq) \longrightarrow CaCO_3(s) + 2NH_4^+(aq) + 2Cl^-(aq)$

$$CO_3^{2-}(aq) + Ca^{2+}(aq) \longrightarrow CaCO_3(s)$$

(c) $2Ag^+(aq) + 2ClO_4^-(aq) + 2Na^+(aq) + CrO_4^{2-}(aq) \longrightarrow Ag_2CrO_4(s) + 2Na^+(aq) + 2ClO_4^-(aq)$

$$2Ag^+(aq) + CrO_4^{2-}(aq) \longrightarrow Ag_2CrO_4(s)$$

6-50 (a) $CuCl_2(aq) + Na_2CO_3(aq) \longrightarrow CuCO_3(s) + 2NaCl(aq)$ Filter the solid $CuCO_3$.

(b) $(NH_4)_2SO_4(aq) + Pb(NO_3)_2(aq) \longrightarrow PbSO_4(s) + 2NH_4NO_3(aq)$
Filter the solid $PbSO_4$.

(c) $2KI(aq) + Hg_2(NO_3)_2(aq) \longrightarrow Hg_2I_2(s) + 2KNO_3(aq)$
Filter the solid Hg_2I_2.

(d) $NH_4Cl(aq) + AgNO_3(aq) \longrightarrow AgCl(s) + NH_4NO_3(aq)$
Filter the solid AgCl; the desired product remains after water is removed by boiling.

(e) $Ca(C_2H_3O_2)_2(aq) + K_2CO_3(aq) \longrightarrow CaCO_3(s) + 2KC_2H_3O_2(aq)$
Filter the solid $CaCO_3$; the desired product remains after the water is removed by boiling.

6-51   (b) HF

6-53   (c) $Al(OH)_3$

6-55   (a) $HI(aq) + CsOH(aq) \longrightarrow CsI(aq) + H_2O(l)$

(b) $2HNO3(aq) + Ca(OH)_2(aq) \longrightarrow Ca(NO3)2(aq) + 2H_2O(l)$

(c) $H_2SO_4(aq) + Sr(OH)_2(aq) \longrightarrow SrSO_4(s) + 2H_2O(l)$

(d) $HNO_3(aq) + NaHCO_3(s) \rightarrow NaNO_3(aq) + CO_2(g) + H_2O(l)$

6-57   (a) $H^+(aq) + I^-(aq) + Cs^+(aq) + OH^-(aq) \longrightarrow Cs^+(aq) + I^-(aq) + H_2O(l)$
   $H^+(aq) + OH^-(aq) \longrightarrow H_2O(l)$

(b) $2H^+(aq) + 2NO3^-(aq) + Ca^{2+}(aq) + 2OH^-(aq) \longrightarrow Ca^{2+}(aq) + 2NO3^-(aq) + 2H_2O(l)$
   $H^+(aq) + OH^-(aq) \longrightarrow H_2O(l)$

(c) $2H^+(aq) + SO_4^{2-}(aq) + Sr^{2+}(aq) + 2OH^-(aq) \longrightarrow SrSO_4(s) + 2H_2O(l)$

Net ionic equation is the same as the total ionic equation since $SrSO_4$ precipitates.

(d) $H^+(aq) + NO_3^-(aq) + Na^+(aq) + HCO_3^-(s) \rightarrow Na^+(aq) + NO_3^-(aq) + CO_2(g) + H_2O(l)$

   $H^+(aq) + HCO_3^-(s) \rightarrow CO_2(g) + H_2O(l)$

6-59 $Mg(OH)_2(s) + 2HCl(aq) \longrightarrow MgCl_2(aq) + 2H_2O(l)$
   $Mg(OH)_2(s) + 2H^+(aq) + 2Cl^-(aq) \longrightarrow Mg^{2+}(aq) + 2Cl^-(aq) + 2H_2O(l)$
   $Mg(OH)_2(s) + 2H^+(aq) \longrightarrow Mg^{2+}(aq) + 2H_2O(l)$

6-61 $C_3H_8(g) + 5O_2(g) \longrightarrow 3CO_2(g) + 4H_2O(l)$

$2C_4H_{10}(g) + 13O_2(g) \longrightarrow 8CO_2(g) + 10H_2O(l)$

$2C_8H_{18}(l) + 25O_2(g) \longrightarrow 16CO_2(g) + 18H_2O(l)$

$C_2H_5OH(l) + 3O_2(g) \longrightarrow 2CO_2(g) + 3H_2O(l)$

6-62   In both of these reactions, the reactants change from neutral atoms to cations and anions in the products. Ions are formed from neutral atoms from the loss or gain of electrons.

6-64   (a) $Ba(s) + I_2(s) \longrightarrow BaI_2(s)$

(b) $HBr(aq) + RbOH(aq) \longrightarrow RbBr(aq) + H_2O(l)$

(c) $Ca(s) + 2HNO_3(aq) \longrightarrow Ca(NO_3)_2(aq) + H_2(g)$

(d) $C_{10}H_8(s) + 12O_2(g) \longrightarrow 10CO_2(g) + 4H_2O(l)$

(e) $(NH_4)_2CrO_4(aq) + BaBr_2(aq) \longrightarrow BaCrO_4(s) + 2NH_4Br(aq)$

(f) $2Al(OH)_3(s) \longrightarrow Al_2O_3(s) + 3H_2O(g)$

6-65   (b) $H^+(aq) + Br^-(aq) + Rb^+(aq) + OH^-(aq) \longrightarrow Rb^+(aq) + Br^-(aq) + H_2O(l)$

$H^+(aq) + OH^-(aq) \longrightarrow H_2O(l)$

(c) $Ca(s) + 2H^+(aq) + 2NO_3^-(aq) \longrightarrow Ca^{2+}(aq) + 2NO_3^-(aq) + H_2(g)$

$Ca(s) + 2H^+(aq) \longrightarrow Ca^{2+}(aq) + H_2(g)$

(e) $2NH_4^+(aq) + CrO_4^{2-}(aq) + Ba^{2+}(aq) + 2Br^-(aq) \longrightarrow BaCrO_4(s) + 2NH_4^+(aq) + 2Br^-(aq)$

$CrO_4^{2-}(aq) + Ba^{2+}(aq) \longrightarrow BaCrO_4(s)$

6-68 $H^+(aq) + OH^-(aq) \longrightarrow H_2O(l)$ $Ba^{2+}(aq) + SO_4^{2-}(aq) \longrightarrow BaSO_4(s)$

6-70 $Fe^{3+}(aq) + PO_4^{3-}(aq) \longrightarrow FePO_4(s)$ $2Fe^{3+}(aq) + 3S^{2-}(aq) \longrightarrow Fe_2S_3(s)$

$Pb^{2+}(aq) + 2I^-(aq) \longrightarrow PbI_2(s)$ $3Pb^{2+}(aq) + 2PO_4^{3-}(aq) \longrightarrow Pb_3(PO_4)_2(s)$

$Pb^{2+}(aq) + S^{2-}(aq) \longrightarrow PbS(s)$

116

7

Quantitative Relationships in Chemistry

Review of Part A *Mass Relationships in Chemical Reactions*

OBJECTIVES AND DETAILED TABLE OF CONTENTS

7-1   Stoichiometry

OBJECTIVE *Perform stoichiometry calculations using mole ratios from balanced equations.*

7-2 Limiting Reactant

OBJECTIVE *Given the masses of two different reactants, determine the limiting reactant and calculate the yield of product.*

   7-2.1 The Definition of Limiting Reactant
   7-2.2 Solving Limiting Reactant Problems

7-3 Percent Yield

OBJECTIVE *Calculate the percent yield of a reaction from the measured actual yield and the calculated theoretical yield.*

SUMMARY OF PART A

A critical skill to be learned in chemistry is the ability to carry out **stoichiometry** calculations. This concerns conversions between the amount (moles, mass, or number) of one reactant or product and that of another reactant or product. The key conversion factor for this endeavor relates moles of all reactants and products by the coefficients in the balanced equation. Thus in the balanced equation

$$Si_2H_6 + 8H_2O \longrightarrow 2Si(OH)_4 + 7H_2$$

conversion factors (**mole ratios**) between H_2O and H_2 are

$$\frac{8 \text{ mol } H_2O}{7 \text{ mol } H_2} \quad \text{and} \quad \frac{7 \text{ mol } H_2}{8 \text{ mol } H_2O}$$

The first ratio is used to convert moles of H_2 to H_2O and the second (which is simply the reciprocal of the first) is used to convert moles of H_2O to H_2. All other conversions (e.g., mass to moles, moles to number) were studied earlier in this chapter. The conversions are represented in the following general scheme. The origin of the conversion factor for each step is indicated next to the arrow.

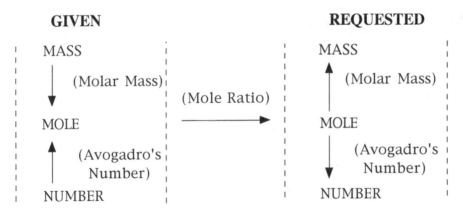

The following two examples illustrate stoichiometry problems.

Example A-1 Mole to Mole Conversions

How many moles of H_2 are produced from 36.5 moles of H_2O according to the following balanced equation?

$$Si_2H_6 + 8H_2O \longrightarrow 2Si(OH)_4 + 7H_2$$

PROCEDURE

From the balanced equation, the mole ratio that converts moles to what's given (H_2O) to moles of what's requested (H_2) is

$$\frac{7 \text{ mol } H_2}{8 \text{ mol } H_2O}$$

SOLUTION

Given x mole ratio = Requested

$$36.5 \text{ mol } H_2O \text{ x } \frac{7 \text{ mol } H_2}{8 \text{ mol } H_2O} = \underline{31.9 \text{ mol } H_2}$$

Example A-2   Mass to Mass Conversions

What mass of H_2 is produced from 165 g of Si_2H_6? (Use the balanced equation from problem A-1.)

PROCEDURE

In this case, the following conversions are necessary:

1. Convert mass of Si_2H_6 (Given) to moles of Si_2H_6 using molar mass of Si_2H_6.

2. Convert moles of Si_2H_6 to moles of H_2 using the appropriate mole ratio.

3. Convert moles of H_2 to grams of H_2 (Requested) using the molar mass of H_2.

| **Given** | **1** | **2** | **3** | **Requested** |

$$165 \text{ g } Si_2H_6 \text{ x } \frac{1 \text{ mol } Si_2H_6}{62.23 \text{ g } Si_2H_6} \text{ x } \frac{7 \text{ mol } H_2}{1 \text{ mol } Si_2H_6} \text{ x } \frac{2.016 \text{ g } H_2}{\text{mol } H_2} = \underline{37.4 \text{ g } H_2}$$

Also, we must realize that reactants are not always mixed in exact proportions so that all are completely consumed (this would be a stoichiometric mixture). In that case, the reactant that is completely consumed limits the yield of products and is called the **limiting reactant**. A quantity of the other reactant (or reactants) is left over and is present in excess. This is illustrated by the following example.

Example A-3 The Limiting Reactant

In the reaction between Si_2H_6 and H_2O, a 50.0-g quantity of Si_2H_6 is mixed with 100 g of H_2O. How many moles of $Si(OH)_4$ are formed?

PROCEDURE

Find the yield in moles from each of the two reactants. The one that produces the smaller yield is the limiting reactant.

$$\text{mass} \begin{bmatrix} H_2O \\ Si_2H_6 \end{bmatrix} \longrightarrow \text{moles} \begin{bmatrix} H_2O \\ Si_2H_6 \end{bmatrix} \longrightarrow \text{moles } Si(OH)_4$$

SOLUTION

H_2O: $100 \text{ g } H_2O \times \dfrac{1 \text{ mol } H_2O}{18.02 \text{ g } H_2O} \times \dfrac{2 \text{ mol } Si(OH)_4}{8 \text{ mol } H_2O} = 1.39 \text{ mol } Si(OH)_4$

Si_2H_6: $50.0 \text{ g } Si_2H_6 \times \dfrac{1 \text{ mol } Si_2H_6}{62.23 \text{ g } Si_2H_6} \times \dfrac{2 \text{ mol } Si(OH)_4}{1 \text{ mol } Si_2H_6} = 1.61 \text{ mol } Si(OH)_4$

The limiting reactant is H_2O and the yield is <u>1.39 mol $Si(OH)_4$</u>.

There are many examples of chemical reactions that do not go to completion (100% to the right). There are two major reasons for this: (1) a reaction may be a reversible reaction where reactants and products reach a point of equilibrium, and (2) other competing reactions may occur between the same reactants. In these cases, it is convenient to express the **percent yield** of a product. The percent yield is obtained by dividing the **actual yield** (which is the measured yield) by the **theoretical yield** (which is the calculated yield) times 100%.

$$\text{Percent yield} = \frac{\text{Actual yield (measured)}}{\text{Theoretical yield (calculated)}} \times 100\%$$

An example of this type of problem follows.

Example A-4 Percent Yield

In the reaction of HCl with O_2, 46.0 g of Cl_2 is formed. If all of the HCl had reacted, 49.5 g of Cl_2 would form. What is the percent yield for this reaction?

PROCEDURE

Substitute the following into the equation used to calculate percent yield.

actual yield $= 46.0$ g   theoretical yield $= 49.5$ g

SOLUTION

percent yield $= \dfrac{46.0 \text{ g}}{49.5 \text{ g}} \times 100\% = \underline{92.9\% \text{ yield}}$

ASSESSMENT OF OBJECTIVES

A-1 Problems

1. Given the following balanced equation illustrating the combustion of butane (C_4H_{10}):

$$2C_4H_{10}(l) + 13O_2(g) \longrightarrow 8CO_2(g) + 10H_2O(l)$$

(a) Write the mole ratio that converts moles of O_2 to moles of H_2O.

(b) Write the mole ratio that converts moles of CO_2 to moles of H_2O.

(c) Write the two conversion factors that are needed to convert grams of C_4H_{10} to moles of O_2.

(d) Write the two conversion factors that are needed to convert moles of C_4H_{10} to number of molecules of CO_2.

(e) Write the three conversion factors that convert grams of H_2O to grams of O_2.

(f) How many moles of CO_2 are produced from 0.115 moles of O_2?

(g) What mass of CO_2 is produced from 8.67 moles of C_4H_{10}?

(h) What mass of O_2 completely reacts with 155 g of C_4H_{10}?

(i) How many individual H_2O molecules are produced from 6.64×10^{-6} g of O_2?

2. Given the following balanced equation:

$$4NH_3(g) + 5O_2 \longrightarrow 4NO(g) + 6H_2O(g)$$

What mass of H_2O is produced from a mixture of 100 g of NH_3 and 200 g of O_2?

3. When a 100-kg quantity of C_4H_{10} is burned in a closed system, a 280 kg quantity of CO_2 is formed. What is the percent yield of the reaction? Use the balanced equation from problem A-1 in this section.

4. An impure sample containing solid gold(III) oxide is heated. The gold(III) oxide decomposes to form metallic gold and gaseous oxygen. If the original sample has a mass of 365 g and a total of 146 g of gold is produced, what percent of the original sample was composed of gold(III) oxide?

Review of Part B *Energy Relationships in Chemical Reactions*

OBJECTIVE AND DETAILED TABLE OF CONTENTS

7-4   Heat Energy in Chemical Reactions

OBJECTIVE *Given the change in enthalpy for a reaction, calculate the amount of heat gained or released by a given mass of reactant.*

SUMMARY OF PART B

The study of **chemical thermodynamics** includes the heat energy that is involved in chemical changes. When heat energy is included, the equation is known as a **thermochemical equation**. In endothermic reactions the reactants absorb energy from the surroundings when a reaction occurs. The origin of this energy is in the difference in the potential energy (in the form of chemical energy) of the reactants and products involved.

An exothermic thermochemical reaction can be written in two ways. In the first, the heat energy is represented as a product. In the second, it is shown as a negative value for **ΔH** [the change in heat content (**enthalpy**)].

$$(1) \quad 2CO\ (g) + O_2\ (g) \longrightarrow 2CO_2\ (g) + 568\ kJ$$

$$(2) \quad 2CO\ (g) + O_2\ (g) \longrightarrow 2CO_2\ (g) \quad \Delta H = -568\ kJ$$

ASSESSMENT OF OBJECTIVE

B-1 Problem

1. A 235-kJ quantity of heat must be supplied to decompose sodium bicarbonate to one mole each of $H_2O(g)$, $CO_2(g)$, and $Na_2CO_3(s)$. Write the thermochemical equation in both ways discussed. What mass of sodium bicarbonate is decomposed by the input of 175 kJ of heat energy?

Chapter Summary Assessment

S-1 Problems

1. The combustion of liquid methyl alcohol (CH_4O), a possible additive in gasoline, produces carbon dioxide and water. Write the balanced equation and determine what mass of water would be produced from the complete combustion of 964 g of methyl alcohol.

2. In the preceding problem, it was found that only 1.15 kg of CO_2 was formed. What is the percent yield of CO_2? The remainder of the alcohol that did not form CO_2 burns to form CO and water. Write the balanced equation illustrating this. What mass of CO is formed in the reaction?

3. Solid potassium superoxide (KO_2) is used in space vehicles to remove carbon dioxide gas and to regenerate oxygen gas. In the reaction KO_2 reacts with carbon dioxide to produce solid potassium carbonate and oxygen gas. Write the balanced equation and determine what mass of oxygen is produced from 83.0 g of CO_2.

4. Solid zinc sulfide reacts with oxygen to form solid zinc oxide and gaseous sulfur dioxide. Write the balanced equation and determine what mass of SO_2 is produced from a mixture of 85.0 g of zinc sulfide and 60.0 g of oxygen.

5. In the preceding problem, it is found that 396 kJ of heat energy is released in the reaction. What is the value of ΔH for the reaction?

Answers to Assessments of Objectives

A-1 Problems

1. $2C_4H_{10} + 13O_2 \longrightarrow 8CO_2 + 10H_2O$

(a) $\dfrac{10 \text{ mol } H_2O}{13 \text{ mol } O_2}$

(b) $\dfrac{10 \text{ mol } H_2O}{8 \text{ mol } CO_2}$

(c) $\dfrac{1 \text{ mol } C_4H_{10}}{58.12 \text{ g } C_4H_{10}}$; $\dfrac{10 \text{ mol } O_2}{2 \text{ mol } C_4H_{10}}$

(d) $\dfrac{8 \text{ mol } CO_2}{2 \text{ mol } C_4H_{10}}$; $\dfrac{6.022 \times 10^{23} \text{ molecules } CO_2}{\text{mol } CO_2}$

(e) $\dfrac{1 \text{ mol } H_2O}{18.02 \text{ g } H_2O}$; $\dfrac{13 \text{ mol } O_2}{10 \text{ mol } H_2O}$; $\dfrac{32.00 \text{ g } O_2}{\text{mol } O_2}$

(f) Convert moles of O_2 (Given) to moles of CO_2 (Requested).

using the mole ratio $\dfrac{8 \text{ mol } CO_2}{13 \text{ mol } CO_2}$

$$0.115 \text{ mol } O_2 \ \times \ \dfrac{8 \text{ mol } CO_2}{13 \text{ mol } O_2} = \underline{0.0708 \text{ mol } CO_2}$$

(g) (1) Convert moles of C_4H_{10} (Given) to moles CO_2 using

the mole ratio $\dfrac{8 \text{ mol } CO_2}{2 \text{ mol } C_4H_{10}}$.

(2) Convert moles of CO_2 to mass of CO_2 (Requested).

$$\overset{(1)}{} \qquad \overset{(2)}{}$$

$$8.67 \text{ mol } C_4H_{10} \ \times \ \dfrac{8 \text{ mol } CO_2}{2 \text{ mol } C_4H_{10}} \ \times \ \dfrac{44.01 \text{ g } CO_2}{\text{mol } CO_2} = 1.53 \times 10^3 \text{ g } CO_2$$

(h) (1) Convert mass of C_4H_{10} (Given) to moles of C_4H_{10}.

(2) Convert moles of C_4H_{10} to moles of O_2 using $\dfrac{13 \text{ mol } O_2}{2 \text{ mol } C_4H_{10}}$.

(3) Convert moles of O_2 to mass of O_2 (Requested).

$$\overset{(1)}{} \qquad \overset{(2)}{} \qquad \overset{(3)}{}$$

$$155 \text{ g } C_4H_{10} \times \dfrac{1 \text{ mol } C_4H_{10}}{58.12 \text{ g } C_4H_{10}} \ \times \ \dfrac{13 \text{ mol } O_2}{2 \text{ mol } C_4H_{10}} \ \times \ \dfrac{32.00 \text{ g } O_2}{\text{mol } O_2} = \underline{555 \text{ g } O_2}$$

(i) (1) Convert mass of O_2 (Given) to moles of O_2.

(2) Convert moles of O_2 to moles of H_2O using $\dfrac{10 \text{ mol } H_2O}{13 \text{ mol } O_2}$.

(3) Convert moles of H_2O to molecules of H_2O.

$$\overset{(1)}{6.64 \times 10^{-6} \ \cancel{g\ O_2}} \ \times \ \overset{(1)}{\frac{1 \ \cancel{mol\ O_2}}{32.00 \ \cancel{g\ O_2}}} \ \times \ \overset{(2)}{\frac{10 \ \cancel{mol\ H_2O}}{13 \ \cancel{mol\ O_2}}}$$

$$\overset{(3)}{\times \ \frac{6.022 \times 10^{23} \ molecules \ H_2O}{\cancel{mol\ H_2O}}} \ = \underline{9.61 \times 10^{16} \ molecules \ H_2O}$$

2.   (1) Convert mass of NH_3 to moles of NH_3 then to moles of H_2O.
    (2) Convert mass of O_2 to moles of O_2 then to moles of H_2O.
    (3) Convert the smaller of the two above to mass of H_2O.

(1) NH_3: $100 \ \cancel{g\ NH_3} \ \times \ \dfrac{1 \ \cancel{mol\ NH_3}}{17.03 \ \cancel{g\ NH_3}} \ \times \ \dfrac{6 \ mol \ H_2O}{4 \ \cancel{mol\ NH_3}} = 8.81 \ mol \ H_2O$

(2) O_2: $200 \ \cancel{g\ O_2} \ \times \ \dfrac{1 \ \cancel{mol\ O_2}}{32.00 \ \cancel{g\ O_2}} \ \times \ \dfrac{6 \ mol \ H_2O}{5 \ \cancel{mol\ O_2}} = 7.50 \ mol \ H_2O$

(3) O_2 is the limiting reactant, so $7.50 \ \cancel{mol\ H_2O} \ \times \ \dfrac{18.02 \ g \ H_2O}{\cancel{mol\ H_2O}} = \underline{135 \ g \ H_2O.}$

3.   (1) Convert kg C_4H_{10} to g C_4H_{10}, then to moles C_4H_{10}.
    (2) Convert moles of C_4H_{10} to moles of CO_2.
    (3) Convert moles of CO_2 to g of CO_2, then to kg of CO_2.
       This is the theoretical yield.

$$100 \ \cancel{kg\ C_4H_{10}} \ \times \ \frac{1 \ \cancel{g\ C_4H_{10}}}{10^{-3} \ \cancel{kg\ C_4H_{10}}} \ \times \ \frac{1 \ \cancel{mol\ C_4H_{10}}}{58.12 \ \cancel{g\ C_4H_{10}}}$$

$$\times \ \frac{8 \ \cancel{mol\ CO_2}}{2 \ \cancel{mol\ C_4H_{10}}} \ \times \ \frac{44.01 \ \cancel{g\ CO_2}}{\cancel{mol\ CO_2}} \ \times \ \frac{10^{-3} \ kg \ CO_2}{\cancel{g\ CO_2}} = 303 \ kg \ CO_2$$

$$\% \ yield \ = \ \frac{280 \ kg}{303 \ kg} \ \times \ 100\% \ = \underline{92.4 \ \%}$$

4. $2Au_2O_3(s) \longrightarrow 4\,Au(s) + 3O_2(g)$

(1) Convert mass of Au to moles of Au.

(2) Convert moles of Au to moles of Au_2O_3 using $\dfrac{2 \text{ mol } Au_2O_3}{4 \text{ mol } Au}$.

(3) Convert moles of Au_2O_3 to mass of Au_2O_3.

(4) Convert mass of Au_2O_3 to percent of original sample.

$$146\ \cancel{g\ Au}\ \times\ \overset{(1)}{\dfrac{1\ \cancel{mol\ Au}}{197.0\ \cancel{g\ Au}}}\ \times\ \overset{(2)}{\dfrac{2\ \cancel{mol\ Au_2O_3}}{4\ \cancel{mol\ Au}}}\ \times\ \overset{(3)}{\dfrac{442.0\ g\ Au_2O_3}{\cancel{mol\ Au_2O_3}}}\ =\ 164\ g\ Au_2O_3$$

$$\overset{(4)}{\dfrac{164\ g}{365\ g}}\ \times\ 100\%\ =\ \underline{44.9\%\ pure}$$

B-1 Problem

1. $2NaHCO_3(s) + 235\ kJ \longrightarrow H_2O(g) + CO_2(g) + Na_2CO_3(s)$

$2NaCO_3(s) \longrightarrow H_2O(g) + CO_2(g) + Na_2CO_3(s) \quad \Delta H = 235\ kJ$

$$\mathbf{kJ} \longrightarrow \mathbf{mol\ NaHCO_3} \longrightarrow \mathbf{g\ NaHCO_3}$$

$$175\ \cancel{kJ}\ \times\ \dfrac{2\ \cancel{mol\ NaHCO_3}}{235\ \cancel{kJ}}\ \times\ \dfrac{84.01\ g\ NaHCO_3}{\cancel{mol\ NaHCO_3}}\ =\ \underline{125\ g\ NaHCO_3}.$$

S-1 Problems

1. $2CH_4O(1) + 3O_2(g) \longrightarrow 2CO_2(g) + 4H_2O(1)$

(1) Convert mass of CH_4O to moles of CH_4O.

(2) Convert moles of CH_4O to moles of H_2O using $\dfrac{4 \text{ mol } H_2O}{2 \text{ mol } CH_4O}$.

(3) Convert moles of H_2O to mass of H_2O.

$$964\ \cancel{g\ CH_4O}\ \times\ \overset{(1)}{\dfrac{1\ \cancel{mol\ CH_4O}}{32.04\ \cancel{g\ CH_4O}}}\ \times\ \overset{(2)}{\dfrac{4\ \cancel{mol\ H_2O}}{2\ \cancel{mol\ CH_4O}}}\ \times\ \overset{(3)}{\dfrac{18.02\ g\ H_2O}{\cancel{mol\ H_2O}}}\ =\ 1.08 \times 10^3\ g\ H_2O$$

2.   (1) Convert mass of CH_4O to moles of CH_4O.

(2) Convert moles of CH_4O to moles of CO_2 using $\dfrac{2 \text{ mol } CO_2}{2 \text{ mol } CH_4O}$

(3) Convert moles of CO_2 to mass of CO_2 (the theoretical yield).

$$964 \text{ g CH}_4\text{O} \times \underset{(1)}{\frac{1 \text{ mol CH}_4\text{O}}{32.04 \text{ g CH}_4\text{O}}} \times \underset{(2)}{\frac{2 \text{ mol CO}_2}{2 \text{ mol CH}_4\text{O}}} \times \underset{(3)}{\frac{44.01 \text{ g CO}_2}{\text{mol CO}_2}} = 1.32 \times 10^3 \text{ g CO}_2$$

Theoretical yield = 1.32×10^3 g = 1.32 kg

Actual yield = 1.15 kg   Percent yield = $\frac{1.15 \text{ kg}}{1.32 \text{ kg}} \times 100\% = 87.1\%$

$$CH_4O(l) + O_2(g) \longrightarrow CO(g) + 2H_2O(l)$$

If 86.5% of the CH_4O is converted to CO_2, the remainder or 13.5% is converted to CO. Therefore, 0.135 x 964 g = 130 g CH_4O forms CO.

g CH_4O $\longrightarrow$ mol CH_4O $\longrightarrow$ mol CO $\longrightarrow$ g CO

$$130 \text{ g CH}_4\text{O} \times \frac{1 \text{ mol CH}_4\text{O}}{32.04 \text{ g CH}_4\text{O}} \times \frac{1 \text{ mol CO}}{1 \text{ mol CH}_4\text{O}} \times \frac{28.01 \text{ g CO}}{\text{mol CO}} = \underline{114 \text{ g CO}}$$

3. $4KO_2(s) + 2CO_2(g) \longrightarrow 2K_2CO_3(s) + 3O_2(g)$

(1) Convert mass of CO_2 to moles of CO_2.

(2) Convert moles of CO_2 to moles of O_2 using $\frac{3 \text{ mol O}_2}{2 \text{ mol CO}_2}$.

(3) Convert moles of O_2 to mass of O_2.

$$83.0 \text{ g CO}_2 \times \underset{(1)}{\frac{1 \text{ mol CO}_2}{44.01 \text{ g CO}_2}} \times \underset{(2)}{\frac{3 \text{ mol O}_2}{2 \text{ mol CO}_2}} \times \underset{(3)}{\frac{32.00 \text{ g O}_2}{\text{mol O}_2}} = \underline{90.5 \text{ g O}_2}$$

4. $2ZnS(s) + 3O_2(g) \longrightarrow 2ZnO(s) + 2SO_2(g)$

(1) Convert mass of ZnS to moles of ZnS then to moles of SO_2.
(2) Convert mass of O_2 to moles of O_2 then to moles of SO_2.
(3) Convert the smaller of the two above to mass of SO_2.

(1) ZnS: $85.0 \text{ g ZnS} \times \frac{1 \text{ mol ZnS}}{97.46 \text{ g ZnS}} \times \frac{2 \text{ mol SO}_2}{2 \text{ mol ZnS}} = 0.872 \text{ mol SO}_2$

(2) O_2: $60.0 \text{ g O}_2 \times \frac{1 \text{ mol O}_2}{32.00 \text{ g O}_2} \times \frac{2 \text{ mol SO}_2}{3 \text{ mol O}_2} = 1.25 \text{ mol SO}_2$

(3) <u>ZnS is the limiting reactant</u> $0.872 \; \cancel{mol \; SO_2} \; \times \; \dfrac{64.07 \; g \; SO_2}{\cancel{mol \; SO_2}}$

$= \underline{55.9 \; g \; SO_2}.$

5. Since ZnS is the limiting reactant, 55.9 g (0.872 mol) of SO_2 forms.

Therefore, $\dfrac{396 \; kJ}{0.872 \; mol \; SO_2} = 454 \; kJ/mol \; SO_2$

Since the equation calls for two mol of SO_2 and heat is evolved,

$$H = 2 \; \cancel{mol \; SO_2} \; \times \; \dfrac{-454 \; kJ}{\cancel{mol \; SO_2}} = \underline{-908 \; kJ}$$

Answers and Solutions to Green Text Problems

7-1 (a) $\dfrac{1 \; mol \; H_2}{1 \; mol \; Mg}$ (b) $\dfrac{2 \; mol \; HCl}{1 \; mol \; Mg}$

(c) $\dfrac{1 \; mol \; H_2}{2 \; mol \; HCl}$ (d) $\dfrac{2 \; mol \; HCl}{1 \; mol \; MgCl_2}$

7-2 (a) $\dfrac{2 \; mol \; C_4H_{10}}{8 \; mol \; CO_2}$ (b) $\dfrac{2 \; mol \; C_4H_{10}}{13 \; mol \; O_2}$

(c) $\dfrac{13 \; mol \; O_2}{8 \; mol \; CO_2}$ (d) $\dfrac{10 \; mol \; H_2O}{13 \; mol \; O_2}$

7-4 (a)

10.0 $\cancel{mol \; Al}$
$\begin{cases} \times \; \dfrac{1 \; mol \; Al_2O_3}{3 \; \cancel{mol \; Al}} = \underline{3.33 \; mol \; Al_2O_3} \\[2mm] \times \; \dfrac{1 \; mol \; AlCl_3}{3 \; \cancel{mol \; Al}} = \underline{3.33 \; mol \; AlCl_3} \\[2mm] \times \; \dfrac{3 \; mol \; NO}{3 \; \cancel{mol \; Al}} = \underline{10.0 \; mol \; NO} \\[2mm] \times \; \dfrac{6 \; mol \; H_2O}{3 \; \cancel{mol \; Al}} = \underline{20.0 \; mol \; H_2O} \end{cases}$

(b)

3.00 $\cancel{mol \; NH_4ClO_4}$
$\begin{cases} \times \; \dfrac{1 \; mol \; Al_2O_3}{3 \; \cancel{mol \; NH_4ClO_4}} = \underline{1.00 \; mol \; Al_2O_3} \\[2mm] \times \; \dfrac{1 \; mol \; AlCl_3}{3 \; \cancel{mol \; NH_4ClO_4}} = \underline{1.00 \; mol \; AlCl_3} \\[2mm] \times \; \dfrac{3 \; mol \; NO}{3 \; \cancel{mol \; NH_4ClO_4}} = \underline{3.00 \; mol \; NO} \\[2mm] \times \; \dfrac{6 \; mol \; H_2O}{3 \; \cancel{mol \; NH_4ClO_4}} = \underline{6.00 \; mol \; H_2O} \end{cases}$

7-6 (a) $10.0 \text{ mol NH}_3 \times \dfrac{3 \text{ mol O}_2}{2 \text{ mol NH}_3} = \underline{15.0 \text{ mol O}_2}$

$10.0 \text{ mol NH}_3 \times \dfrac{2 \text{ mol CH}_4}{2 \text{ mol NH}_3} = \underline{10.0 \text{ mol CH}_4}$

(b) $10.0 \text{ mol O}_2 \times \dfrac{2 \text{ mol HCN}}{3 \text{ mol O}_2} = \underline{6.67 \text{ mol HCN}}$

$10.0 \text{ mol O}_2 \times \dfrac{6 \text{ mol H}_2\text{O}}{3 \text{ mol O}_2} = \underline{20.0 \text{ mol H}_2\text{O}}$

7-8 (a) $4.86 \text{ mol HF} \times \dfrac{1 \text{ mol SiF}_4}{4 \text{ mol HF}} \times \dfrac{104.1 \text{ g SiF}_4}{\text{mol SiF}_4} = \underline{126 \text{ g SiF}_4}$

(b) $4.86 \text{ mol HF} \times \dfrac{2 \text{ mol H}_2\text{O}}{4 \text{ mol HF}} \times \dfrac{18.02 \text{ g H}_2\text{O}}{\text{mol H}_2\text{O}} = \underline{43.8 \text{ g H}_2\text{O}}$

(c) $4.86 \text{ mol HF} \times \dfrac{1 \text{ mol SiO}_2}{4 \text{ mol HF}} \times \dfrac{60.09 \text{ g SiO}_2}{\text{mol SiO}_2} = \underline{73.0 \text{ g SiO}_2}$

7-9 (a) **mol H$_2$O → mol H$_2$**

$0.400 \text{ mol H}_2\text{O} \times \dfrac{2 \text{ mol H}_2}{2 \text{ mol H}_2\text{O}} = \underline{0.400 \text{ mol H}_2}$

(b) **g O$_2$ → mol O$_2$ → mol H$_2$O → g H$_2$O**

$0.640 \text{ g O}_2 \times \dfrac{1 \text{ mol O}_2}{32.00 \text{ g O}_2} \times \dfrac{2 \text{ mol H}_2\text{O}}{1 \text{ mol O}_2} \times \dfrac{18.02 \text{ g H}_2\text{O}}{\text{mol H}_2\text{O}} = \underline{0.721 \text{ g H}_2\text{O}}$

(c) **g O$_2$ → mol O$_2$ → mol H$_2$ → g H$_2$**

$0.032 \text{ g O}_2 \times \dfrac{1 \text{ mol O}_2}{32.00 \text{ g O}_2} \times \dfrac{2 \text{ mol H}_2}{1 \text{ mol O}_2} \times \dfrac{2.016 \text{ g H}_2}{\text{mol H}_2} = \underline{0.0040 \text{ g H}_2}$

(d) **g H$_2$ → mol H$_2$ → mol H$_2$O → g H$_2$O**

$0.400 \text{ g H}_2 \times \dfrac{1 \text{ mol H}_2}{2.016 \text{ g H}_2} \times \dfrac{2 \text{ mol H}_2\text{O}}{2 \text{ mol H}_2} \times \dfrac{18.02 \text{ g H}_2\text{O}}{\text{mol H}_2\text{O}} = \underline{3.58 \text{ g H}_2\text{O}}$

7-10 (a)

$$0.450 \;\cancel{\text{mol } C_3H_8} \begin{cases} \times \dfrac{3 \text{ mol } CO_2}{1 \;\cancel{\text{mol } C_3H_8}} = \underline{1.35 \text{ mol } CO_2} \\[2mm] \times \dfrac{4 \text{ mol } H_2O}{1 \;\cancel{\text{mol } C_3H_8}} = \underline{1.80 \text{ mol } H_2O} \\[2mm] \times \dfrac{5 \text{ mol } O_2}{1 \;\cancel{\text{mol } C_3H_8}} = \underline{2.25 \text{ mol } O_2} \end{cases}$$

(b) $\mathbf{g\ CO_2} \rightarrow \mathbf{mol\ CO_2} \rightarrow \mathbf{mol\ H_2O} \rightarrow \mathbf{g\ H_2O}$

$$8.80 \;\cancel{\text{g } CO_2} \times \frac{1 \;\cancel{\text{mol } CO_2}}{44.01 \;\cancel{\text{g } CO_2}} \times \frac{4 \;\cancel{\text{mol } H_2O}}{3 \;\cancel{\text{mol } CO_2}} \times \frac{18.02 \text{ g } H_2O}{\cancel{\text{mol } H_2O}} = \underline{4.80 \text{ g } H_2O}$$

(c) $\mathbf{g\ H_2O} \rightarrow \mathbf{mol\ H_2O} \rightarrow \mathbf{mol\ C_3H_8} \rightarrow \mathbf{g\ C_3H_8}$

$$1.80 \;\cancel{\text{g } H_2O} \times \frac{1 \;\cancel{\text{mol } H_2O}}{18.02 \;\cancel{\text{g } H_2O}} \times \frac{1 \;\cancel{\text{mol } C_3H_8}}{4 \;\cancel{\text{mol } H_2O}} \times \frac{44.09 \text{ g } C_3H_8}{\cancel{\text{mol } C_3H_8}} = \underline{1.10 \text{ g } C_3H_8}$$

(d) $\mathbf{g\ O_2} \rightarrow \mathbf{mol\ O_2} \rightarrow \mathbf{mol\ C_3H_8} \rightarrow \mathbf{g\ C_3H_8}$

$$160 \;\cancel{\text{g } O_2} \times \frac{1 \;\cancel{\text{mol } O_2}}{32.00 \;\cancel{\text{g } O_2}} \times \frac{1 \;\cancel{\text{mol } C_3H_8}}{5 \;\cancel{\text{mol } O_2}} \times \frac{44.09 \text{ g } C_3H_8}{\cancel{\text{mol } C_3H_8}} = \underline{44.1 \text{ g } C_3H_8}$$

(e) $\mathbf{g\ O_2} \rightarrow \mathbf{mol\ O_2} \rightarrow \mathbf{mol\ CO_2} \rightarrow \mathbf{g\ CO_2}$

$$6.38 \;\cancel{\text{g } O_2} \times \frac{1 \;\cancel{\text{mol } O_2}}{32.00 \;\cancel{\text{g } O_2}} \times \frac{3 \;\cancel{\text{mol } CO_2}}{5 \;\cancel{\text{mol } O_2}} \times \frac{44.01 \text{ g } CO_2}{\cancel{\text{mol } CO_2}} = \underline{5.26 \text{ g } CO_2}$$

(f) $\mathbf{molecules\ CO_2} \rightarrow \mathbf{mol\ CO_2} \rightarrow \mathbf{mol\ H_2O}$

$$4.50 \times 10^{22} \;\cancel{\text{molecules}} \times \frac{1 \;\cancel{\text{mol } CO_2}}{6.022 \times 10^{23} \;\cancel{\text{molecules}}} \times \frac{4 \text{ mol } H_2O}{3 \;\cancel{\text{mol } CO_2}} = \underline{0.0996 \text{ mol } H_2O}$$

7-12
$$N_2 + O_2 \rightarrow 2NO$$
$$2NO + O_2 \rightarrow 2NO_2$$
$$N_2 + 2O_2 \rightarrow 2NO_2 \quad \text{(total reaction)}$$

$$\mathbf{g\ NO_2} \rightarrow \mathbf{mol\ NO_2} \rightarrow \mathbf{mol\ N_2} \rightarrow \mathbf{g\ N_2}$$

$$155 \;\cancel{\text{g } NO_2} \times \frac{1 \;\cancel{\text{mol } NO_2}}{46.01 \;\cancel{\text{g } NO_2}} \times \frac{1 \;\cancel{\text{mol } N_2}}{2 \;\cancel{\text{mol } NO_2}} \times \frac{28.02 \text{ g } N_2}{\cancel{\text{mol } N_2}} = \underline{47.2 \text{ g } N_2}$$

131

7-14   g CaCO$_3$ → mol CaCO$_3$ → mol HCl → g HCl

$$1.00 \text{ g CaCO}_3 \times \frac{1 \text{ mol CaCO}_3}{100.1 \text{ g CaCO}_3} \times \frac{2 \text{ mol HCl}}{1 \text{ mol CaCO}_3} \times \frac{36.46 \text{ g HCl}}{\text{mol HCl}} = \underline{0.728 \text{ g HCl}}$$

7-16   g FeS$_2$ → mol FeS$_2$ → mol H$_2$S → g H$_2$S

$$62.4 \text{ g FeS}_2 \times \frac{1 \text{ mol FeS}_2}{120.0 \text{ g FeS}_2} \times \frac{1 \text{ mol H}_2\text{S}}{1 \text{ mol FeS}_2} \times \frac{34.09 \text{ g H}_2\text{S}}{\text{mol H}_2\text{S}} = \underline{17.7 \text{ g H}_2\text{S}}$$

7-17   kg NO$_2$ → g NO$_2$ → mol NO$_2$ → mol HNO$_3$ → g HNO$_3$

$$18.5 \text{ kg NO}_2 \times \frac{10^3 \text{ g NO}_2}{\text{kg NO}_2} \times \frac{1 \text{ mol NO}_2}{46.01 \text{ g NO}_2} \times$$

$$\frac{2 \text{ mol HNO}_3}{3 \text{ mol NO}_2} \times \frac{63.02 \text{ g HNO}_3}{\text{mol HNO}_3} = \underline{16,900 \text{ g (16.9 kg) HNO}_3}$$

7-19   g sugar → mol sugar → mol alcohol → g alcohol

$$451 \text{ g C}_6\text{H}_{12}\text{O}_6 \times \frac{1 \text{ mol C}_6\text{H}_{12}\text{O}_6}{180.2 \text{ g C}_6\text{H}_{12}\text{O}_6} \times \frac{2 \text{ mol C}_2\text{H}_5\text{OH}}{1 \text{ mol C}_6\text{H}_{12}\text{O}_6} \times \frac{46.07 \text{ g C}_2\text{H}_5\text{OH}}{\text{mol C}_2\text{H}_5\text{OH}}$$
$$= \underline{231 \text{ g C}_2\text{H}_5\text{OH}}$$

7-20   molecules CH$_4$ → mol CH$_4$ → mol CO → g CO

$$8.75 \times 10^{25} \text{ molecules CH}_4 \times \frac{1 \text{ mol CH}_4}{6.022 \times 10^{23} \text{ molecules CH}_4}$$

$$\times \frac{2 \text{ mol CO}}{1 \text{ mol CH}_4} \times \frac{28.01 \text{ g CO}}{\text{mol CO}} = \underline{8140 \text{ g CO} = 8.14 \text{ kg CO}}$$

7-21   (a) $3.00 \text{ mol CuO} \times \frac{1 \text{ mol N}_2}{3 \text{ mol CuO}} = 1.00 \text{ mol N}_2$ - limiting reactant

$3.00 \text{ mol NH}_3 \times \frac{1 \text{ mol N}_2}{2 \text{ mol NH}_3} = 1.50 \text{ mol N}_2$

(b) Stoichiometric mixture producing 1.00 mol N$_2$

(c) $1.00 \text{ mol NH}_3 \times \frac{1 \text{ mol N}_2}{2 \text{ mol NH}_3} = 0.500 \text{ mol N}_2$ - limiting reactant

(d) $0.628 \text{ mol CuO} \times \dfrac{1 \text{ mol N}_2}{3 \text{ mol CuO}} = 0.209 \text{ mol N}_2$ - limiting reactant

$0.430 \text{ mol NH}_3 \times \dfrac{1 \text{ mol N}_2}{2 \text{ mol NH}_3} = 0.215 \text{ mol N}_2$

(e) $5.44 \text{ mol CuO} \times \dfrac{1 \text{ mol N}_2}{3 \text{ mol CuO}} = 1.81 \text{ mol N}_2$

$3.50 \text{ mol NH}_3 \times \dfrac{1 \text{ mol N}_2}{2 \text{ mol NH}_3} = 1.75 \text{ mol N}_2$ - limiting reactant

7-22   (a) $3.00 \text{ mol CuO} \times \dfrac{2 \text{ mol NH}_3}{3 \text{ mol CuO}} = 2.00 \text{ mol NH}_3$ used

$3.00 - 2.00 = 1.00 \text{ mol NH}_3$ in excess

(c) 1.50 mol CuO in excess

7-25   First, find the limiting reactant:

$0.800 \text{ mol Al} \times \dfrac{3 \text{ mol H}_2}{2 \text{ mol Al}} = 1.20 \text{ mol H}_2$

$1.00 \text{ mol H}_2\text{SO}_4 \times \dfrac{3 \text{ mol H}_2}{3 \text{ mol H}_2\text{SO}_4} = 1.00 \text{ mol H}_2$

Therefore, H_2SO_4 is the limiting reactant and the yield of H_2 is <u>1.00 mole.</u>

Now convert moles of H_2SO_4 to moles of Al.

$1.00 \text{ mol H}_2\text{SO}_4 \times \dfrac{2 \text{ mol Al}}{3 \text{ mol H}_2\text{SO}_4} = 0.667 \text{ mol of Al used}$

$0.800 - 0.667 = $ <u>0.133 mol Al remaining</u>

7-26 $3.44 \text{ mol C}_5\text{H}_6 \times \dfrac{10 \text{ mol CO}_2}{2 \text{ mol C}_5\text{H}_6} = 17.2 \text{ mol CO}_2$

$20.6 \text{ mol O}_2 \times \dfrac{10 \text{ mol CO}_2}{13 \text{ mol O}_2} = 15.8 \text{ mol CO}_2$

Since O_2 is the limiting reactant: $15.8 \text{ mol CO}_2 \times \dfrac{44.01 \text{ g CO}_2}{\text{mol CO}_2} = $ <u>695 g CO$_2$</u>

7-28

$g\,O_2 \;\rightarrow\; mol\,O_2$

$mol\,NH_3$

$mol\,N_2$

$40.0 \text{ g O}_2 \times \dfrac{1 \text{ mol O}_2}{32.00 \text{ g O}_2} \times \dfrac{2 \text{ mol N}_2}{3 \text{ mol O}_2} = 0.833 \text{ mol N}_2$

$1.50 \text{ mol NH}_3 \times \dfrac{2 \text{ mol N}_2}{4 \text{ mol NH}_3} = 0.750 \text{ mol N}_2$

Since NH_3 produces the least N_2, it is the limiting reactant and the yield of N_2 is <u>0.750 mol</u>.

7-29 g $AgNO_3$ → mol $AgNO_3$

mol AgCl → g AgCl

g $CaCl_2$ → mol $CaCl_2$

$$20.0 \text{ g AgNO}_3 \times \frac{1 \text{ mol AgNO}_3}{169.9 \text{ g AgNO}_3} \times \frac{2 \text{ mol AgCl}}{2 \text{ mol AgNO}_3} = 0.118 \text{ mol AgCl}$$

$$10.0 \text{ g CaCl}_2 \times \frac{1 \text{ mol CaCl}_2}{111.0 \text{ g CaCl}_2} \times \frac{2 \text{ mol AgCl}}{1 \text{ mol CaCl}_2} = 0.180 \text{ mol AgCl}$$

Since $AgNO_3$ produces the least AgCl, it is the limiting reactant.

$$0.118 \text{ mol AgCl} \times \frac{143.4 \text{ g AgCl}}{\text{mol AgCl}} = \underline{16.9 \text{ g AgCl}}$$

Convert moles of AgCl (the limiting reactant) to grams of $CaCl_2$ used.

$$0.118 \text{ mol AgCl} \times \frac{1 \text{ mol CaCl}_2}{2 \text{ mol AgCl}} \times \frac{111.0 \text{ g CaCl}_2}{\text{mol CaCl}_2} = 6.55 \text{ g CaCl}_2 \text{ used}$$

$$10.0 \text{ g} - 6.55 \text{ g} = 3.5 \text{ g CaCl}_2 \text{ remaining}$$

7-31 g HNO_3 → mol HNO_3

mol H_2O → g H_2O

g H_2S → mol H_2S

$$10.0 \text{ g HNO}_3 \times \frac{1 \text{ mol HNO}_3}{63.02 \text{ g HNO}_3} \times \frac{4 \text{ mol H}_2O}{2 \text{ mol HNO}_3} = 0.317 \text{ mol H}_2O$$

$$5.00 \text{ g H}_2S \times \frac{1 \text{ mol H}_2S}{34.09 \text{ g H}_2S} \times \frac{4 \text{ mol H}_2O}{3 \text{ mol H}_2S} = 0.196 \text{ mol H}_2O \text{ (limiting reactant)}$$

$$0.196 \text{ mol H}_2O \times \frac{18.02 \text{ g H}_2O}{\text{mol H}_2O} = \underline{3.53 \text{ g H}_2O}$$

Convert moles H_2O to grams of other two products.

$$0.196 \text{ mol H}_2O \times \frac{3 \text{ mol S}}{4 \text{ mol H}_2O} \times \frac{32.07 \text{ g S}}{\text{mol S}} = \underline{4.71 \text{ g S}}$$

$$0.196 \text{ mol H}_2O \times \frac{2 \text{ mol NO}}{4 \text{ mol H}_2O} \times \frac{30.01 \text{ g NO}}{\text{mol NO}} = \underline{2.94 \text{ g NO}}$$

Convert mol H_2O to grams of excess reactant used.

$$0.196 \text{ mol } H_2O \times \frac{2 \text{ mol } HNO_3}{4 \text{ mol } H_2O} \times \frac{63.02 \text{ g } HNO_3}{\text{mol } HNO_3} = 6.18 \text{ g } HNO_3$$

$$10.0 \text{ g} - 6.18 \text{ g} = 3.8 \text{ g } HNO_3 \text{ remaining}$$

7-32 $\quad g\,SO_2 \;\rightarrow\; mol\,SO_2 \;\rightarrow\; mol\,SO_3 \;\rightarrow\; g\,SO_3$

$$24.0 \text{ g } SO_2 \times \frac{1 \text{ mol } SO_2}{64.07 \text{ g } SO_2} \times \frac{2 \text{ mol } SO_3}{2 \text{ mol } SO_2} \times \frac{80.07 \text{ g } SO_3}{\text{mol } SO_3} = 30.0 \text{ g } SO_3 \text{ (theoretical yield)}$$

$$\text{percent yield} = \frac{21.2 \text{ g}}{30.0 \text{ g}} \times 100\% = \underline{70.7\%}$$

7-35 $\qquad g\,C_8H_{18} \;\rightarrow\; mol\,C_8H_{18} \;\rightarrow\; mol\,CO_2 \;\rightarrow\; g\,CO_2$

$$57.0 \text{ g } C_8H_{18} \times \frac{1 \text{ mol } C_8H_{18}}{114.2 \text{ g } C_8H_{18}} \times \frac{16 \text{ mol } CO_2}{2 \text{ mol } C_8H_{18}} \times \frac{44.01 \text{ g } CO_2}{\text{mol } CO_2} = 176 \text{ g } CO_2$$

$$\frac{152 \text{ g}}{176 \text{ g}} \times 100\% = \underline{86.4\%}$$

7-36 If 86.4% is converted to CO_2, the remainder (13.6%) is converted to CO. Thus, $0.136 \times 57.0 \text{ g} = 7.75 \text{ g of } C_8H_{18}$ is converted to CO. Notice that 1 mole of C_8H_{18} forms 8 moles of CO (because of the eight carbons in C_8H_{18}). Thus

$$g\,C_8H_{18} \;\rightarrow\; mol\,C_8H_{18} \;\rightarrow\; mol\,CO \;\rightarrow\; g\,CO$$

$$7.75 \text{ g } C_8H_{18} \times \frac{1 \text{ mol } C_8H_{18}}{114.2 \text{ g } C_8H_{18}} \times \frac{8 \text{ mol } CO}{1 \text{ mol } C_8H_{18}} \times \frac{28.01 \text{ g } CO}{\text{mol } CO} = \underline{15.2 \text{ g } CO}$$

7-37 Theoretical yield $\times 0.700 = 250 \text{ g}$ (actual yield) Theoretical yield $= 250 \text{ g}/0.700 = 357 \text{ g } N_2$

$$g\,N_2 \;\rightarrow\; mol\,N_2 \;\rightarrow\; mol\,H_2 \;\rightarrow\; g\,H_2$$

$$357 \text{ g } N_2 \times \frac{1 \text{ mol } N_2}{28.02 \text{ g } N_2} \times \frac{4 \text{ mol } H_2}{1 \text{ mol } N_2} \times \frac{2.016 \text{ g } H_2}{\text{mol } H_2} = \underline{103 \text{ g } H_2}$$

7-40 $\quad 2Mg(s) + O_2 \;\rightarrow\; 2MgO(s) + 1204 \text{ kJ}$

$$2Mg(s) + O_2(g) \;\rightarrow\; 2MgO(s) \quad \Delta H = -1204 \text{ kJ}$$

7-42 $\quad CaCO_3(s) + 176 \text{ kJ} \;\rightarrow\; CaO(s) + CO_2(g)$

$$CaCO_3(s) \;\rightarrow\; CaO(s) + CO_2(g) \quad \Delta H = 176 \text{ kJ}$$

7-43 $g\ C_8H_{18}$ → $mol\ C_8H_{18}$ → kJ

$$1.00\ \cancel{g\ C_8H_{18}} \times \frac{1\ \cancel{mol\ C_8H_{18}}}{114.2\ \cancel{g\ C_8H_{18}}} \times \frac{5480\ kJ}{\cancel{mol\ C_8H_{18}}} = \underline{48.0\ kJ}$$

$$1.00\ \cancel{g\ CH_4} \times \frac{1\ \cancel{mol\ CH_4}}{16.04\ \cancel{g\ CH_4}} \times \frac{890\ kJ}{\cancel{mol\ CH_4}} = \underline{55.6\ kJ}$$

7-45 $2Al(s) + Fe_2O_3(s)$ → $Al_2O_3(s) + Fe(l)$ $\Delta H = -850\ kJ$

kJ → mol Al → g Al

$$35.8\ \cancel{kJ} \times \frac{2\ \cancel{mol\ Al}}{850\ \cancel{kJ}} \times \frac{26.99\ g\ Al}{\cancel{mol\ Al}} = \underline{2.27\ g\ Al}$$

7-46   kJ → $mol\ C_6H_{12}O_6$ → $g\ C_6H_{12}O_6$

$$975\ \cancel{kJ} \times \frac{1\ \cancel{mol\ C_6H_{12}O_6}}{2519\ \cancel{kJ}} \times \frac{180.2\ g\ C_6H_{12}O_6}{\cancel{mol\ C_6H_{12}O_6}} = 69.7\ g\ C_6H_{12}O_6$$

7-48 $g\ Fe_2O_3$ → $mol\ Fe_2O_3$ → $mol\ Fe_3O_4$ →

$mol\ FeO$ → $mol\ Fe$ → $g\ Fe$

$$125\ \cancel{g\ Fe_2O_3} \times \frac{1\ \cancel{mol\ Fe_2O_3}}{159.7\ \cancel{g\ Fe_2O_3}} \times \frac{2\ \cancel{mol\ Fe_3O_4}}{3\ \cancel{mol\ Fe_2O_3}} \times \frac{3\ \cancel{mol\ FeO}}{1\ \cancel{mol\ Fe_3O_4}}$$

$$\times \frac{1\ \cancel{mol\ Fe}}{1\ \cancel{mol\ FeO}} \times \frac{55.85\ g\ Fe}{\cancel{mol\ Fe}} = \underline{87.4\ g\ Fe}$$

7-49 $2KClO_3$ → $2KCl + 3O_2$
Find the mass of $KClO_3$ needed to produce 12.0 g O_2.

$g\ O_2$ → $mol\ O_2$ → $mol\ KClO_3$ → $g\ KClO_3$

$$12.0\ \cancel{g\ O_2} \times \frac{1\ \cancel{mol\ O_2}}{32.00\ \cancel{g\ O_2}} \times \frac{2\ \cancel{mol\ KClO_3}}{3\ \cancel{mol\ O_2}} \times \frac{122.6\ g\ KClO_3}{\cancel{mol\ KClO_3}} = 30.7\ g\ KClO_3$$

$$\text{percent purity} = \frac{30.7\ g}{50.0\ g} \times 100\% = \underline{61.4\%}$$

7-50 Convert g of SO_2 to g of FeS_2

$$g\,SO_2 \rightarrow mol\,SO_2 \rightarrow mol\,FeS_2 \rightarrow g\,FeS_2$$

$$312\,\cancel{g\,SO_2} \times \frac{1\,\cancel{mol\,SO_2}}{64.07\,\cancel{g\,SO_2}} \times \frac{4\,\cancel{mol\,FeS_2}}{8\,\cancel{mol\,SO_2}} \times \frac{120.0\,g\,FeS_2}{\cancel{mol\,FeS_2}} = 292\,g\,FeS_2$$

$$\frac{292\,g}{6500\,g} \times 100\% = \underline{4.49\%\,FeS_2}$$

7-52 (1) Find the limiting reactant.

$$g\,NH_3 \rightarrow mol\,NH_3$$

$$\searrow$$

$$mol\,NO \rightarrow g\,NO$$

$$g\,O_2 \rightarrow mol\,O_2 \nearrow$$

$$80.0\,\cancel{g\,NH_3} \times \frac{1\,\cancel{mol\,NH_3}}{17.03\,\cancel{g\,NH_3}} \times \frac{4\,mol\,NO}{4\,\cancel{mol\,NH_3}} = 4.70\,mol\,NO\,(limiting\,reactant)$$

$$200\,\cancel{g\,O_2} \times \frac{1\,\cancel{mol\,O_2}}{32.00\,\cancel{g\,O_2}} \times \frac{4\,mol\,NO}{5\,\cancel{mol\,O_2}} = 5.00\,mol\,NO$$

(2) Find the theoretical yield based on NH_3.

$$4.70\,\cancel{mol\,NO} \times \frac{30.01\,g\,NO}{\cancel{mol\,NO}} = 141\,g\,NO\,(theoretical\,yield) \qquad \frac{40.0\,g}{141\,g} \times 100\% = \underline{28.4\%\,yield}$$

7-54 $0.250\,\cancel{mol\,H_2O} \times \frac{1\,mol\,CaCl_2 \cdot 6H_2O}{5\,\cancel{mol\,H_2O}} = 0.0500\,mol\,CaCl_2 \cdot 6H_2O$ - limiting reactant

$$9.50 \times 10^{22}\,\cancel{molecules\,HCl} \times \frac{1\,\cancel{mol\,HCl}}{6.022 \times 10^{23}\,\cancel{molecules\,HCl}} \times$$

$$\frac{1\,mol\,CaCl_2 \cdot 6H_2O}{2\,\cancel{mol\,HCl}} = 0.0789\,mol\,CaCl_2 \cdot 6H_2O$$

$$15.0\,\cancel{g\,CaCO_3} \times \frac{1\,\cancel{mol\,CaCO_3}}{100.1\,\cancel{g\,CaCO_3}} \times \frac{1\,mol\,CaCl_2 \cdot 6H_2O}{1\,\cancel{mol\,CaCO_3}} = 0.150\,mol\,CaCl_2 \cdot 6H_2O$$

$$0.050\,\cancel{mol\,CaCl_2 \cdot 6H_2O} \times \frac{219.1\,g\,CaCl_2 \cdot 6H_2O}{\cancel{mol\,CaCl_2 \cdot 6H_2O}} = \underline{11.0\,g\,CaCl_2 \cdot 6H_2O}$$

7-55 C: $14.1\,\cancel{g\,C} \times \frac{1\,mol\,C}{12.01\,\cancel{g\,C}} = 1.17\,mol\,C$

H: $2.35\,\cancel{g\,H} \times \frac{1\,mol\,H}{1.008\,\cancel{g\,H}} = 2.33\,mol\,H$

Cl: $83.5\,\cancel{g\,Cl} \times \frac{1\,mol\,Cl}{35.45\,\cancel{g\,Cl}} = 2.36\,mol\,Cl$

C: $\frac{1.17}{1.17} = 1.0$ H: $\frac{2.33}{1.17} = 2.0$ Cl: $\frac{2.36}{1.17} = 2$. Molecular formula $= \underline{CH_2Cl_2}$

$$CH_4(g) + 2Cl_2(g) \longrightarrow CH_2Cl_2(l) + 2HCl(g)$$

$$2.85\,\cancel{g\,CH_4} \times \frac{1\,\cancel{mol\,CH_4}}{16.04\,\cancel{g\,CH_4}} \times \frac{1\,mol\,CH_2Cl_2}{1\,\cancel{mol\,CH_4}} = 0.178\,mol\,CH_2Cl_2$$

$15.0 \text{ g } Cl_2 \times \dfrac{1 \text{ mol } Cl_2}{70.90 \text{ g } Cl_2} \times \dfrac{1 \text{ mol } CH_2Cl_2}{2 \text{ mol } Cl_2} = 0.106 \text{ mol } CH_2Cl_2 \text{ - limiting reactant}$

$0.106 \text{ mol } CH_2Cl_2 \times \dfrac{84.93 \text{ g } CH_2Cl_2}{\text{mol } CH_2Cl_2} = \underline{9.00 \text{ g } CH_2Cl_2}$

8

Modern Atomic Theory

Review of Part A   *The Energy of the Electron in the Atom*

OBJECTIVES AND DETAILED TABLE OF CONTENTS

8-1 The Emission Spectra of the Elements and Bohr's Model

OBJECTIVE *Describe how Bohr's model of the atom accounts for observed wavelengths and energies of emission spectra.*

8-2 Modern Atomic Theory: A Closer Look at Energy Levels

OBJECTIVE *Describe the electronic structure of an atom, including shells, subshells, and various types of orbitals.*

8-2.1 The Wave Mechanical Model
8-2.2 The First Shell and *s* Orbitals
8-2.3 The Second Shell and *p* Orbitals
8-2.4 Outer Shells with *d* and *f* Orbitals

SUMMARY OF PART A

Light is a form of energy that interacts with matter. The light coming to us from the sun contains all of the colors of the rainbow and is thus known as a **continuous spectrum**. Light from a hot, glowing, gaseous element, however, emits specific colors of light, which is known as a **discrete spectrum**. The light that we perceive in the **visible spectrum** is bordered by light with **wavelengths** slightly longer than red (the **infrared**) and shorter than violet (the **ultraviolet**). It was the orderly spectrum of hydrogen, the lightest element, which attracted the most curiosity of scientists in the early part of the 20th century, however.

In 1913, Niels Bohr explained the origin of the discrete spectrum of hydrogen by describing the energy states available to the one electron in hydrogen. His model was analogous to the planets revolving around the sun, except that the electron can exist in only definite (**quantized**) **energy levels** with each level designated by a **principal quantum number (n)**. When energy in the form of heat is supplied to the atom, the electron jumps from the lowest energy state (the **ground state**) to a higher energy state (an **excited state**). When the electron falls back down to a lower energy state, it re-emits the energy in the form of light. Since the difference in energy between the higher and lower energy levels is discrete, the emitted light has a discrete energy. In the visible region of the spectrum, hydrogen emits four colors: violet, indigo, green, and red. Even though modern theories use a more complicated model of the atom, Bohr's model remains useful to the chemist because it presents an easily visualized picture.

The modern view of the atom, called the **wave mechanical model**, provides more detail about the energy and other properties of the electron in the hydrogen atom. In this model, the electron is not viewed as a hard particle with a definite energy and position but instead as having a probability of existing in a region of space known as an **orbital**. Different types of orbitals exist with different shapes. For example, *s* type orbitals have spherical shapes that include most of the electron density. In the second shell (n = 2), *p* type orbitals exist which have two "lobes" of electron density lying along a three-coordinate axis. There are three *p* type orbitals in the second and successive shells. The third and higher energy shells also have *d* type orbitals and the fourth and higher shells have *f* type orbitals. The latter two types have complex shapes. If the principal energy level is known as a **shell**, the orbitals of each type within a shell together make up what is known as a **subshell**. For example the second shell (n = 2) contains two subshells, the 2*s* (one orbital) and the 2*p* (three orbitals). The n = 3 shell has "*s*", "*p*", and "*d*" subshells, which are listed in order of increasing energy. This information is summarized in the following table. Notice that the capacity of each shell is $2n^2$ electrons.

Table 8-A									
Shell	1	2		3			4		
Subshell	s	s	p	s	p	d	s	p	d	f
Subshell capacity	2	2	6	2	6	10	2	6	10	14
Shell capacity	2	8		18			32			

ASSESSMENT OF OBJECTIVES

A-1 Multiple Choice

_____ 1. Which of the following is an important postulate of the Bohr model of the atom?

(a) electrons exist in indefinite orbits around the nucleus
(b) the energy of the light is directly proportional to its wavelength
(c) atoms emit light when electrons jump to higher energy orbits
(d) the discrete spectrum of hydrogen is a property of the nucleus
(e) atoms emit discrete wavelengths of light because there is a discrete difference in energy between orbits

_____ 2. How many orbitals are in the $4f$ subshell?

(a) 14   (b) 7   (c) 6   (d) 10   (e) 4

_____ 3. How many total orbitals are in the $n = 2$ shell?

(a) 2   (b) 3   (c) 4   (d) 8   (e) 6

_____ 4. How many "lobes" does one p orbital have?

(a) one   (b) none   (c) two   (d) four   (e) three

_____ 5. What is the electron capacity of the $n = 3$ shell?

(a) 2   (b) 8   (c) 18   (d) 10   (e) 32

_____ 6. What subshells are in the $n = 3$ shell?

(a) s, p, d (b) s, p (c) s, p, f (d) p, d, f (e) s, d, f

_____ 7. What is the electron capacity of the d subshell?

(a) 2   (b) 6   (c) 10   (d) 8   (e) 14

_____ 8. Which of the following subshells does not exist?

(a) $7s$ (b) $2d$ (c) $2p$ (d) $5f$ (e) $6d$

141

_____ 9. Which of the following subshells is the highest in energy?

(a) 6*p*   (b) 5*d*   (c) 6*s*   (d) 4*f*   (e) 5*f*

A-2. Matching

_____ Orbital

_____ Excited state

_____ Ultraviolet light

_____ Shell

_____ Discrete spectra

_____ Subshell

_____ Wavelength

(a) When an electron occupies an energy level higher than the lowest available energy level

(b) A principal energy level in an atom

(c) Light that is lower in energy than the visible

(d) The rainbow

(e) A region of space in which one or two electrons are found

(f) The distance between the lowest and highest point on a wave

(g) Bands of light of definite energies

(h) When all electrons are in the lowest available energy states

(i) All of the orbitals of one type within a shell

(j) Light with shorter wavelengths than the visible

(k) The distance between two equivalent points on a wave

A-3. Orbital Shape

1. Sketch the shapes of all orbitals in the $n = 2$ shell.

8-3   Electron Configurations of the Elements

OBJECTIVE *Using the periodic table, write the outer electron configuration of the atoms of a specific element.*

8-3.1 The Aufbau Principle and Electron Configuration
8-3.2 Assignment of Electrons into Shells and Subshells
8-3.3 Exceptions to the Normal Order of Filling
8-3.4 Using the Periodic Table to Determine Electron Configuration
8-3.5 Electron Configurations of Specific Groups

8-4   Orbital Diagrams of the Elements

OBJECTIVE *Using orbital diagrams, determine the distribution of electrons in an atom.*

8-4.1 The Pauli Exclusion Principle and Orbital Diagrams
8-4.2 Hund's Rule

8-5   Periodic Trends

OBJECTIVE *Using the periodic table, predict trends in atomic and ionic radii, ionization energy and electron affinity.*

8-5.1 Atomic Radius
8-5.2 Ionization Energy
8-5.3 Electron Affinity
8-5.4 Electron Configurations of Ions
8-5.5 Ionic Radii

SUMMARY OF PART B

We are now ready to discuss the **electron configuration** of the elements. This summarizes the location of all the electrons in the atoms of an element (shell, subshell, and number of electrons in each subshell). The standard method for electron configuration is as follows.

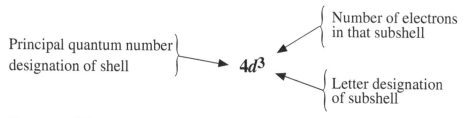

Electrons fill the subshells in the order of the lowest energy levels first. This is known as the **Aufbau principle**. The electron configurations of a few sample elements are as follows.

Element	Electron Configuration	Comments
Li	$1s^2 2s^1$	Two electrons fill the first shell, the third is in the next lowest level which is the $2s$ subshell.
Ne	$1s^2 2s^2 2p^6$	This element completes the filling of the two subshells of the $n = 2$ shell.
K	$[Ar]\, 4s^1$	The [Ar] shorthand notation represents all of the electrons in the noble gas Ar. Notice that the $4s$ is lower in energy than the $3d$ so the $4s$ fills first.
Zn	$[Ar]\, 4s^2 3d^{10}$	The $3d$ fills after the $4s$. This completes the filling of the $n = 3$ shell.
Kr	$[Ar]\, 4s^2 3d^{10} 4p^6$	This completes the filling of the $4p$ subshell. Notice that noble gases have filled outer ns and np subshells (except He which doesn't have a $1p$ subshell).
La	$[Xe]\, 6s^2 5d^1$	The periodic table tells us that this element has a d^1 configuration since it comes under Y which has $4d^1$.
Ce	$[Xe]\, 6s^2 5d^1 4f^1$	The next element after La has one electron in the $4f$ subshell.

When the orbitals in a subshell are represented as boxes, the illustration is known as an **orbital diagram** or **box diagram**. The orbitals present in the various subshells are thus illustrated as follows.

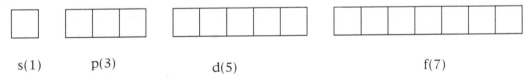

s(1)   p(3)   d(5)   f(7)

Electrons in an atom have one additional important property known as electronic "spin." Spin arises because the electron has properties of a spinning particle. The spin of an electron is represented as an arrow that points either up or down illustrating the two possible spins. According to the **Pauli exclusion principle**, if two electrons are present in the same orbital, they must have opposite or "paired" spins. The two "paired" electrons in an orbital are represented as one arrow pointing up and the other down, indicating opposite spins.

According to **Hund's rule**, electrons occupy separate orbitals with their spins parallel (given a choice). Thus the orbital diagram of a nitrogen atom is represented as follows. Notice that, according to Hund's rule, the three electrons are in separate p orbitals with parallel spins, leaving nitrogen with three unpaired electrons.

$1s^2$ $2s^2$ 2p

N   [↓↑]   [↓↑]   [↑ | ↑ | ↑]

Many years after the periodic table was first displayed, we find that the vertical columns have the same subshell configuration but successively higher shells. Thus electron configuration is called a *periodic* property. We can put this to good use in establishing the electron configurations of the elements by locating the element on the table and then writing the electron configuration expected for an element in that position. The four categories of elements can then be illustrated by their general electron configurations as follows. [NG] represents a noble gas configuration.

Representative elements: [NG] ns^x or [NG] ns^2np^y
     e.g.,   IA (alkali metals) [NG] ns^1 Rb: [Kr] $5s^1$
              IVA [NG] ns^2np^2 Si: [Ne] $3s^23p^2$

Noble gases: [NG] ns^2np^6 e.g., Ar: [Ne] $3s^23p^6$

Transition metals: [NG] $ns^2 (n-1)d^x$ e.g., Zr: [Kr] $5s^24d^2$

Inner transition metals: [NG] $ns^2 (n-1)d^1 (n-2)f^x$ e.g., Ce: [Xe] $6s^25d^14f^1$

Besides electron configuration, there are other properties of the atoms of the elements that vary in a periodic manner. The first property discussed that shows definite trends is the **atomic radius** of the atoms. The radius is the distance from the nucleus to the outermost electrons. There is a trend across a period and a trend down a group.

To understand these trends, one must appreciate how one electron affects the charge of the nucleus felt by another electron. Electrons in the same subshell do not *shield* each other effectively. As a result, the more electrons there are in an outer subshell, the more the nuclear charge is felt by all of these outer electrons and the smaller the atom. Thus elements generally become smaller as a subshell is filling. Going down a group, the outer electrons lie in a shell farther from the nucleus so atoms are generally larger.

Notice that the trend toward large atomic radii (down and to the left in the periodic table) is the same as the trend toward metallic character. This arrangement is no coincidence, since **ionization energy** (the energy required to remove an electron from a gaseous atom) is related to the trends in the atomic radii. The smaller the atom, the more tightly the outer electrons are held and the harder they are to remove. Thus large atoms like metals lose an electron easily compared to smaller atoms like the nonmetals. This is a very important difference in the properties of the metals and nonmetals. We also noticed that it is comparatively very difficult to remove electrons from noble gas structures. For example, the energy required to form Mg^{3+} is very large compared to the energy required to form Mg^{2+}. This is because Mg has two outer electrons beyond a noble gas (i.e., [Ne] $3s^2$). Notice that the first two electrons removed lie outside the neon core. The third electron, however, must be removed from the neon core and is much more firmly attached.

The electron configuration of representative metal ions tend to be **isoelectronic** with noble gases. Thus we find ions such as Se^{2-}, Br^-, Rb^+, and Sr^{2+} are all isoelectronic with the noble gas krypton. **Ionic radii** are a function of the charge. Anions are larger and cations are smaller than their parent atoms. The higher the charge on the ion, the larger is the anion and the smaller is the cation.

ASSESSMENT OF OBJECTIVES

B-1 Multiple Choice

_____ 1. Which of the following groups (vertical columns) in the periodic table have no unpaired electrons?

(a) IIB (b) IA (c) IIIA (d) IVA (e) IIIB

_____ 2. How many unpaired electrons are in the atoms of group IVB?

(a) none (b) one (c) two (d) three (e) four

_____ 3. Which of the following orbital diagrams is excluded by the Pauli exclusion principle?

(a)

(b)

(c)

(d)

_____ 4. Which of the following orbital diagrams is excluded by Hund's rule?

(a)

(b)

(c)

(d)

_____ 5. Which of the following elements has the general electron configuration, ns^2np^5 (n is the principal quantum number)?

(a) P (b) Ar (c) S (d) Mn (e) Cl

_____ 6. From the periodic table, predict which of the following elements has the smallest radius:

(a) Ga (b) P (c) O (d) S (e) N

_____ 7. From the periodic table, predict which of the following elements has the lowest ionization energy:

(a) K (b) Br (c) Ca (d) Rb (e) I

_____ 8. From the periodic table, predict which of the following cations would require the largest amount of energy to form:

(a) Ba^{3+} (b) Ga^{3+} (c) Cs^+ (d) Sr^{2+} (e) Ca^+

146

___ 9. Which of the following would have the smallest radius?

(a) Cl^- (b) S^{2-} (c) S   (d) Na   (e) Na^+

B-2 Matching

I.

_____ Aufbau principle

_____ Hund's rule

_____ Unpaired electron

_____ Pauli exclusion principle

(a) Electrons occupy excited states first.

(b) Present when an orbital is singly occupied

(c) Electrons occupy the lowest possible energy state.

(d) Electrons occupy separate orbitals of the same energy with parallel spin.

(e) Present when an orbital is doubly occupied.

(f) Electrons cannot occupy the same orbital with the same spin.

II. If any of the following ions are energetically feasible (comparably easy to form), put a check. If not, select the reason from the possibilities listed.

_____ Ba^{2+}

_____ O^{2+}

_____ Na^{2+}

_____ F^{2-}

_____ Mg^{2-}

(a) Nonmetals do not easily form cations.

(b) An electron would have to be removed from a noble gas core.

(c) Metals do not easily form anions.

(d) The outer p subshell cannot accommodate all the electrons.

B-3 Problems

1. Write the orbital diagram showing the arrangement of all electrons beyond the previous noble gas for the following. For example, the orbital diagram for Ca is

$$4s^2$$

$$\boxed{\uparrow\downarrow}$$

   (a) V

   (b) Cl

   (c) Ce

   (d) Zn

2. Using only the periodic table, write the electron configuration for:

   (a) N _____   (b) Ga _____

   (c) I _____   (d) Pm _____

   (e) Ta _____

3. Using only the periodic table, write the symbol of the element whose atoms have the following electron configurations:

   ____ (a) [Ne] $3s^2$ ____ (c) [Xe] $6s^2 4f^{14} 5d^{10}$

   ____ (b) [Ar] $4s^2 3d^5$ ____ (d) [Rn] $7s^2 6d^1 5f^3$

4. Indicate the symbol of the appropriate element.

   ____ (a) the first element with a d electron

   ____ (b) the first element with an f electron

   ____ (c) the first element to have a filled d subshell

   ____ (d) the first element with a d electron and a filled $4f$

   ____ (e) the first element that has the general configuration
   $$ns^2 (n-1)d^{10} np^3$$

   ____ (f) the atomic number of the first element with a filled $6d$ subshell

   ____ (g) the first element with a filled $5p$ subshell

Chapter Summary Assessment

S-1 Match the Elements

Using the periodic table, write the symbol of the appropriate element. The elements that correspond to a particular statement are listed below.

_____ 1. The element that has one electron in addition to a complete $4s$ subshell

_____ 2. The element that completes the filling of the $n = 3$ shell

_____ 3. Two elements with two unpaired electrons in the $4p$ subshell

_____ 4. The element with four electrons more than the noble gas xenon

_____ 5. The element with the electron configuration: [Kr] $5s^2 4d^{10} 5p^3$

_____ 6. The element with the electron configuration: [Ar] $4s^1 3d^5$

_____ 7. The element with one electron in a p subshell and a full $4d$ subshell

_____ 8. The first element with p electrons that also has no unpaired electrons

_____ 9. The element that starts the filling of the $n = 4$ shell

_____ 10. The first element that has six filled orbitals

_____ 11. The first element that has three unpaired electrons in orbitals that lie at a $90°$ angle from each other

_____ 12. The first element with paired electrons in a $3d$ orbital

Elements: (a) Cr   (b) Ge   (c) Ne   (d) Mg   (e) Fe   (f) Sb   (g) Se   (h) Cu   (i) K
(j) Ce   (k) N   (l) In   (m) Sc

Answers to Assessments of Objectives

A-1 Multiple Choice

1. **e**   Atoms emit discrete wavelengths of light.

2. **b**   7

3. **c**   4 (one s and three p)

4. **c**   two lobes on either side of a nucleus

5. **c** $2n^2 = 18$

6. **a** s, p, d

7. **c**   10

8. **b** $2d$

9. **e** $5f$

A-2 Matching

e   An **orbital** is a region of space in which one or two electrons are found.

a   An electron is in an **excited state** when it occupies an energy level higher than the lowest available energy level.

j   **Ultraviolet light** has wavelengths shorter than light in the visible.

b   A **shell** is a principal energy level in an atom.

g   **Discrete spectra** are bands of light of definite energies.

i   A **subshell** contains all of the orbitals of one type within a shell.

l   **Wavelength** of light is the distance between two equivalent points on a wave.

A-3 Orbital Shape

The shapes of the $n = 2$ orbitals (s and p)

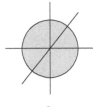 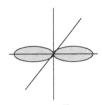 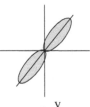 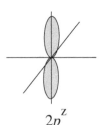

$2s$ $2p^x$ $2p^y$ $2p^z$

B-1 Multiple Choice

1. **a**   IIB

2. **c**   two unpaired electrons in d orbitals

3. **b**

4. **d**

5. **e**   Cl   6. **c**   O   7. **d**   Rb   8. **a** Ba^{3+} 9. **e** Na^{+}

150

B-2 Matching

I.   **c**   The **Aufbau principle** states that electrons occupy the lowest possible energy state.

d   **Hund's rule** states that electrons occupy separate orbitals of the same energy with parallel spins.

b   **Unpaired electron**s are present when an orbital is singly occupied.

f   The **Pauli exclusion principle** states that electrons in the same orbital must have opposite spins.

II. Ba^{2+} is OK.

O^{2+} **(a)** does not form because nonmetals do not form cations.

Na^{2+} **(b)** does not form because an electron would have to be removed from a noble gas core.

F^{2-} **(d)** does not form because the outer p subshell in fluorine cannot accommodate two extra electrons.

S^{2-} is OK.

Mg^{2-} **(c)** does not form because metals do not easily form anions.

B-3 Problems

1.   (a) V

$4s^2$ $3d^3$

(b) Cl

$3s^2$ $3p^5$

(c) Ce

$6s^2$ $5d^1$ $4f^1$

(d) Zn

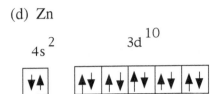

$4s^2$ $3d^{10}$

2. Electron configuration

(a) N [He] $2s^22p^3$ (d) Pm [Xe] $6s^25d^14f^4$

(b) Ga [Ar] $4s^23d^{10}4p^1$ (e) Ta [Xe] $6s^24f^{14}5d^3$

(c) I [Kr] $5s^24d^{10}5p^5$

3. Locate the elements

(a) [Ne] $3s^2$ Mg   (c) [Xe] $6s^24f^{14}5d^{10}$ Hg

(b) [Ar] $4s^23d^5$ Mn   (d) [Rn] $7s^26d^15f^3$ U

4. Indicate the element

(a) The first element with a d electron is **Sc (#21)**.

(b) The first element with an f electron is **Ce (#58)**.

(c) The first element to have a filled d subshell is **Cu (#29)**.

(d) The first element with a d electron and a filled $4f$ is **Lu (#71).**

(e) The first element that has the general configuration
$ns^2(n-1)d^{10}np^3$ is **As (#33)**.

(f) The atomic number of the first element with a filled $6d$ subshell
is element **#112** or **Rg (#111)** if $7s^16d^{10}$.

(g) The first element with a filled $5p$ subshell is **Xe (#54)**.

S-1 Match the Elements

1. Sc   7. In
2. Cu   8. Ne
3. Si and S   9. K
4. Ce   10. Mg
5. Sb   11. N
6. Cr   12. Fe

Answers and Solutions to Green Text Problems

8-1   Ultraviolet light has shorter wavelengths but higher energy than visible light. Ultraviolet light can damage living cells in tissues, thus causing a burn.

8-2   Since these two shells are close in energy, transitions of electrons from these two levels to the $n = 1$ shell have similar energy. Thus, the wavelengths of light from the two transitions are very close together.

8-3   Since these two shells are comparatively far apart in energy, transitions from these two levels to the $n = 1$ shell have comparatively different energies. Thus, the wavelengths of light from the two transitions are quite different. (The $n = 3$ to $n = 1$ transition has a shorter wavelength and higher energy than the $n = 2$ to the $n = 1$ transition.)

8-5   1p and 3f

8-8   A 4p orbital is shaped roughly like a two-sided baseball bat with two "lobes" lying along one of the three axes. This shape represents the region of highest probability of finding the 4p electrons.

8-9   The 3s orbital is spherical in shape. There is an equal probability of finding the electron regardless of the orientation from the nucleus. (In fact, the probability lies in three concentric spheres with the highest probability in the sphere farthest from the nucleus.) The highest probability of finding the electron lies farther from the nucleus in the 3s than in the 2s.

8-11   3s (one), 3p (three), 3d (five). Total (nine)

8-13   (a) 3p – three   (b) 4d - five   (c) 6s – one

8-15 $n = 3$ $2(3)^2 = \underline{18}$ $n = 2$ $2(2)^2 = \underline{8}$

8-18 $n = 5$ $2(5)^2 = \underline{50}$

8-19   The 5s, 5p, 5d, and 5f subshells hold a total of 32 electrons. Therefore, the 5g subshell must hold 50 - 32 = 18 electrons.

8-20   Since each orbital holds two electrons, 18 electrons means <u>nine orbitals</u> are present.

8-21   The 1s subshell always fills first.

8-24   (a) 6s   (b) 5p   (c) 4p   (d) 4d

8-26   4s, 4p, 5s, 4d, 5p, 6s, 4f

8-27   (a) Mg: $1s^2, 2s^2 2p^6 3s^2$
           (b) Ge: $1s^2 2s^2 2p^6 3s^2 3p^6 4s^2 3d^{10} 4p^2$
           (c) Cd: $1s^2 2s^2 2p^6 3s^2 3p^6 4s^2 3d^{10} 4p^6 5s^2 4d^{10}$
           (d) Si: $1s^2 2s^2 2p^6 3s^2 3p^2$

8-29   Ar: $1s^2 2s^2 2p^6 3s^2 3p^6$

8-31   (a) S: [Ne] $3s^2\,3p^4$ (c) Hf: [Xe]$6s^2 5d^2 4f^1$

         (b) Zn: [Ar] $4s^2\,3d^{10}$ (d) I: [Kr] $5s^2 4d^{10}\,5p^5$

8-33   (a) In   (b) Y   (c) Ce   (d) Ar

8-35   (a) 3p   (b) 5p   (c) 5d   (d) 5s

8-37   Both have three valence electrons. The outer electron in IIIA is in a p subshell; in IIIB it is in a d subshell.

8-39   (a) F   (b) Ga   (c) Ba   (d) Gd   (e) Cu

8-41   (a) VIIA   (b) IIIA   (c) IIA   (e) IB

8-42   VA [Noble gas] $ns^2\,np^3$ VB [Noble gas] $ns^2\,(n-1)d^3$

8-44   (b) and (e) - Group IVA

8-46   (a) IVA   (b) VIIIA   (c) IB   (d) Pr and Pa

8-47   (a) [NG] ns^2 (b) [NG] $ns^2(n-1)d^{10}$

         (c) [NG] ns^2np^4 or [NG] $ns^2(n-1)d^{10}np^4$ (d) [NG] $ns^2(n-1)d^2$

8-48   Helium does not have a filled p subshell. (There is no 1p subshell.)

8-50   [Ne] $3s^2\,3p^6$

8-52   The theoretical order of filling is 6d, 7p, 8s, 5g. The 6d is completed at element #112. The 7p and 8s fill at element #120. Thus, element #121 would theoretically begin the filling of the 5g subshell. This assumes the normal order of filling.

8-54 $7s^2\,6d^{10}\,5f^{14}\,7p^6$ Total 32

8-55 $8s^2\,5g^{18}\,7d^{10}\,6f^{14}\,8p^6$ Total 50

8-56   (a) transition   (b) representative   (c) noble gas   (d) inner transition

8-58   Element number 118 (under Rn)

8-60   (a) This is excluded by Hund's rule since electrons are not shown in separate orbitals of the same subshell with parallel spins.

         (b) This is correct.

         (c) This is excluded by the Aufbau principle because the 2s subshell fills before the 2p.

         (d) This is excluded by the Pauli exclusion principle since the two electrons in the 2s orbital cannot have the same spin.

8-61   (a) S

         3s   3p

(b) V

(c) Br

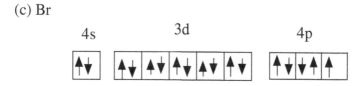

(d) Pm

8-63 IIB, none; VB, three; VIA, two; VIIA, one; [ns^1(n-1)d^5], six; Pm, five

8-65 IVA and VIA

8-67 (a) As (b) Ru (c) Ba (d) I

8-69 Cr - 117 pm, Nb - 134 pm

8-71 (a) V (b) Cl (c) Mg (d) Fe (e) B

8-72 In - 558 kJ/mol, Ge - 762 kJ/mol

8-73 Te - 869 kJ/mol, Br - 1140 kJ/mol

8-75 (a) Cs$^+$ easiest, (b) Rb^{2+} hardest

8-77 (d) I

8-79 Ba^{2+} = [Xe], S^{2-} = [Ar], Ga^{3+} = [Ar]3d^{10}, I- = [Xe], Cu$^+$ [Ar]3d^{10}

8-81 (d) Mg^{2+} (e) K$^+$ (c) S (a) Mg (b) S^{2-} (f) Se^{2-}

8-83 The outer electron in Hf is in a shell higher in energy than Zr. This alone would make Hf a larger atom. However, in between Zr and Hf lie several subshells including the long 4f subshell (Ce through Lu). The filling of these subshells, especially the 4f, causes a gradual contraction that offsets the higher shell for Hf.

8-84 C$^+$ (1086 kJ/mol), C^{2+} (3439 kJ/mol), C^{3+} (8059 kJ/mol, C^{4+} (14,282 kJ/mol), C^{5+} (52,112 kJ/mol). Notice that the energy required to form C$^+$ (1086 kJ) is about twice the energy required to form Ga$^+$ (579 kJ), which is a metal. Thus, it is apparent that metal cations form easier than nonmetal cations.

8-85 (a) B (b) Kr (c) K, Cr, and Cu (d) Ga (e) Hf

8-87 (a) B (b) Br (c) Sb (d) Si

8-89 (a) Sr, metal, representative element, IIA
(b) Pt, metal, transition metal, VIIIB
(c) Br, nonmetal, representative element, VIIA

8-91 P

8-93 Z = Sn, X = Zr

8-97 (a) Ca (b) Br⁻ (c) S   (d) S^{2-} (e) Na^+

8-99

s^1	s^2		d^1	d^2	d^3	d^4	d^5	d^6	d^7	d^8		p^1	p^2	p^3	p^4
1															2
3	4											5	6	7	8
9	10											11	12	13	14
15	16		17	18	19	20	21	22	23	24		25	26	27	28
29	30		31	32	33	34	35	36	37	38		39	40	41	42
43	44		45*												

* 46-50 would be in a 4f subshell

(a) six in second period, 14 in fourth

(b) third period - #14; fourth period - #28

(c) first inner transition element is #46 (assuming an order of filling like that on Earth)

(d) Elements #11 and #17 are most likely to be metals.

(e) Element #12 would have the larger radius in both cases.

(f) Element #7 would have the higher ionization energy in all three cases.

(g) The ions that would be reasonable are 16^{2+} (metal cation), 13⁻ (nonmetal anion), 15⁺ (metal cation), and 1⁻ (nonmetal anion). The 9^{2+} ion is not likely because the second electron would come from a filled inner subshell. The 7⁺ ion is not likely because it would be a nonmetal cation. The 17^{4+} ion is not likely because #17 has only three electrons in the outer shell.

9

The Chemical Bond

Review of Part A   *Chemical Bonds and the Nature of Ionic Compounds*

OBJECTIVES AND DETAILED TABLE OF CONTENTS

9-1   Bond Formation and Representative Elements

OBJECTIVE *Write the Lewis dot symbols of atoms and monatomic ions by following the octet rule.*

9-2   Formation of Ions and Ionic Compounds

OBJECTIVE *Using the Lewis dot structure and the octet rule, predict the charges on the ions of representative elements and the formulas of binary ionic compounds.*

SUMMARY OF PART A

One of the most fundamental topics of chemistry is how the atoms of one element bond to those of another. From a discussion of the structure of atoms in Chapter 2, we progressed to the arrangements of the electrons in those atoms in Chapter 8. That discussion gave us the theoretical basis for the periodic table, which was first introduced in Chapter 4. With this background, we are now ready to discuss why the elements combine in the ratios and in the manner in which they do.

At the end of Chapter 8, we learned that metals can lose electrons comparatively easily and that nonmetals have a tendency to gain electrons. To illustrate these losses and gains of electrons, we make use of **Lewis dot symbols** where outer s and p electrons (known as **valence electrons**) are represented as dots. By the use of Lewis symbols and through the knowledge of the electron configuration of the ions, we can see that many ions of the representative elements form as predicted by the octet rule. The **octet rule** states that these atoms gain, lose or share electrons so as to have access to eight valence electrons. Binary ionic compounds thus form between metals and nonmetals. The ions in the solid exist in a geometric arrangement known as a **lattice**, which is held together by **ionic bonds**. The formula of the compound is determined by the charges on the ions. A formula represents the ratio of cations to anions present so that there is no net charge, illustrated for the compound Tl_2O_3 [thallium (III) oxide] as shown on the following page.

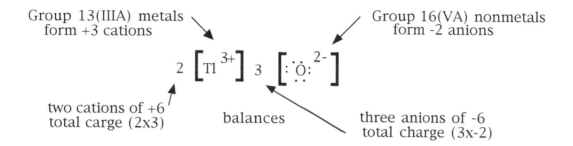

Group 13(IIIA) metals form +3 cations

Group 16(VA) nonmetals form -2 anions

$$2 \left[Tl^{3+} \right] \quad 3 \left[:\ddot{O}: {}^{2-} \right]$$

two cations of +6 total carge (2x3)

balances

three anions of -6 total charge (3x-2)

ASSESSMENT OF OBJECTIVES

A-1 Multiple Choice

___ 1. What is the expected charge on a Te ion?

   (a) +2   (b) -2   (c) -1   (d) +3   (e) +6

___ 2. Which of the following pairs of atoms would not be expected to combine to form a binary ionic compound?

   (a) B and H   (b) Na and H   (c) Li and Cl
   (d) Rb and O   (e) K and S

___ 3. What is the expected charge on an alkaline earth ion?

   (a) +1   (b) +6   (c) +2   (d) -2   (e) -3

___ 4. Based on the charge on the metal ion, which of the following metal-nonmetal compounds would not be expected to be ionic?

   (a) LiBr   (b) CaO   (c) Rb_2S (d) CrO   (e) Mn_2O_7

___ 5. Which of the following ionic compounds would not be expected to exist based on the octet rule?

   (a) LiI   (b) CaS   (c) BeF_2 (d) BaCl   (e) Mg_3N_2

159

A-2 Matching

_____ A nonmetal ion with the same electron configuration as Xe

_____ A metal ion that does not follow the octet rule

_____ A metal ion with the same electron configuration as Xe

_____ A metal ion with the same electron configuration as Xe

_____ A possible -3 ion

_____ An ion predicted for Group IIIA

_____ A nonmetal ion that violates the octet rule

(a) Te^-

(b) Sb^{3-}

(c) OH^-

(d) Al^{3+}

(e) I^-

(f) Pb^{2+}

(g) B^{3+}

(h) La^{3+}

(i) N^{3-}

(j) Br^-

(k) Si^{4+}

(l) S^{2-}

A-3 Problems

1. Write the ion expected to form from the following representative elements:

(a) Sr   (b) Ga

(c) H   (d) Se

(e) I

2. Write the formula of the ionic compound formed between each pair:

(a) Al and Se _____

(b) Be and F _____

(c) Ba and O _____

(d) K and N _____

Review of Part B *Chemical Bonds and the Nature of Molecular Compounds*

OBJECTIVES AND DETAILED TABLE OF CONTENTS

9-3   The Covalent Bond

OBJECTIVE *Apply the octet rule to determine the number of bonds formed in simple compounds.*

9-4   Writing Lewis Structures

OBJECTIVE *Draw Lewis structures of a number of molecular compounds and polyatomic ions.*

9-5   Resonance Structures

OBJECTIVE *Write multiple Lewis structures for molecules capable of resonance.*

9-6   Formal Charge

OBJECTIVE *Determine the validity of a Lewis structure based on formal charge considerations.*

SUMMARY OF PART B

We now turn the discussion to compounds where atoms share rather than gain or lose electrons. Such compounds are generally, but not always, between two nonmetals where neither element loses electrons easily. Bonds of this nature are known as **covalent bonds**. The octet rule predicts that representative elements (other than H) achieve an octet by having access to eight valence electrons. The octet is established through a combination of unshared electrons and shared electrons in a bond. This process is illustrated on the next page with the **Lewis structure** of Br_2. (Two dots represent an **unshared pair** of outer electrons, a dash represents a shared pair of electrons in a covalent bond.)

Each Br has six unshared electrons but counts the two shared electrons in the bond for a total of eight.

Notice that there are a total of fourteen valence electrons (seven from each Br).

$$: \ddot{Br} \!-\! \ddot{Br} :$$

$6 + 2 = 8 \qquad\qquad 2 + 6 = 8$

In many of the representative elements, especially C, N, O, and S compounds, two pairs (a **double bond**) or three pairs (a **triple bond**) of electrons are shared between two atoms. For example, the Lewis structures of CO_2 and CO illustrate this as follows:

Double bond

$$\ddot{O} \!=\! C \!=\! \ddot{O}$$

C has access to four shared pairs for a total of eight. Each O has two unshared pairs and access to two shared pairs for a total of eight.

Triple bond

$$: C \!\equiv\! O :$$

Both C and O have an unshared pair and access to three shared pairs for a total of eight for each.

The charge and formula of a polyatomic ion can also be rationalized by a Lewis structure and the octet rule. For example, the structure of the NO_2^- ion is illustrated as follows. Notice that the NO_2^- ion has 18 valence electrons (five from N, six each from the two Os, and one extra electron to account for the -1 charge).

N has access to one unshared pair and three shared pairs

This O has two unshared pairs and two shared pairs.

$$\left[\ddot{O} \!=\! \overset{..}{N} \!-\! \ddot{O} \right]^-$$

This O has three unshared pairs and one shared pair.

The octet rule is not an unbreakable law. Some stable molecules exist where an atom has access to less than eight electrons. In other cases, we can write Lewis structures that follow the octet rule (e.g., O_2 and BF_3) but we know that these structures do not agree with experimental observations. Also, nonmetals of the third period and down form many compounds where an atom has access to more than eight electrons. These latter compounds are not emphasized in this text.

Writing Lewis structures of molecules and ions is an important and fundamental endeavor of general chemistry and should be mastered. Structures not only allow us to understand the ratio of atoms in a molecule or an ion but, in some cases, allow us to predict possible formulas. From the Lewis structure, it is also possible to predict the geometry of the molecule or ion, which is discussed in Section C in this chapter.

Most students have very little difficulty writing correct Lewis structures with knowledge of the guidelines listed in the text and a generous amount of practice. We will run through the rules for C_2H_4 (ethylene) as an example.

1. H and C are both nonmetals, so this is a molecular compound.

2. The number of valence or outer electrons is:

atom	number of atoms		number of valence electrons		total
C	2	x	4	=	8
H	4	x	1	=	4
					12

3. Arrange the atoms symmetrically.

H   H
   C   C
H   H

4. Form bonds between atoms
   (There is always one less bond than there are atoms.)

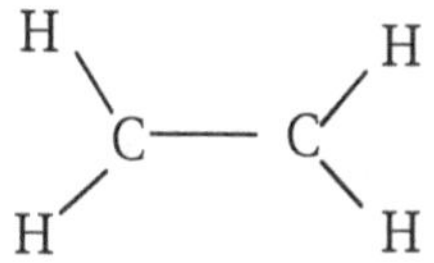

5. Add the remaining electrons
   $(12 - 10 = 2)$ to atoms with less
   than eight remaining.

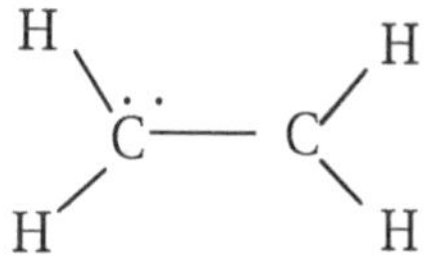

6. Check octets (two for Hs).
   Notice one C has only six electrons.
   By making a double bond with the
   other C, both Cs have access to eight.

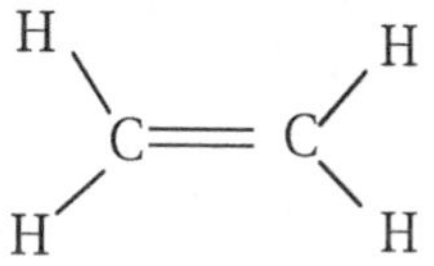

It is possible that a Lewis structure may present an incomplete picture. For example, consider one of the structures of the nitrite ion shown below. Each structure implies that one N-O bond is different from the other. In fact, they are both identical. To remedy this misconception, we can write the two equally possible Lewis structures, which are called **resonance structures**. The actual structure is a **resonance hybrid** of the two shown. Both N-O bonds are thus predicted to be identical and between a double and a single bond. This corresponds to experimental observation.

163

In the example of the nitrite ion, the two resonance structures are identical except for the location of the N-O bone. They both contribute equally to the hybrid structure. In some resonance structures there are some obvious differences. How can we tell if a certain structure is important? By calculating the **formal charges** on the atoms in the resonance structure, we can establish their relevance. Formal charge is calculated by assuming all electrons in a bond are shared equally. (As we will see in the next section, this is rarely the case.) It is calculated by subtracting the number of unshared electrons plus half of the shared electrons from the group number. Structures with the least amount of formal charges are usually the more relevant structures. This will be illustrated with a problem that follows. In some areas of chemistry, such as organic and biochemistry, some sulfur and phosphorus compounds are shown that violate the octet rule but do have favorable formal charge distributions.

ASSESSMENT OF OBJECTIVES

B-1 Multiple Choice

_____ 1. Which of the following two elements would be expected to form a covalent bond?

(a) Na, O   (b) H, S   (c) K, Te
(d) Mn, F   (e) Ba, S

_____ 2. Based on the octet rule, which of the following compounds would not be expected to exist?

(a) NH_4 (b) HI   (c) OF_2 (d) CF_4 (e) H_2Te

_____ 3. How many total valence electrons are present in the H_2CO molecule?

(a) 16   (b) 24   (c) 12   (d) 14   (e) 18

_____ 4. How many valence electrons are present in the NO^+ ion?

(a) 14   (b) 15   (c) 10   (d) 11   (e) 12

_____ 5. The N_2 molecule contains which of the following bonds?

(a) an ionic bond   (b) a single bond   (c) a double bond
(d) a triple bond   (e) a quadruple bond

_____ 6. Which of the following is not a resonance structure for $C_2O_4{}^{2-}$?

_____ 7. Which of the following has a carbon-oxygen bond halfway between a single and a double bond?

(a) $CO_3{}^{2-}$ (b) $H_3C\text{-}CO_2{}^-$ (c) CO_2 (d) CO

___ 8. What is the formal charge on the sulfur in SO_2 that follows the octet rule?

(a) 0   (b) +1   (c) +2   (d) –1   (e) -2

B-2 Problems

1. Write the simplest formulas of covalent compounds formed between the following two elements (single bonds only):

    (a) P and H   _____

    (b) F and I   _____

    (c) O and Br   _____

    (d) Se and H   _____

2. Write Lewis structures for the following. Where appropriate, write equivalent resonance structures.

    (a) $AsCl_3$

    (b) TeO_2

    (c) Na_2SO_3

    (d) HNO_3

    (e) C_2H_2

3. Write resonance structures for the cyanate ion (NCO^-) where C is the central atom. By means of formal charge, decide whether any of the structures do not contribute to the hybrid structure. What do the legitimate resonance structures tell us about the N – C and C – O bonds?

Review of Part C *The Distribution of Charge in Chemical Bonds*

OBJECTIVES AND DETAILED TABLE OF CONTENTS

9-7   Electronegativity and Polarity of Bonds

OBJECTIVE *Classify a bond as being nonpolar, polar, or ionic.*

9-8   Geometry of Simple Molecules

OBJECTIVE *Determine the bond angles and geometries present in simple molecules or ions from the Lewis structure.*

9-9   Polarity of Molecules

OBJECTIVE *Classify a molecule as polar or nonpolar based on geometry and electronegativity.*

SUMMARY OF PART C

Electronegativity is a periodic property of elements that is a measure of the attraction an atom has for electrons in a chemical bond. Nonmetals are more electronegative than metals with the most electronegative element, fluorine, having an electronegativity of 4.0. When different elements form a bond, the more electronegative atom acquires a partial negative charge, leaving the other atom with a partial positive charge. The bond then contains a **dipole** (two poles) and is thus a **polar bond**. This is illustrated as follows:

$$\overset{\delta^+ \qquad \delta^-}{\underset{:\overset{..}{\underset{..}{Br}} \quad :\overset{..}{\underset{..}{F}}:}{\longrightarrow}}$$

Indicates direction of polarity from δ^+ to δ^-.

Pair of electrons in the bond is drawn closer to the F thus giving the F a partial negative charge (indicted by δ^-).

The existence of polar covalent bonds indicates that the dividing line between purely covalent bonds (equal sharing between elements of the same electronegativity) and ionic (no sharing between elements with a difference in electronegativity of 1.8 or greater) is not exact. The more polar a bond, the more ionic character it has. Thus a polar covalent bond represents the intermediate ground between the two extremes of an ionic bond and a **nonpolar** covalent bond (equal sharing) illustrated as follows:

$$Na^+ \quad :\ddot{\underset{\cdot\cdot}{F}}: \quad ^- \qquad\qquad :\ddot{\underset{\cdot\cdot}{F}}\!-\!\ddot{\underset{\cdot\cdot}{F}}:$$

<u>Ionic</u>   <u>Nonpolar Covalent</u>

Large difference in electro-   No difference in electronegativity between Na and F.   negativity between two Fs.
Complete electron exchange.   *Equal sharing of electrons.*

Even though a molecule may contain polar bonds, the molecule itself may not be polar. The polarity of the molecule depends on the geometry of the molecule. The **electronic geometry**, which includes the bonds and the unshared pairs on the atom in question, can be determined from its Lewis structure and the application of **VSEPR theory**. This simple theory simply tells us that electron pairs on a central atom, either bonded or as lone pairs, tend to repel each other to the maximum extent. The three symmetrical geometries assumed by molecules containing no lone pairs are linear (two bonds), trigonal planar (three bonds), and tetrahedral (four bonds). When the molecule has lone pairs in place of bonded pairs, the **molecular geometries** are described as V-shaped (two bonds and one lone pair), trigonal pyramidal (three bonds and one lone pair), and V-shaped (two bonds and two lone pairs).

A polar bond causes a bond dipole that is like a force with both magnitude and direction. If the forces are equal and are oriented in space so as to exactly oppose each other, the dipoles cancel. If a molecule has one of the three symmetrical geometries and all terminal atoms are the same, the molecule is nonpolar. An example of a nonpolar molecule with three polar bonds is as follows:

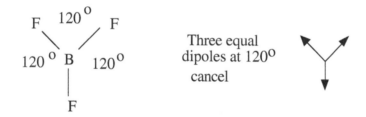

Because H_2O is a V-shaped molecule, it has a net **molecular dipole**. This very important fact accounts for many of the properties of water that will be discussed in Chapter 11.

ASSESSMENT OF OBJECTIVES

C-1 Multiple Choice

____ 1. Which of the following has the highest electronegativity?

(a) N   (b) I   (c) C   (d) H   (e) Na

____ 2. Which of the following two elements would have the greatest difference in electronegativity?

(a) Al, I   (b) K, N   (c) K, S   (d) B, H   (e) C, H

167

___ 3. Which of the following contains a polar covalent bond?

(a) H_2 (b) N_2 (c) KBr   (d) H_2S (e) KF

___ 4. Which of the following elements always has the negative dipole in any polar covalent bond?

(a) Fr   (b) At   (c) F   (d) H   (e) O

___ 5. Which of the following molecules is polar?

(a) SO_3 (b) O_3 (c) N_2 (d) BCl_3 (e) SO_2

Chapter Summary Assessment

S-1 Formulas

Write the formula of the simplest compound formed by each pair of elements. Indicate whether the compound is primarily ionic or covalent.

	Formula	Ionic or Covalent
(a) Si and H	_____	_____
(b) Sr and Cl	_____	_____
(c) S and F	_____	_____
(d) Ga and O	_____	_____
(e) Mg and N	_____	_____
(f) B and S	_____	_____

S-2 Matching

___ A formula unit with both ionic and covalent bonds

___ A formula unit with only ionic bonds

___ A molecule that cannot follow the octet rule

___ A molecule containing the most electronegative element

___ A molecule with a nonpolar covalent bond

___ A molecule with resonance structures

(a) NO

(b) Cl_2O

(c) SO_2

(d) Rb_2Se

(e) N_2

(f) $KBrO_4$

(g) CCl_4

(h) SiF_4

(i) CO

S-3 Bonding

Given the following five compounds with three atoms in one formula unit:

$$OF_2, K_2O, SeO_2, CO_2, \text{ and } SrF_2$$

(a) The two formula units that contain ions   _____ and _____

(b) The formula unit that has V-shaped molecules with an angle of about 120°   _____

(c) The formula unit that has nonpolar molecules   _____

(d) The formula unit that has V-shaped molecules with an angle of about 109°   _____

Answers to Assessments of Objectives

A-1 Multiple Choice

1. **b**   -2, Te is in Group VIA, which can gain two electrons.

2. **a**   B and H. Both of these elements are nonmetals.

3. **c**   +2. Alkaline earth metals can lose two electrons to form an octet.

4. **e** Mn_2O_7. The Mn would be a +7 ion, which is too high for an ion.

5. **d**   BaCl. Either the Ba would have a +1 charge or the Cl a -2 charge, neither of which has an octet of electrons.

A-2 Matching

e I^- is a nonmetal ion with the same electron configuration as Xe.

d Pb^{2+} is a metal ion that does not follow the octet rule.

g La^{3+} is a metal ion with the same electron configuration as Xe.

h N^{3-} is a possible -3 ion. (Sb is a metal.)

d Al^{3+} is an ion predicted for Group IIIA. (B is a nonmetal.)

a Te^- is a nonmetal ion that violates the octet rule.

A-3 Problems

1. (a) Sr^{2+} [Group IIA] (b) Ga^{3+} [Group IIIA]

   (c) H^- (d) Se^{2-} [Group VIA]

   (e) I^- [Group VIIA]

2. (a) Al^{3+} and Se^{2-} = Al_2Se_3 (b) Be^{2+} and F^{1-} = BeF_2

   (c) Ba^{2+} and O^{2-} = BaO (d) K^{1+} and N^{3-} = K_3N

B-1 Multiple Choice

1. **b** H, S Both elements are nonmetals.

2. **a** NH_4 There would be nine electrons on the nitrogen.

3. **c** 12 [4 + 6 + (2 x 1) = 12]

4. **c** 10 (6 + 5 – 1 = 10)

5. **d** a triple bond

6. **d** This is not a correct Lewis structure.

7. **b** $H_3C-CO_2^-$ The two resonance structures are

8. **c** The formal charge is 6 – 2 (unshared electrons) – 3 (1/2 of shared electrons) = +1

B-2 Problems

1.

(a) PH_3

(b) IF

(c) Br_2O

(d) H_2S

2. (a) $AsCl_3$ Valence electrons
As 1 x 5 =   5
Cl 3 x 7 =   21
              $\overline{26}$

(b) TeO_2 Valence electrons
Te 1 x 6 =   6
O   2 x 6 =   12
              $\overline{18}$ Two resonance structures

(c) Na_2SO_3 This contains $2(Na^+)$ ions and a SO_3^{2-} ion.
For SO_3^{2-} Valence electrons
S 1 x 6 =   6
O 3 x 6 =   18
Charge   =   2
              $\overline{26}$

(d) HNO_3 Valence electrons
H 1 x 1 =   1
N 1 x 5 =   5
O 3 x 6 =   18   (The N is the central atom, and
              $\overline{24}$ the H is attached to O.)
              It has two resonance structures.

(e) C_2H_2 Valence electrons
H 2 x 1 =   2
C 2 x 4 =   8
              $\overline{10}$

H—C≡C—H

171

3. NCO^-

Valence electrons
C 1 x 4 = 4
N 1 x 5 = 5
O 1 x 6 = 6
Charge = 1
――――――――
16

Three resonance structures can be written as shown below.

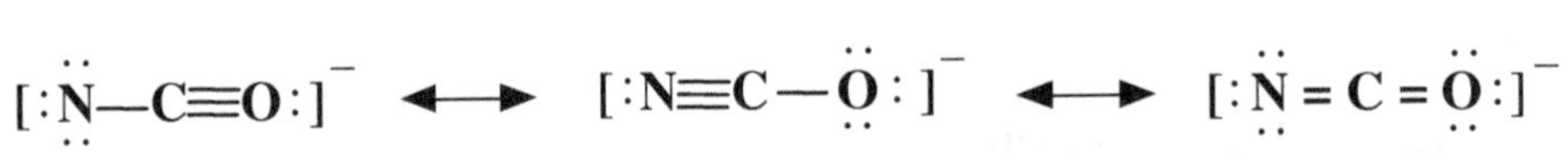

$$[:\ddot{N}-C\equiv O:]^- \longleftrightarrow [:N\equiv C-\ddot{O}:]^- \longleftrightarrow [:\ddot{N}=C=\ddot{O}:]^-$$

The first structure on the left does not contribute to the hybrid structure. It has a −2 formal charge on the nitrogen and a +1 on the oxygen. The middle structure has only a −1 formal charge on the oxygen. The structure on the right has only a −1 formal charge on the nitrogen. The hybrid structure of the ion would be represented by the two resonance structures on the right. The C − O bond would be predicted to be between a double and a single bond. The C − N bond would be predicted to be between a triple and a double bond.

C-1 Multiple Choice

1. **a**   N

2. **b**   K, N (0.8 and 3.0)

3. **d** H_2S

4. **c**   F It is the most electronegative element.

5. **e** SO_2 The bond dipoles do not cancel because the molecule is V-shaped.

S-1 Formulas

(a) **SiH_4**   Covalent

(b) **$SrCl_2$**   Ionic

(c) **SF_2**   Covalent

(d) **Ga_3O_2**   Ionic

(e) **Mg_3N_2**   Ionic by metal-nonmetal criterion but molecular by electronegativity difference.

(f) **B_2S_3**   Covalent

S-2 Matching

f   **$KBrO_4$** contains both ionic and covalent bonds (i.e., $K^+BrO_4^-$).

d   **Rb_2Se** contains only an ionic bond (i.e., Rb^+ and Se^{2-}).

a   **NO** cannot follow the octet rule because it has 11 valence electrons.

h SiF_4 is a molecule containing the most electronegative element fluorine.

e N_2 is a molecule with a nonpolar covalent bond.

c SO_2 is a molecule with two resonance structures.

S-3 Bonding

(a) **K_2O** ($2K^+, O^{2-}$) and SrF_2 ($Sr^{2+}, 2F^-$) contain only ions.

(b) **SeO_2** has one lone pair on the central atom so it is V-shaped with an angle of about $120°$.

(c) **CO_2** is linear so it is nonpolar.

(d) **OF_2** has two lone pairs on the central atom so it is V-shaped with an angle about $109°$.

Answers and Solutions to Green Text Problems

9-1   (a) $Ca\cdot$ (b) $\cdot \dot{S}b \cdot$ (c) $\cdot \dot{S}n \cdot$ (d) $:\dot{I}\cdot$ (e) $:\dot{N}e:$ (f) $\cdot \dot{B}i \cdot$ (g) $:\dot{V}IA\cdot$

9-2   (a) Group IIIA   (b) Group VA   (c) Group IIA

9-4   The electrons from filled inner subshells are not involved in bonding.

9-6   (b) Sr and S   (c) H and K   (d) Al and F

9-8   (b) S^- (c) Cr^{2+} (e) In^+ (f) Pb^{2+} (h) Tl^{3+}

9-10   They have the noble gas configuration of He, which requires only two electrons.

9-11   (a) K^+ (b) $:\dot{O}\cdot^-$ (c) $:\dot{I}:^-$ (d) $:\dot{P}:^{3-}$ (e) Ba^+ (f) $:\dot{X}e\cdot^+$ (g) Sc^{3+}

9-13   (b) O^- (e) Ba^+ (f) Xe^+

9-14   (a) Mg^{2+} (b) Ga^{3+} (pseudo-noble gas config.)   (c) Br^- (d) S^{2-} (e) P^{3-}

9-17 $Se^{2-}, Br^-, Rb^+, Sr^{2+}, Y^{3+}$

9-18 Tl^{3+} has a full 5d subshell. This is a pseudo-noble gas configuration.

9-19 $Cs_2S, Cs_3N, BaBr_2, BaS, Ba_3N_2, InBr_3, In_2S_3, InN$

9-21   (a) CaI_2 (b) CaO (c) Ca_3N_2 (d) $CaTe$ (e) CaF_2

9-23   (a) Cr^{3+} (b) Fe^{3+} (c) Mn^{2+} (d) Co^{2+} (e) Ni^{2+} (f) V^{3+}

9-26 (a) H_2Se (b) GeH_4 (c) ClF (d) Cl_2O (e) NCl_3 (f) CBr_4

9-28 (b) Cl_3, (d) NBr_4, and (e) H_3O are impossible.

9-30 (a) SO_4^{2-} (b) IO_3^- (c) SeO_4^{2-} (d) $H_2PO_4^-$ (e) $S_2O_3^{2-}$

9-32 (a)
```
      H   H
      |   |
  H — C — C — H
      |   |
      H   H
```
(b)
```
  H — Ö — Ö — H
```
(c)
```
  F — N̈ — F
      |
      F
```
(d)
```
  :C̈l — S̈ — C̈l:
```
(e)
```
      H   H
      |   |
  H — C — C — Ö — H
      |   |
      H   H
```
```
      H       H
      |       |
  H — C — Ö — C — H
      |       |
      H       H
```

9-34 (a) :C ≡ O:

(b)
```
      :O:
      ‖
      S
    ／   ＼
  :Ö:    :Ö:
```
(c) $K^+[:C \equiv N:]^-$

(d)
```
          :O:
          ‖
  H — Ö — S — Ö — H
          |
          :O:
```

9-35 (a) $\ddot{\text{N}} = \text{C} = \ddot{\text{N}}^{2-}$

(b)
```
          Ö.⁻
        ∥
  H — C
        ＼
          .Ö.
```
(c)
```
              Ö.⁻
            ∥
  H — Ö — C
            ＼
              .Ö.
```

(d) $:\ddot{\text{C}}\text{l} — \ddot{\text{C}} = \ddot{\text{O}}.^-$

9-36 (a) $.\ddot{\text{N}} = \text{N} = \ddot{\text{O}}.$

(b)
```
              N̈
            ∥        ⁻
  Ca²⁺ 2  .Ö.    .Ö.
```
(c)
```
  :C̈l — As — C̈l:
        |
       :Cl:
```
(d)
```
         S̈
       ／   ＼
      H     H
```

(e)
$$:\ddot{C}l - \underset{\underset{H}{|}}{\overset{\overset{H}{|}}{C}} - \ddot{C}l:$$

(f)
$$H - \underset{\underset{H}{|}}{\overset{\overset{H}{|}}{N}} - H \quad +$$

9-37 (a) $:\ddot{C}l - \ddot{O} - \ddot{C}l:$

(b) $:\ddot{O} - \underset{\underset{:\ddot{O}:}{|}}{S} - \ddot{O}:^{2-}$

(c)
$$\overset{H}{\underset{H}{>}}C = C\overset{H}{\underset{H}{<}}$$

(d)
$$\overset{H}{\underset{H}{>}}C = \ddot{O}:$$

(e)
$$\underset{:\ddot{F} \qquad \ddot{F}:}{\overset{:\ddot{F}:}{\underset{|}{B}}}$$

(f) $[:N \equiv O:]^+$

9-40 (a)
$$\underset{:\ddot{O}. \qquad .\ddot{O}:}{\overset{\dot{O}\cdot}{\underset{\|}{S}}} \quad \longleftrightarrow \quad \underset{:\ddot{O}. \qquad .\dot{O}}{\overset{:\ddot{O}:}{\underset{|}{S}}} \quad \longleftrightarrow \quad \underset{:\ddot{O}. \qquad .\ddot{O}:}{\overset{:\ddot{O}:}{\underset{\|}{S}}}$$

(b)
$$\underset{:\ddot{O}. \qquad .\ddot{O}}{\overset{\ddot{N}}{}} \quad - \quad \longleftrightarrow \quad \underset{:\ddot{O}. \qquad .\ddot{O}:}{\overset{\ddot{N}}{}} \quad -$$

(c) $:\ddot{O} - \underset{\underset{:\ddot{O}:}{|}}{\ddot{S}} - \ddot{O}:^{2-}$ (only one structure)

9-42
$$H - \underset{\underset{H}{|}}{\overset{\overset{H}{|}}{B}} - C\overset{\overset{\dot{O}}{\|} \, ^{2-}}{\underset{.\ddot{O}.}{\diagdown}} \quad \longleftrightarrow \quad H - \underset{\underset{H}{|}}{\overset{\overset{H}{|}}{B}} - C\overset{\overset{\ddot{O}:}{\diagup} \, ^{2-}}{\underset{\ddot{O}}{\|}}$$

9-43 A resonance hybrid is the actual structure of a molecule or ion that is implied by the various resonance structures. Each resonance structure contributes a portion of the actual structure. For example, the two structures shown in Problem 9-42 imply that both C – O bonds have properties that are halfway between those of a single and a double bond.

9-45   no formal charge

$$\overset{-1}{:\ddot{O}} - \overset{+1}{\dot{N}} = \overset{..}{\dot{O}.} \qquad\qquad :\overset{..}{\dot{O}} - \overset{..}{\ddot{N}} = \overset{..}{\dot{O}.}$$

Formal charge indicates that the unpaired electron is on the oxygen. Two nitrogen dioxide molecules bond through the nitrogen, however.

9-47 **(a)** C = -1; O = +1   **(b)** S = +2; single bonded Os = -1

      (c) K = +1; C = -1; N = 0   **(d)** Hs = 0; S = +1; O in H—O—S = 0, other O = -1

9-49 $\overset{-1}{:\ddot{O}} - \overset{+2}{\ddot{Cl}} - \overset{-1}{\ddot{O}:}^{-}$ $.\overset{..}{O} = \overset{..}{\ddot{Cl}} = \overset{..}{O}.^{-}$

             | |

           $:\ddot{O}:$ -1 $:\ddot{O}:$ -1

9-51 $\overset{-1}{.\ddot{N}} = \overset{+1}{N} = \overset{..}{\dot{O}.} \longleftrightarrow :N \equiv N - \overset{..}{\dot{O}:} \qquad \overset{-1}{.\dot{N}} = \overset{+2}{O} = \overset{-1}{\dot{N}.}$

Both resonance structures for NNO have much less formal charge than the NON structure.

9-52 Cs, Ba, Be, B, C, Cl, O, F

9-53 (a) $\overset{\delta^-}{N} — \overset{\delta^+}{H}$ (b) $\overset{\delta^+}{B} — \overset{\delta^-}{H}$ (c) $\overset{\delta^+}{Li} — \overset{\delta^-}{H}$ (d) $\overset{\delta^-}{F} — \overset{\delta^+}{O}$ (e) $\overset{\delta^-}{O} — \overset{\delta^+}{Cl}$
        ←   →   →   ←   ←

        (f) $\overset{\delta^-}{S} — \overset{\delta^+}{Se}$ (g) $\overset{\delta^-}{C} — \overset{\delta^+}{B}$ (h) $\overset{\delta^+}{Cs} — \overset{\delta^-}{N}$ (i) C — S (essentially nonpolar)
        ←   ←   →

9-54 (i) nonpolar, (b) = (f), (d) = (e) = (g), (a), (c), (h)

9-55 (a) CF polar covalent   (b) AlF ionic   (c) FF nonpolar

9-58 (b) I – I in nonpolar, C-H is nearly nonpolar.

9-60 (a) $\overset{..}{\dot{S}}$ V-shaped   (b) $.\dot{S} = C = \dot{S}.$ linear

            /   \

      :F   F:

(c) 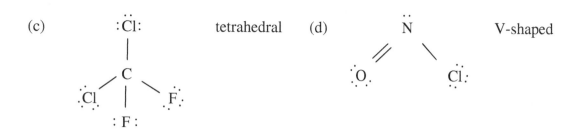 tetrahedral   (d)   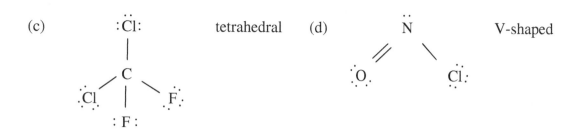 V-shaped

(e)   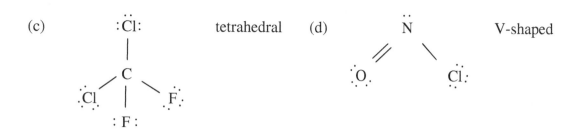 V-shaped

9-62   (a) No angle between two points   (b) 180°   (c) 120°

9-64   (a)   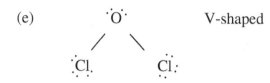 N - trigonal pyramidal, O- V-shaped

(b)   O - V-shaped, C - linear

(c)   H — C — C ≡ N: C - tetrahedral, C - linear

(d) See problem 9-42 for Lewis structure. B- tetrahedral, C- trigonal planar

9-66   If the molecule has a symmetrical geometry and all bonds are the same, the bond dipoles cancel and the molecule is nonpolar.

9-67   (a) polar - bond dipoles do not cancel   (b) nonpolar - equal bond dipoles cancel
(c) polar - bond dipoles not equal   (d) polar - unequal bond dipoles
(e) polar - bond dipoles do not cancel

9-69

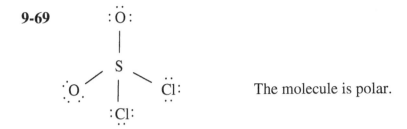

The molecule is polar.

9-70 Since the H-S bond is much less polar than the H-O bond, the resultant molecular dipole is much less. The H_2S molecule is less polar than the H_2O molecule.

9-72 The CHF_3 molecule is more polar than the $CHCl_3$ molecule. The C-F bond is more polar than the C-Cl bond, which means that the resultant molecular dipole is larger for CHF_3.

9-74

$$ \overset{\cdot\cdot}{\underset{\cdot}{O}} \quad \text{All bond dipoles in} $$
$$ \parallel \quad SO_3 \text{ cancel.} $$
$$ \overset{}{\underset{\cdot\cdot}{O}} \diagdown\!\!\!\!\diagup S \diagdown \overset{}{\underset{\cdot\cdot}{O}} $$

All bond dipoles in SO_3 cancel.

$$ \overset{\cdot\cdot}{S} $$
$$ \diagup\!\!\diagup \quad \diagdown $$
$$ \overset{\cdot\cdot}{\underset{\cdot}{O}} \qquad \overset{\cdot\cdot}{\underset{\cdot}{O}} $$

Bond dipoles in SO_2 do not cancel.

9-76 (a) e.g., RbCl   (b) e.g., SrO   (c) e.g., AlN

9-77

$$ \underset{:O:}{\overset{\cdot\cdot}{Xe}} $$
$$ \overset{\cdot\cdot}{\underset{\cdot}{O}} \diagup \mid \diagdown \overset{\cdot\cdot}{\underset{\cdot}{O}} $$

Trigonal pyramidal - polar

9-79

$$ H $$
$$ \diagdown $$
$$ H - B - C \equiv O: $$
$$ \diagup $$
$$ H $$

No resonance structures
Geometry around B - tetrahedral
Geometry around C - linear

9-81

$$ \diagup\!\!\diagup \overset{}{N} - \overset{\cdot\cdot}{N} \diagdown\!\!\diagdown $$
$$ \overset{\cdot}{\underset{\cdot\cdot}{O}} \qquad \overset{\cdot}{\underset{\cdot\cdot}{O}} $$

Geometry is angular for this structure.

Other resonance structures:

$$ \overset{\cdot\cdot}{\underset{\cdot}{O}}=N=\overset{\cdot\cdot}{N}-\overset{\cdot\cdot}{\underset{\cdot\cdot}{O}}: \quad\longleftrightarrow\quad :\overset{\cdot\cdot}{\underset{\cdot\cdot}{O}}-N\equiv N-\overset{\cdot\cdot}{\underset{\cdot\cdot}{O}}: \quad\longleftrightarrow\quad :\overset{\cdot\cdot}{\underset{\cdot\cdot}{O}}-\overset{\cdot\cdot}{N}-N\equiv O: $$

In fact, none of these last three resonance structures contribute to the actual structure. They all have too much formal charge compared to the first structure.

9-84 Valence electrons for neutral $PO_3 = 5(P) + (3 \times 6)(Os) = 23$. Since the species has 26 electrons, it must be an anion with a -3 charge (i.e., $PO_3{}^{3-}$).

$$:\ddot{O} - \overset{..}{P} - \ddot{O}:\qquad^{3-}$$
$$|$$
$$:\ddot{O}:$$

The geometry of the anion is trigonal pyramid al with a O-P-O angle of about 109°.

9-85 Valence electrons for neutral $ClI_2 = 3 \times 7 = 21$. Since the species has 20 valence electrons it must be a cation with a +1 charge (i.e., $ClI_2{}^+$).

$$\overset{..~..}{Cl}\qquad^{+}$$
$$\diagup\quad\diagdown$$
$$\ddot{I}.\qquad\qquad.\ddot{I}.$$

The geometry of the cation is V-shaped with an angle of about 109°.

9-87 K_3N - potassium nitride $KN_3 = K^+ N_3{}^{\cdot}$ Resonance structures:

$$\overset{..}{.N} = N = \ddot{N}:{}^{-} \quad \longleftrightarrow \quad :N \equiv N - \ddot{N}:{}^{-} \quad \longleftrightarrow \quad \overset{.}{N} - N \equiv N:{}^{-}$$

In all resonance structures, the geometry around the central N is linear. The last two structures together say the same as the first. That is, both bonds are double bonds.

9-90

$$H$$
$$\diagdown$$
$$\overset{.}{O} - C \equiv N:$$

The angle of 105° indicates that the H-O-C angle is V-shaped with an angle of about 109°. This would involve two bonds and two pairs of electrons on the O. The geometry around the C is linear.

9-91

$$H$$
$$\ddot{S} = C = N:$$

The angle of 116° indicates that the H-S-C angle is V-shaped with an angle of about 120°. This would involve three bonds and one pair of electrons on the S. The geometry around the C is linear.

9-92

$$:\ddot{F} - \ddot{N} = \ddot{N} - \ddot{F}:$$

9-94 Oxygen difluoride, OF_2 Lewis structure:

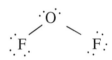

Dioxygen difluroide, O_2F_2 Lewis structure:

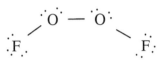

Oxygen is less electronegative (more metallic) than fluorine so is named first. Both molecules are angular, thus they are polar.

9-96 Assume 100 g of compound. There is then 27.4 g Na, 14.3 g C, 57.1 g O, and 1.19 g H per 100 g.

$$27.4 \text{ g Na} \times \frac{1 \text{ mol Na}}{22.99 \text{ g Na}} = 1.19 \text{ mol Na} \qquad 14.3 \text{ g C} \times \frac{1 \text{ mol C}}{12.01 \text{ g C}} = 1.19 \text{ mol C}$$

$$57.1 \text{ g O} \times \frac{1 \text{ mol O}}{16.00 \text{ g O}} = 3.57 \text{ mol O} \qquad 1.19 \text{ g H} \times \frac{1 \text{ mol H}}{1.008 \text{ g H}} = 1.18 \text{ mol H}$$

O: $\frac{3.57}{1.18} = 3.0$ Formula = NaHCO3 = $\underline{Na^+ \ HCO3^-}$

$$H - \ddot{O} - C \Big\langle{\overset{\displaystyle \ddot{O}:^-}{\underset{\displaystyle \ddot{O}:}{}}}$$

The geometry around the C is trigonal planar with the approximate H-O-C angle of 120°.

9-98 In 100 g of compound there is 19.8 g Ca, 1.00 g H, 31.7 g S, and 47.5 g O.

$$19.8 \text{ g Ca} \times \frac{1 \text{ mol Ca}}{40.08 \text{ g Ca}} = 0.494 \text{ mol Ca} \qquad 1.00 \text{ g H} \times \frac{1 \text{ mol H}}{1.008 \text{ g H}} = 0.992 \text{ mol H}$$

$$31.7 \text{ g S} \times \frac{1 \text{ mol S}}{32.07 \text{ g S}} = 0.988 \text{ mol S} \qquad 47.5 \text{ g O} \times \frac{1 \text{ mol O}}{16.00 \text{ g O}} = 2.97 \text{ mol O}$$

H: $\frac{0.992}{0.494} = 2.0$ S: $\frac{0.988}{0.494} = 2.0$ O: $\frac{2.97}{0.494} = 6.0$

Formula = $CaH_2S_2O_6$ = $\underline{Ca(HSO_3)_2}$ calcium bisulfite or calcium hydrogen sulfite

$$H - \overset{..}{\underset{..}{O}} - \overset{..}{S} - \overset{..}{\underset{..}{O}}:^-$$
$$\overset{|}{\underset{..}{:O:}}$$

The geometry around the S is trigonal pyramid. The H-O-S angle is about 109°.

9-99 (1) (a) 17   (b) 31   (c) 51_3 (d) 97   (e) 7(13)   (f) $10(13)_2$ (g) 67_2 (h) 3_26

(2) On Zerk, six electrons fill the outer s and p orbitals to make a noble gas configuration. Therefore, we have a "sextet" rule on Zerk.

(a) $1 - \overset{..}{\underset{.}{7}}:$ (b) $3^+ 1:^-$ (ionic)   (c) $1 - \overset{}{\underset{\underset{1}{|}}{5}} - 1$

(d) $9^+ :\overset{..}{\underset{..}{7}}:^-$ (ionic)   (e) $\overset{.}{\underset{.}{7}} - 13\overset{.}{\underset{.}{}}$ (f) $10^{2+} (\overset{..}{\underset{..}{13}}:^-)_2$ (ionic)

(g) $\overset{.}{\underset{.}{7}} - 6 - \overset{..}{\underset{.}{7}}$ (h) $(3^+)_2 :\overset{..}{6}:^{2-}$ (ionic)

181

10

The Gaseous State

Review of Part A *The Nature of the Gaseous State and the Effects of Changing Conditions*

OBJECTIVES AND DETAILED TABLE OF CONTENTS

10-1 The Nature of Gaseous State and the Kinetic Molecular Theory

OBJECTIVE *List the general properties of gases based on the postulates of Kinetic Molecular Theory.*

10-2 The Pressure of a Gas

OBJECTIVE *Calculate the effect changing pressure has on the volume of a gas (Boyle's Law).*

10-3 Charles's, Gay-Lussac's, and Avogadro's Laws

OBJECTIVE *Using the gas laws, perform calculations involving relationships between volume, pressure, and temperature.*

SUMMARY OF PART A

The gaseous state is unique compared to the other two states in many respects. Unlike solids and liquids, gases are compressible, have low densities, fill containers uniformly, mix rapidly and thoroughly, and exert pressure evenly in all directions. These and other properties, including what are known as "the gas laws," become obvious from an understanding of the **kinetic molecular theory** applied to gases. The major assumptions of this theory are:

(1) Gas molecules have negligible volume.

(2) Gas molecules undergo random collisions with each other and with the walls of the container, thus exerting pressure.

(3) The total energy of all of the collisions is conserved.

(4) Gas molecules have negligible interactions with each other.

(5) The average kinetic energy of the molecules is proportional to the temperature.

An immediate consequence of the last point of kinetic theory is that the average velocity of a gas is related to its formula weight (or molar mass) since K.E. $= 1/2mv^2$. The extension of this principle to the **effusion** and **diffusion** of gases is known as **Graham's law**. The lower the molar mass of a gas, the higher is its average velocity as well as rates of effusion and diffusion.

It seems obvious now, but the understanding of the nature of gases began with the demonstration that the atmosphere exerts **pressure** that can be measured with a **barometer**. Pressure is the weight of the gas (the **force**) applied per unit area. The pressure of a gas is measured in many different units but all can be compared to the standard unit, which is the average pressure of the atmosphere at sea level (**one atmosphere**). The most common unit of pressure in chemistry calculations besides atmosphere is **torr**, which is the same as mm of mercury. One atmosphere is the pressure that supports a column of Hg 760 mm high, which is thus 760 torr.

The gas laws discussed in this part of the review relate the pressure, temperature, and number of moles to the volume of a gas. The laws can be used to calculate how one parameter changes as the others are varied. Examples of these laws are as follows.

Example A-1 Boyle's Law (V and P)

If the volume of a gas is 32.5 L at a pressure of 4.25 atm, what is the volume if the pressure is increased to 5.70 atm?

PROCEDURE

The appropriate expression of Boyle's law under two conditions is

$$P_1V_1 = P_2V_2 \text{ or solving for } V_2, \quad V_2 = V_1 \times \frac{P_1}{P_2}$$

The problem can be solved by substitution or, better yet, by reason. (That way you don't have to remember the formula.)

$$V_{final} = V_{initial} \times P_{corr}$$

P_{corr} symbolizes the factor that converts the given volume to the final volume. In this problem, the pressure increases. According to Boyle's law, an *increase* in pressure results in a *decrease* in volume. Therefore, we reason that P_{corr} is a factor less than one in order to convert $V_{initial}$ to a smaller value. In summary, P goes *up*, therefore V goes *down*, and the P_{corr} is less than one.

SOLUTION

$$V = 32.5 \text{ L} \times \frac{4.25 \text{ atm}}{5.70 \text{ atm}} = \underline{24.2 \text{ L}}$$

Example A-2 Charles's Law (V and T)

If the volume of a gas is 32.5 L at a temperature of 28°C, what is the volume at 112°C?

PROCEDURE

$$V_{final} = V_{initial} \times T_{corr}$$

T goes *up*, therefore V goes *up*, and the T_{corr} is greater than one. In these calculations, the *Kelvin* temperature scale is used, which begins at *absolute zero*.

SOLUTION

$$V = 32.5 \text{ L} \times \frac{(112 + 273)\text{K}}{(28 + 273)\text{K}} = \underline{41.6 \text{ L}}$$

Example A-3 Gay-Lussac's Law (P and T)

If the pressure on a gas is 485 torr and the temperature is 162°C, what is the pressure if the temperature is increased to 215°C?

    PROCEDURE

$$P_{final} = P_{initial} \times T_{corr}$$

    T goes *up*, therefore P goes *up* and the T_{corr} is greater than one.

    SOLUTION

$$P = 485 \text{ torr} \times \frac{(215 + 273)\ \cancel{K}}{(162 + 273)\ \cancel{K}} = \underline{544 \text{ torr}}$$

       One additional law follows from the previous three laws. This law relates a quantity of gas under two sets of conditions and is known as the **combined gas law**.

Example A-4 Combined Gas Law (P, V, and T)

If the volume of a gas is 285 mL at a temperature of 16°C and a pressure of 685 torr, what is the volume at a temperature of 116°C and a pressure of 842 torr?

    PROCEDURE

$$V_{final} = V_{initial} \times P_{corr} \times T_{corr}$$

    P goes *up*, therefore V goes *down* and P_{corr} is less than one.
    T goes *up*, therefore V goes *up* and T_{corr} is greater than one.

    SOLUTION

$$V = 285 \text{ mL} \times \frac{685\ \cancel{torr}}{842\ \cancel{torr}} \times \frac{(116 + 273)\cancel{K}}{(16 + 273)\cancel{K}} = \underline{312 \text{ mL}}$$

Example A-5 Avogadro's Law (V and n)

If 1.51 moles of N_2 has a volume of 67.7 L, what is the volume of 1.96 moles of N_2 under the same conditions?

    PROCEDURE

$$V_{final} = V_{initial} \times n_{corr}$$

    n goes *up*; therefore V goes *up* and n_{corr} is greater than one.

    SOLUTION

$$V = 67.7 \text{ L} \times \frac{1.96\ \cancel{mol}}{1.51\ \cancel{mol}} = \underline{87.9 \text{ L}}$$

ASSESSMENT OF OBJECTIVES

A-1 Multiple Choice

___ 1. Which of the following is NOT characteristic of gases?

    (a) They fill a container uniformly.
    (b) Different gases mix rapidly.
    (c) Gases have a low density.
    (d) Gases are virtually incompressible.

___ 2. Which of the following is false about the kinetic theory of gases?

    (a) Gas molecules are in random motion.
    (b) The volume of a gas is composed mostly of matter.
    (c) Gas molecules have no attraction to each other.
    (d) Collisions between molecules are elastic.
    (e) Gas molecules collide with each other and with the sides of the container.

___ 3. Which of the following is false?

    (a) At the same temperature, light molecules move faster on the average than heavy molecules.
    (b) At the same temperature, all molecules have the same average kinetic energy.
    (c) Gases diffuse faster at higher temperatures.
    (d) Gases are compressible.
    (e) All gas molecules have the same average speed at the same temperature.

___ 4. Which of the following is not a unit of pressure?

    (a) in. of Hg   (b) pascals   (c) torr
    (d) kelvins   (e) bars

___ 5. One atmosphere at sea level supports a column of mercury

    (a) 76.0 cm high   (b) 14.7 in. high   (c) 760 cm high
    (d) 76.0 mm high   (e) 29.9 cm high

___ 6. When the pressure on a gas is doubled at constant temperature, the volume

    (a) is doubled   (b) stays the same
    (c) is halved   (d) is quartered

___ 7. The temperature of a volume of gas is increased from 0 °C to 20°C. The volume

    (a) increases by a factor of 20   (b) decreases by a factor of 20
    (c) decreases by a factor of 20/273   (d) increases by a factor of 293/273
    (e) decreases by a factor of 273/293

___ 8. What is -50°C on the Kelvin scale?

    (a) -323   (b) 323   (c) -223   (d) 223   (e) 273

___ 9. Which of the following is a representation of Gay-Lussac's law?

    (a) $PV = k$ (b) $P \propto T$ (c) $T(^{\circ}C) \propto P$ (d) $V/T = k$

_____ 10. Standard temperature and pressure (STP) is:

(a) one atm pressure and $25^{\circ}C$ (b) 760 torr and 273 K
(c) 760 atm and $0^{\circ}C$ (d) one atm and $273^{\circ}C$
(e) one atm and $760^{\circ}C$

_____ 11. When the temperature of a volume of gas increases,

(a) the pressure decreases
(b) the molecules collide more frequently with the sides of the container
(c) the molecules move more slowly
(d) the molecules become heavier
(e) the volume decreases

_____ 12. If one mole of a gas occupies 25 L at a certain temperature, what would be the volume of 6.02×10^{21} molecules of the gas under the same conditions?

(a) 2.5 L (b) 250 L (c) 0.25 L (d) 75 L (e) 6.25 L

A-2 Problem

1. A 575-mL quantity of gas is contained at a pressure of 855 torr. What is the volume if the pressure is decreased to 725 torr?

2. A quantity of gas has a volume of 12.5 L at a temperature of $-16^{\circ}C$. What is the temperature if the gas expands to 14.6 L at a constant pressure?

3. A certain quantity of gas in a pressurized can (constant volume) has a pressure of 1.22 atm at a temperature of $20^{\circ}C$. What temperature is required to raise the pressure to 3.75 atm (which causes the can to explode)?

4. A quantity of gas has a volume of 4350 mL at a temperature of 298 K and a pressure of 0.862 atm. What is the volume if the temperature is changed to $0^{\circ}C$ and the pressure to 887 torr?

5. A 525-mL quantity of gas under pressure at $25^{\circ}C$ expands to 3.62 L when the pressure is released to one atmosphere. If the temperature of the expanded gas is $19^{\circ}C$, what was the original pressure in the container in atmosphere?

6 . The volume of 50.0 g of O_2 is 37.1 L at a certain temperature and pressure. How many molecules must be *added* to this amount of oxygen to cause the volume to increase to 43.8 L at the same temperature and pressure?

Review of Part B *Relationships Among Quantities of Gases, Conditions, and Chemical Reactions*

OBJECTIVES AND DETAILED TABLE OF CONTENTS

10-4 The Ideal Gas Law

OBJECTIVE *Using the ideal gas law, calculate one condition of a gas given the other stated conditions.*

10-5 Dalton's Law of Partial Pressures

OBJECTIVE *Using Dalton's law, calculate the partial pressure of a gas.*

10-6 The Molar Volume and Density of a Gas

OBJECTIVE *Convert between volume, moles and density of a gas measured at STP.*

10-7 Stoichiometry Involving Gases

OBJECTIVE *Using the molar volume at STP, perform stoichiometric calculations involving gas volumes.*

SUMMARY OF PART B

The four gas laws discussed above can be joined into one general law. This law, known as the **ideal gas law**, relates Boyle's, Charles's and Avogadro's laws. With this law, one property (temperature, pressure, volume, or moles) can be calculated if the other three are known. An example follows.

Example B-1 Ideal Gas Law

If 0.743 mole of a gas occupies 18.6 L at a pressure of 0.831 atm, what is the temperature (in $^{\circ}C$)?

PROCEDURE

$$PV = nRT \qquad T = \frac{PV}{nR}$$

SOLUTION

$$T = \frac{0.831 \text{ atm} \times 18.6 \text{ L}}{0.743 \text{ mol} \times 0.0821 \frac{\text{L} \cdot \text{atm}}{\text{K} \cdot \text{mol}}} = 253 \text{ K} \qquad (253 - 273) = \underline{-20^{\circ}C}$$

According to kinetic molecular theory, gas molecules do not interact. Thus properties such as pressure and volume depend only on the amount of gas present and not on its identity. **Dalton's law** is a statement of this observation. It states that the pressure of a mixture of gases is the sum of the partial pressures of the component gases. A sample calculation involving this law follows.

Example B-2 Dalton's Law

If a mixture of gases exerts a pressure of 1250 torr and is composed of 2.35 moles of O_2, 1.76 moles of N_2, and 0.85 moles of CO_2, what is the partial pressure due to N_2?

PROCEDURE

Calculate the decimal fraction of moles of N_2 then multiply that by the total pressure.

SOLUTION

$$\frac{1.76}{1.76 + 2.35 + 0.85} = 0.355$$

$$P(N_2) = 0.355 \times 1250 \text{ torr} = \underline{444 \text{ torr}}$$

Since all gases act the same (physically), one mole of either a pure gas or a mixture has the same volume. The **molar volume** of a gas is 22.4 L, which is the volume occupied by 1.00 mole of a gas at standard temperature (0 $^{\circ}C$) and standard pressure (1 atm). The density of a gas is usually expressed in grams per liter at STP. It is obtained by dividing the molar mass by the molar volume.

$$\frac{\text{g/mol}}{\text{L/mol}} = \text{g/L}$$

The molar volume relationship or the ideal gas law allows us a straightforward way to incorporate gases into the general scheme of stoichiometry introduced in Chapter 7 because this law converts volume at a specified temperature and pressure to moles.

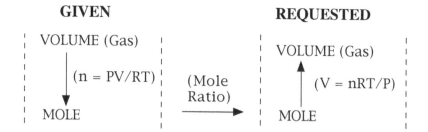

GIVEN		REQUESTED	
VOLUME (Gas)		VOLUME (Gas)	
↓ (n = PV/RT)	(Mole Ratio) →	↑ (V = nRT/P)	
MOLE		MOLE	

A sample problem follows.

Example B-3 Stoichiometry and Gases

What volume of O_2 is produced from 188 g of KO_2 if the O_2 is measured at 25°C and a pressure of 745 torr?

$$4KO_2(s) + 2CO_2(g) \longrightarrow 2K_2CO_3(s) + 3O_2(g)$$

PROCEDURE

1. Convert mass of KO_2 (Given) to moles of KO_2.

2. Convert moles of KO_2 to moles of O_2.

3. Convert moles of O_2 to volume of O_2 (Requested) in a separate calculation.

SOLUTION

$$\underset{1}{188 \text{ g } KO_2} \times \frac{1 \text{ mol } KO_2}{71.10 \text{ g } KO_2} \times \underset{2}{\frac{3 \text{ mol } O_2}{4 \text{ mol } KO_2}} = 1.98 \text{ mol } O_2$$

$$3 \quad V = \frac{nRT}{P} = \frac{1.98 \text{ mol} \times 0.0821 \frac{L \cdot atm}{K \cdot mol} \times 298 \text{ K}}{\frac{745 \text{ torr}}{760 \text{ torr/atm}}} = \underline{49.4 \text{ L}}$$

ASSESSMENT OF OBJECTIVES

B-1 Multiple Choice

___ 1. Which of the following is a correct value of the gas constant?

(a) $62.4 \dfrac{L \cdot atm}{mol \cdot {}^{o}C}$ (b) $82.1 \dfrac{mL \cdot atm}{K \cdot mol}$

(c) $0.0821 \dfrac{L \cdot atm}{{}^{o}C \cdot mol}$ (d) $62.4 \dfrac{mL \cdot torr}{K \cdot mol}$

___ 2. Under what conditions can there be serious deviation from the ideal gas law?

(a) high pressure and temperature
(b) low pressure and temperature
(c) high pressure and low temperature
(d) low pressure and high temperature

___ 3. What is the total pressure of a mixture of gases if 60.0% of the gas is N_2 and the partial pressure of N_2 is 300 torr?

(a) 240 torr   (b) 500 torr   (c) 1000 torr
(d) 180 torr   (e) 360 torr

___ 4. A quantity of oxygen is collected over water. The pressure of the gas above the water is 755 torr. The pressure that the oxygen alone would exert is 728 torr. The vapor pressure of water at this temperature is

(a) 728 torr   (b) 755 torr
(c) 782 torr   (d) 27 torr

___ 5. A 224-L quantity of N_2 measured at STP

(a) contains one mole of N_2
(b) has a mass of 28.0 g
(c) contains 6.02×10^{24} N_2 molecules
(d) is 0.10 mole
(e) contains 1.20×10^{24} N atoms

___ 6. The density of a gas is 4.1 g/L at STP. The gas is

(a) NO_2 (b) CO_2 (c) N_2O_4 (d) SO_3 (e) CO

B-2 Problems

1. What is the volume of a gas if 0.334 moles of the gas has a temperature of $11^{\circ}C$ and a pressure of 0.900 atm?

2. What mass of SO_2 occupies 47.6 L at a pressure of 745 torr and a temperature of $-6^{\circ}C$?

3. How many molecules are in a volume of 1.00 mL at a temperature of $117^{\circ}C$ and a pressure of 1.65×10^{-4} torr?

4. A mixture of gases is composed of 46.2% N_2, 31.4% SO_2, and the rest CO_2. If the total pressure is 1.08 atm, what is the partial pressure of each gas?

5. A mixture of gases is composed of 132 g of O_2 and 2.72×10^{24} molecules of N_2. If the total pressure is 685 torr, what is the partial pressure of each gas?

6. What mass of CO_2 occupies 54.2 mL at STP?

7. Given the balanced equation:

$$2HNO_3(aq) + 3H_2S(aq) \longrightarrow 2NO(g) + 3S(s) + 4H_2O(l)$$

What mass of H_2S is required to produce 8.62 L of NO measured at STP?

8. What mass of H_2O is produced if 16.3 L of NO measured at $22^{\circ}C$ and 0.954 atm is also formed? (Use the balanced equation in problem 7 above.)

Chapter Summary Assessment

S-1 Matching

___ Combined gas law

___ Ideal gas law

___ Charles's law

___ Dalton's law

___ Boyle's law

___ Gay-Lussac's law

___ Avogadro's law

___ Graham's law

(a) $P_{tot} = P_1 + P_2 +$

(b) $V_1 T_1 = V_2 T_2$

(c) $\dfrac{P_1}{T_1} = \dfrac{P_2}{T_2}$

(d) $P = \dfrac{k}{V}$

(e) $PT = nVT$

(f) $\dfrac{m_1}{m_2} = \sqrt{\dfrac{v_1}{v_2}}$

(g) $V \propto n$

(h) $\dfrac{T_1}{V_1} = \dfrac{T_2}{V_2}$

(i) $P = \dfrac{nRT}{V}$

(j) $V \propto P$

(k) $v_1 = v_2 \sqrt{\dfrac{m_2}{m_1}}$

(l) $P = X_1 P_1$

(m) $V \propto \dfrac{T}{P}$

Answers to Assessments of Objectives

A-1 Multiple Choice

1. **d** Gases are virtually incompressible.

2. **b** The volume of a gas is composed mostly of matter. (It is, in fact, almost entirely empty space.)

3. **e** All gas molecules have the same average speed at the same temperature. (They have the same kinetic energy.)

4. **d** kelvins

5. **a** 76.0 cm

6. **c** The volume is halved.

7. **d** increases by a factor of 293/273

8. **d** 223 (-50 + 273 = 223)

9. **b** $P \propto T$

10. **b** 760 torr and 273 K

11. **b** The molecules collide more frequently with the sides of the container. (This is because they are moving faster.)

12. **c** 0.25 L $(6.02 \times 10^{21} = 0.01$ mol$)$ 25 L $\times \dfrac{0.01 \text{ mol}}{1.00 \text{ mol}} = 0.25$ L

A-2 Problem

1. $V_{initial} = 575$ mL, $P_{in} = 855$ torr $V_{final} = ?$, $P_F = 725$ torr

   Since P goes *down* V goes *up* and P_{corr} is greater than one.

   $V_f = V_{in} \times P_{corr} = 575$ mL $\times \dfrac{855 \text{ torr}}{725 \text{ torr}} = \underline{678 \text{ mL}}$

2. $V_{initial} = 12.5$ L, $T_{In} = (-16 + 273) = 257$ K $V_{final} = 14.6$ L, $T_F = ?$

   Since V goes *up* T goes *up* and V_{corr} is greater than one.

   $T_{final} = T_{in} \times V_{corr} = 257$ K $\times \dfrac{14.6 \text{ L}}{12.5 \text{ L}} = 300$ K $T(^{o}C) = \underline{27^{o}C}$

3. $P_{in} = 1.22$ atm, $T_{in} = (20 + 273) = 293$ K, $P_v = 3.75$ atm, $T_v = ?$

   Since P goes *up* T goes *up* and P_{corr} is greater than one.

   $T = 293$ K $\times \dfrac{3.75 \text{ atm}}{1.22 \text{ atm}} = 901$ K $T(^{o}C) = (901 - 273) = \underline{628^{o}C}$

4. $V_{in} = 4350$ mL, $T_{in} = 298$ K, $P_{in} = 0.862$ atm, $V_f = ?$

   $T_F = 0^{o}C = 273$ K, $P_F = 887$ torr $\times \dfrac{1 \text{ atm}}{760 \text{ torr}} = 1.17$ atm

   Since T goes *down* V goes *down* and T_{Corr} is less than one.

   Since P goes *up* V goes *down* and P_{corr} is less than one.

   $V = 4350$ mL $\times \dfrac{273 \text{ K}}{298 \text{ K}} \times \dfrac{0.862 \text{ atm}}{1.17 \text{ atm}} = \underline{2940 \text{ mL}}$

5. Don't let the wording confuse you. The initial situation should be considered after expansion and the compressed gas regarded as the final condition.

   $V_{in} = 3.62$ L $= 3.62 \times 10^3$ mL, $T_{in} = (19 + 273) = 292$ K,

   $P_{in} = 1.00$ atm, $V_f = 525$ mL, $T_f = (25 + 273) = 298$ K, $P_f = ?$

   Since T goes *up* P goes *up* and T_{corr} is greater than one.

   Since V goes *down* P goes *up* and V_{corr} is greater than one.

   $P_F = 1.00$ atm $\times \dfrac{298 \text{ K}}{292 \text{ K}} \times \dfrac{3.62 \times 10^3 \text{ mL}}{525 \text{ mL}} = \underline{7.04 \text{ atm}}$

6. V_{in} = 37.1 L, n_{in} = 50.0 g O$_2$ x $\dfrac{1\ mol}{32.00\ g\ O_2}$ = 1.56 mol

V_f = 43.8 L, n_f = 1.56 + X (X = number of moles of gas added)

Since V goes **up** n goes **up** and n_{corr} is greater than one.

$V_F = V_{In}$ x n_{corr} 43.8 L = 37.1 L x $\dfrac{(1.56 + X)\ mol}{1.56\ mol}$

438 x 1.56 = 37.1 (1.56 + X)   X = 0.28 mol added

0.28 mol x $\dfrac{6.022 \times 10^{23}\ molecules}{mol}$ = $\underline{1.7 \times 10^{23}\ molecules\ added}$

B-1 Multiple Choice

1.   **b** 82.1 $\dfrac{mL \cdot atm}{K \cdot mol}$

2.   **c** high pressure and low temperature

3.   **b** 500 torr (0.600 x P_T = 300 torr; P_T = 500 torr)

4.   **d** 27 torr

5.   **c** contains 6.02×10^{24} N$_2$ molecules (224 L equals 10 moles at STP)

6.   **c** N$_2$O$_4$ (4.1 $\dfrac{g}{L}$ x 22.4 $\dfrac{L}{mol}$ = 92 g/mol)

B-2 Problems

1. n = 0.334 mol, T = (11 + 273) = 284 K, P = 0.900 atm, V = ?

PV = nRT   V = $\dfrac{nRT}{P}$

V = $\dfrac{0.334\ mol\ x\ 0.0821\ \frac{L \cdot atm}{K \cdot mol}\ x\ 284\ K}{0.900\ atm}$ = $\underline{8.65\ L}$

2. n = ?, T = (-6 + 273) = 267 K, P = 745 torr, V = 47.6 L

PV = nRT   n = $\dfrac{PV}{RT}$ = $\dfrac{\frac{745\ torr}{760\ torr/atm}\ x\ 47.6\ L}{0.0821\ \frac{L \cdot atm}{K \cdot mol}\ x\ 267\ K}$ = 2.13 mol

2.13 mol SO$_2$ x $\dfrac{64.07\ g\ SO_2}{mol\ SO_2}$ = 136 g SO$_2$

3. $n = ?$, $T = (117 + 273) = 390 \text{ K}$, $P = 1.65 \times 10^{-4}$ torr, $V = 1.00 \text{ mL} = 1.00 \times 10^{-3} \text{ L}$

$$n = \frac{PV}{RT} = \frac{\frac{1.65 \times 10^{-4} \text{ torr}}{760 \text{ torr/atm}} \times (1.00 \times 10^{-3} \text{ L})}{390 \text{ K} \times 0.0821 \frac{\text{L} \cdot \text{atm}}{\text{K} \cdot \text{mol}}} = 6.78 \times 10^{-12} \text{ mol}$$

$$6.78 \times 10^{-12} \text{ mol} \times \frac{6.022 \times 10^{23} \text{ molecules}}{\text{mol}} = \underline{4.08 \times 10^{12} \text{ molecules}}$$

4. $P_{Component} = [\frac{\%}{100\%}] \times P_{tot}$

$P(N_2) = 0.462 \times 1.08 \text{ atm} = 0.50 \text{ atm}$ $P(SO_2) = 0.314 \times 1.08 \text{ atm} = 0.34 \text{ atm}$

$P(CO_2) = 1.08 \text{ atm} - (0.50 \text{ atm} + 0.34 \text{ atm}) = 0.24 \text{ atm}$

5. $P(O_2) = [\frac{\% O_2}{100\%}] \times P_{tot}$ $\% O_2 = [\frac{\text{moles} (O_2)}{\text{total moles}}] \times 100\%$

Calculate the total moles of gas present.

$$n(O_2) = 132 \text{ g } O_2 \times \frac{1 \text{ mol}}{32.00 \text{ g } O_2} = 4.12 \text{ mol}$$

$$n(N_2) = 2.72 \times 10^{24} \text{ molecules} \times \frac{1 \text{ mol}}{6.022 \times 10^{23} \text{ molecules}} = 4.52 \text{ mol}$$

$$P(O_2) = [\frac{4.12}{4.12 + 4.52}] \times 685 \text{ torr} = \underline{327 \text{ torr}}$$

$$P(N_2) = P_T - P(O_2) = 685 \text{ torr} - 327 \text{ torr} = \underline{358 \text{ torr}}$$

6. $54.2 \text{ mL} \times \frac{10^{-3} \text{ L}}{\text{mL}} \times \frac{1 \text{ mol}}{22.4 \text{ L (STP)}} = 2.42 \times 10^{-3} \text{ mol}$

$2.42 \times 10^{-3} \text{ mol } CO_2 \times \frac{44.01 \text{ g}}{\text{mol } CO_2} = \underline{0.107 \text{ g}}$

7.   (1) Convert volume of NO to moles of NO using the molar volume relationship since conditions are at STP.
     (2) Convert moles of NO to moles of H_2S.
     (3) Convert moles of H_2S to mass of H_2S.

$$\overset{(1)}{} \qquad \overset{(2)}{} \qquad \overset{(3)}{}$$

$$8.62 \text{ L (STP)} \times \frac{1 \text{ mol NO}}{22.4 \text{ L (STP)}} \times \frac{3 \text{ mol } H_2S}{2 \text{ mol NO}} \times \frac{34.09 \text{ g } H_2S}{\text{mol } H_2S} = \underline{19.7 \text{ g } H_2S}$$

8. (1) Convert volume of NO to moles of NO using the ideal gas law.
(2) Convert moles of NO to moles of H_2O.
(3) Convert moles of H_2O to mass of H_2O.

$$(1)\quad n = \frac{PV}{RT} = \frac{0.954\ \cancel{atm}\ \times\ 16.3\ \cancel{L}}{0.0821\ \dfrac{\cancel{L}\cdot\cancel{atm}}{\cancel{K}\cdot mol}\ \times\ 295\ \cancel{K}} = 0.642\ mol\ NO$$

$$\underset{(2)}{}\qquad\qquad \underset{(3)}{}$$

$$0.642\ \cancel{mol\ NO}\ \times\ \frac{4\ \cancel{mol\ H_2O}}{2\ \cancel{mol\ NO}}\ \times\ \frac{18.02\ g\ H_2O}{\cancel{mol\ H_2O}} = \underline{23.1\ g\ H_2O}$$

S-1 Matching

m   **Combined gas law** is $V \propto \dfrac{T}{P}$ or $\dfrac{PV}{T} = k$

i   **Ideal gas law** $P = \dfrac{nRT}{V}$ or $PV = nRT$

h   **Charles's law** $\dfrac{T_1}{V_1} = \dfrac{T_2}{V_2}$ or $\dfrac{V_1}{T_1} = \dfrac{V_2}{T_2}$

a   **Dalton's law** $P_T = P_1 + P_2 + \cdots$

d   **Boyle's Law** $P = \dfrac{k}{V}$ or $PV = k$

c   **Gay-Lussac's law** $\dfrac{P_1}{T_1} = \dfrac{P_2}{T_2}$

g   **Avogadro's law** $V \propto n$

k   **Graham's law** $v_1 = v_2\sqrt{\dfrac{m_2}{m_1}}$

Answers and Solutions to Green Text Problems

10-1 The molecules of water are closely packed together and thus offer much more resistance. The molecules in a gas are dispersed into what is mostly empty space.

10-3 Since a gas is mostly empty space, more molecules can be added. In a liquid, molecules occupy most of the space so no more can be added.

10-5 Gas molecules are in rapid but random motion. When gas molecules collide with a light dust particle suspended in the air, they impart a random motion to the particle.

10-6 $\dfrac{v_1}{v_2} = \sqrt{\dfrac{m_2}{m_1}}$ $v_2 = 20.0$ miles/hr, $m_2 = 6.00$ kg $= 6000$ g $v_1 = ?$ $m_1 = 1.50$ g

$v_1 = 20.0$ miles/hr x $\sqrt{\dfrac{6000\ g}{1.50\ g}}$ $= \underline{1260\ miles/hr}$

10-7   The molecule with the largest formula weight travels the slowest. SF_6(146.1 amu)
$< SO_2$(64.07 amu) $< N_2O$(44.02 amu) $< CO_2$(44.01 amu) $< N_2$ (28.02 amu)
$< H_2$(2.016 amu)

10-9   When the pressure is high, the gas molecules are forced close together. In a highly compressed gas, the molecules can occupy an appreciable part of the total volume. When the temperature is low, molecules have a lower average velocity. If there is some attraction, they can momentarily stick together when moving slowly.

10-10 (a) 1650 torr x $\dfrac{1\ atm}{760\ torr}$ $= \underline{2.17\ atm}$

(b) 3.50×10^{-5} atm x $\dfrac{760\ torr}{atm}$ $= \underline{0.0266\ torr}$

(c) 18.5 lb/in.^{2} x $\dfrac{1\ atm}{14.7\ lb/in.^2}$ x $\dfrac{760\ torr}{atm}$ $= \underline{9560\ torr}$

(d) 5.65 kPa x $\dfrac{1\ atm}{101.3\ kPa}$ $= \underline{0.0558\ atm}$

(e) 190 torr x $\dfrac{1\ atm}{760\ torr}$ x $\dfrac{14.7\ lb/in.^2}{atm}$ $= \underline{3.68\ lb/in.^2}$

(f) 85 torr x $\dfrac{1\ atm}{760\ torr}$ x $\dfrac{101.3\ kPa}{atm}$ $= \underline{11\ kPa}$

10-11 (a) 30.2 in. Hg x $\dfrac{1\ atm}{29.9\ in.\ Hg}$ x $\dfrac{760\ torr}{atm}$ $= \underline{768\ torr}$

(b) 25.7 kbar x $\dfrac{10^3\ bar}{kbar}$ x $\dfrac{1\ atm}{1.013\ bar}$ $= \underline{2.54 \times 10^4\ atm}$

(c) 57.9 kPa x $\dfrac{1\ atm}{101.3\ kPa}$ x $\dfrac{14.7\ lb/in.^2}{atm}$ $= \underline{8.40\ lb/in.^2}$

(d) 0.025 atm x $\dfrac{760\ torr}{atm}$ $= \underline{19\ torr}$

10-13 10.3 mbar x $\dfrac{10^{-3}\ bar}{mbar}$ x $\dfrac{1\ atm}{1.013\ bar}$ $= \underline{0.0102\ atm}$

10-15 Assume a column of Hg has a 1 cm^2 cross section and is 76.0 cm high.
Weight of Hg = 76.0 cm x 1 cm^2 x 13.6 g/cm^3 = 1030 g.
If water is substituted, 1030 g of water in the column is required.
height x 1 cm^2 x 1.00 g/cm^3 = 1030 g   height = 1030 cm

1030 cm x $\dfrac{1\ in.}{2.54\ cm}$ x $\dfrac{1\ ft}{12\ in.}$ $= \underline{33.8\ ft}$

If a well is 40 ft deep, the water cannot be raised in one stage by suction since 33.8 ft is the theoretical maximum height that is supported by the atmosphere.

10-16 $V_2 = 6.85$ L x $\dfrac{0.650\ atm}{0.435\ atm}$ $= \underline{10.2\ L}$

10-18 $V_2 = 785$ mL $\times \dfrac{760 \text{ torr}}{610 \text{ torr}} = \underline{978 \text{ mL}}$

10-19 $P_2 = 62.5$ torr $\times \dfrac{125 \text{ mL}}{115 \text{ mL}} = \underline{67.9 \text{ torr}}$

10-22 $V_{final}\,(V_f) = 15\,V_{initial}\,(V_i)$ $\dfrac{V_f}{V_i} = \dfrac{P_i}{P_f}$; $\dfrac{15\,V_i}{V_i} = \dfrac{0.950 \text{ atm}}{P_f}$

$P_f = 0.950$ atm $\times\ 15 = \underline{14.3 \text{ atm}}$

10-23

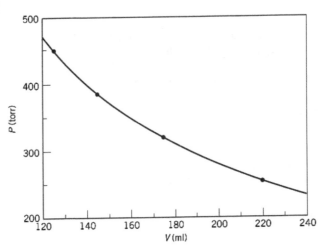

No, the graph is not linear.

10-24 $V_2 = 1.55$ L $\times \dfrac{373 \text{ K}}{298 \text{ K}} = \underline{1.94 \text{ L}}$

10-26 $T_2 = 290$ K $\times \dfrac{392 \text{ mL}}{325 \text{ mL}} = 350$ K 350 K $- 273 = \underline{77\ ^{o}C}$

10-28 $V_2 = 3.66 \times 10^4$ L $\times \dfrac{323 \text{ K}}{455 \text{ K}} = 2.60 \times 10^4$ L

10-29 $V_2 = 1.25\,V_1$ $T_2 = 273$ K $\times \dfrac{1.25\,V_1}{V_1} = 341$ K 341 K $- 273 = \underline{68\ ^{o}C}$

10-31 $P_2 = 2.50$ atm $\times \dfrac{295 \text{ K}}{251 \text{ K}} = \underline{2.94 \text{ atm}}$

10-32 $P_1 = 850$ torr $\times \dfrac{1 \text{ atm}}{760 \text{ torr}} = 1.12$ atm

$T_2 = 328$ K $\times \dfrac{0.652 \text{ atm}}{1.12 \text{ atm}} = 191$ K 191 K $- 273 = \underline{-82\ ^{o}C}$

10-34 $T_2 = 298$ K $\times \dfrac{2.50 \text{ atm}}{1.25 \text{ atm}} = 596$ K 596 K $- 273 = \underline{323\ ^{o}C}$

10-35 $P_2 = 28.0$ lb/in.^{2} $\times \dfrac{313 \text{ K}}{290 \text{ K}} = \underline{30.2 \text{ lb/in.}^2}$

10-38 (a) $PV = kT$ and (d) $\dfrac{P}{T} \propto \dfrac{1}{V}$

10-40 Exp. 1, T increases   Exp. 2, V decreases   Exp. 3, P increases   Exp. 4, T increases

10-42 $P_2 = 0.950 \text{ atm} \times \dfrac{5.50 \text{ L}}{4.75 \text{ L}} \times \dfrac{308 \text{ K}}{273 \text{ K}} = \underline{1.24 \text{ atm}}$

10-43 $V_2 = 17.5 \text{ L} \times \dfrac{6.00 \text{ atm}}{1.00 \text{ atm}} \times \dfrac{273 \text{ K}}{373 \text{ K}} = \underline{76.8 \text{ L}}$

10-45 $T_2 = 223 \text{ K} \times \dfrac{155 \text{ torr}}{78.0 \text{ torr}} \times \dfrac{9.55 \times 10^{-5} \text{ mL}}{4.78 \times 10^{-4} \text{ mL}} = 88.5 \text{ K} = \underline{-185\ ^{\circ}C}$

10-48 $T_2 = 298 \text{ K} \times \dfrac{1.38 \text{ L}}{1.55 \text{ L}} \times \dfrac{1.02 \text{ atm}}{1.05 \text{ atm}} = \underline{258 \text{ K } (-15\ ^{\circ}C)}$

10-49 $V_2 = 35.0 \text{ mL} \times \dfrac{295 \text{ K}}{290 \text{ K}} \times \dfrac{11.5 \text{ atm}}{1.00 \text{ atm}} = \underline{409 \text{ mL}}$

10-50 $V_2 = 2.54 \text{ L} \times \dfrac{0.0750 \text{ mol}}{0.112 \text{ mol}} = \underline{1.70 \text{ L}}$

10-51 Let X = total moles needed in the expanded balloon.

$188 \text{ L} \times \dfrac{X}{8.40 \text{ mol}} = 275 \text{ L} \quad X = 12.3 \text{ mol}$

$12.3 - 8.4 = \underline{3.9 \text{ moles must be added.}}$

10-53 $n_2 = 2.50 \times 10^{-3} \text{ mol} \times \dfrac{164 \text{ mL}}{75.0 \text{ mL}} = 5.47 \times 10^{-3} \text{ mol}$

$(5.47 \times 10^{-3}) - (2.50 \times 10^{-3}) = 2.97 \times 10^{-3} \text{ mol added}$

$2.97 \times 10^{-3} \text{ mol N}_2 \times \dfrac{28.02 \text{ g N}_2}{\text{mol N}_2} = \underline{0.0832 \text{ g N}_2}$

10-55 $T = \dfrac{PV}{nR} = \dfrac{2.25 \text{ atm} \times 4.50 \text{ L}}{0.332 \text{ mol} \times 0.0821 \dfrac{\text{L} \cdot \text{atm}}{\text{K} \cdot \text{mol}}} = 371 \text{ K} \qquad 371 \text{ K} - 273 = \underline{98\ ^{\circ}C}$

10-57 $n = \dfrac{PV}{RT} = \dfrac{0.955 \text{ atm} \times 16.4 \text{ L}}{250 \text{ K} \times 0.0821 \dfrac{\text{L} \cdot \text{atm}}{\text{K} \cdot \text{mol}}} = 0.763 \text{ mol NH}_3$

$0.763 \text{ mol NH}_3 \times \dfrac{17.03 \text{ g NH}_3}{\text{mol NH}_3} = \underline{13.0 \text{ g NH}_3}$

10-58 $n = 0.250 \text{ g O}_2 \times \dfrac{1 \text{ mol O}_2}{32.00 \text{ g O}_2} = 7.81 \times 10^{-3} \text{ mol O}_2$

$P = \dfrac{nRT}{V} = \dfrac{7.81 \times 10^{-3} \text{ mol} \times 0.0821 \dfrac{\text{L} \cdot \text{atm}}{\text{K} \cdot \text{mol}} \times 302 \text{ K}}{0.250 \text{ L}} = 0.775 \text{ atm}$

$0.775 \text{ atm} \times 760 \text{ torr/atm} = \underline{589 \text{ torr}}$

10-60 $n = \dfrac{PV}{RT} = \dfrac{1.15 \text{ atm} \times 3.50 \text{ L}}{0.0821 \dfrac{\text{L} \cdot \text{atm}}{\text{K} \cdot \text{mol}} \times 296 \text{ K}} = 0.166 \text{ mol}$

$0.166 \text{ mol Ne} \times \dfrac{20.18 \text{ g Ne}}{\text{mol Ne}} = \underline{3.35 \text{ g Ne}}$

10-62 $P = \dfrac{780 \text{ torr}}{760 \text{ torr/atm}} = 1.03 \text{ atm}$

$n = \dfrac{PV}{RT} = \dfrac{1.03 \text{ atm} \times 2.5 \times 10^7 \text{ L}}{0.0821 \dfrac{\text{L} \cdot \text{atm}}{\text{K} \cdot \text{mol}} \times 300 \text{ K}} = 1.0 \times 10^6 \text{ mol of gas}$

He: $1.0 \times 10^6 \text{ mol He} \times \dfrac{4.003 \text{ g He}}{\text{mol He}} \times \dfrac{1 \text{ lb}}{453.6 \text{ g}} = \underline{8800 \text{ lb He}}$

Air: $1.0 \times 10^6 \text{ mol air} \times \dfrac{29.0 \text{ g air}}{\text{mol air}} \times \dfrac{1 \text{ lb}}{453.6 \text{ g}} = \underline{64,000 \text{ lb air}}$

Lifting power with He = 64,000 – 8800 = $\underline{55,000 \text{ lb}}$

H_2: $1.0 \times 10^6 \text{ mol H}_2 \times \dfrac{2.016 \text{ g H}_2}{\text{mol H}_2} \times \dfrac{1 \text{ lb}}{453.6 \text{ g}} = \underline{4400 \text{ lb H}_2}$

Lifting power with H_2 = 64,000 - 4000 = $\underline{60,000 \text{ lb}}$
Helium is a noncombustible gas whereas hydrogen forms an explosive mixture with O_2.

10-64 $P_T = P(CO_2) + P(N_2) + P(He) = 250 \text{ torr} + 375 \text{ torr} + 137 \text{ torr} = \underline{762 \text{ torr}}$

10-66 $P(Ar) = \left(\dfrac{Ar\%}{100\%}\right) \times P_T = 0.0090 \times 756 \text{ torr} = \underline{6.8 \text{ torr}}$

10-69 $P(N_2) = 1050 \text{ torr} \times 0.720 = 756 \text{ torr}$ $P(O_2) = 1050 \text{ torr} \times 0.0800 = 84.0 \text{ torr}$

$P(SO_2) = P_T - [P(N_2) + P(O_2)] = 1050 - (756 + 84) = \underline{210 \text{ torr}}$

10-70 $P(O_2) = \left(\dfrac{O_2\%}{100\%}\right) \times P_T$ $256 \text{ torr} = (0.350) \times P_T$ $P_T = \underline{731 \text{ torr}}$

10-72 The partial pressure of each gas must be determined when confined in a 2.00-L volume.
$P(N_2) = 300 \text{ torr}$ $P(O_2) = 85 \text{ torr} \times \dfrac{4.00 \text{ L}}{2.00 \text{ L}} = 170 \text{ torr}$

$P(CO_2) = 450 \text{ torr} \times \dfrac{1.00 \text{ L}}{2.00 \text{ L}} = 225 \text{ torr}$
$P_T = 300 + 170 + 225 = \underline{695 \text{ torr}}$

10-73 First, find the partial pressure of gas A in the 4.00-L container.
$P_A = 0.880 \text{ atm} \times \dfrac{2.50 \text{ L}}{4.00 \text{ L}} = 0.550 \text{ atm}$
$P_B = 0.850 - 0.550 = \underline{0.300 \text{ atm}}$

10-75 $15.0 \text{ g CO}_2 \times \dfrac{1 \text{ mol CO}_2}{44.01 \text{ g CO}_2} \times \dfrac{22.4 \text{ L}}{\text{mol}} = \underline{7.63 \text{ L}}$

10-77 3.01×10^{24} ~~molecules~~ $\times \dfrac{1 \text{ mol } N_2}{6.022 \times 10^{23} \text{ ~~molecules~~}} \times \dfrac{22.4 \text{ L } N_2}{\text{~~mol~~} N_2} = \underline{112 \text{ L } N_2}$

10-79 6.78×10^{-4} ~~L~~ $\times \dfrac{1 \text{ ~~mol~~}}{22.4 \text{ ~~L~~}} \times \dfrac{46.01 \text{ g}}{\text{~~mol~~}} = \underline{1.39 \times 10^{-3} \text{ g}}$

10-80 Molar mass of $B_2H_6 = 27.67$ g/mol $\qquad \dfrac{27.67 \text{ g}}{\text{~~mol~~}} \times \dfrac{1 \text{ ~~mol~~}}{22.4 \text{ L}} = \underline{1.24 \text{ g/L}}$

10-82 $\dfrac{1.52 \text{ g}}{\text{~~L~~}} \times \dfrac{22.4 \text{ ~~L~~}}{\text{mol}} = \underline{34.0 \text{ g/mol}}$

10-84 Find the volume at STP using the combined gas law.

$1.00 \text{ L} \times \dfrac{273 \text{ ~~K~~}}{298 \text{ ~~K~~}} \times \dfrac{1.20 \text{ ~~atm~~}}{1.00 \text{ ~~atm~~}} = 1.10 \text{ L (STP)} \qquad \dfrac{3.60 \text{ g}}{1.10 \text{ L}} = \underline{3.27 \text{ g/L (STP)}}$

10-85 Find moles of N_2 in 1 L at 500 torr and 22 $^{\circ}$C using the ideal gas law.

$P = \dfrac{500 \text{ ~~torr~~}}{760 \text{ ~~torr~~/atm}} = 0.658 \text{ atm} \quad n = \dfrac{PV}{RT} = \dfrac{0.658 \text{ ~~atm~~} \times 1.00 \text{ ~~L~~}}{0.0821 \dfrac{\text{~~L~~} \cdot \text{~~atm~~}}{\text{~~K~~} \cdot \text{mol}} \times 295 \text{ ~~K~~}} = 0.272 \text{ mol } N$

$0.272 \text{ ~~mol~~} N_2 \times \dfrac{28.02 \text{ g } N_2}{\text{~~mol~~} N_2} = 0.762 \text{ g } N_2$

Density $= \underline{0.762 \text{ g/L}}$ (500 torr and 22 $^{\circ}$C)

10-87 g $CaCO_3 \longrightarrow$ mol $CaCO_3 \longrightarrow$ mol $CO_2 \longrightarrow$ Vol. CO_2

$115 \text{ ~~g~~} CaCO_3 \times \dfrac{1 \text{ ~~mol~~ } CaCO_3}{100.1 \text{ ~~g~~ } CaCO_3} \times \dfrac{1 \text{ ~~mol~~ } CO_2}{1 \text{ ~~mol~~ } CaCO_3} \times \dfrac{22.4 \text{ L}}{\text{~~mol~~} CO_2} = \underline{25.7 \text{ L } CO_2 \text{ (STP)}}$

10-88 Vol. $O_2 \longrightarrow$ mol $O_2 \longrightarrow$ mol Mg $\longrightarrow$ g Mg

$5.80 \text{ ~~L~~ } O_2 \times \dfrac{1 \text{ ~~mol~~ } O_2}{22.4 \text{ ~~L~~ } O_2} \times \dfrac{2 \text{ ~~mol~~ Mg}}{1 \text{ ~~mol~~ } O_2} \times \dfrac{24.31 \text{ g Mg}}{\text{~~mol~~ Mg}} = \underline{12.6 \text{ g Mg}}$

10-90 g $H_2O \longrightarrow$ mol $H_2O \longrightarrow$ mol $C_2H_2 \longrightarrow$ Vol C_2H_2

$5.00 \text{ ~~g~~ } H_2O \times \dfrac{1 \text{ ~~mol~~ } H_2O}{18.02 \text{ ~~g~~ } H_2O} \times \dfrac{1 \text{ mol } C_2H_2}{2 \text{ ~~mol~~ } H_2O} = 0.139 \text{ mol } C_2H_2$

$P = \dfrac{745 \text{ ~~torr~~}}{760 \text{ ~~torr~~/atm}} = 0.980 \text{ atm}$

$V = \dfrac{nRT}{P} = \dfrac{0.139 \text{ ~~mol~~} \times 0.0821 \dfrac{\text{L} \cdot \text{~~atm~~}}{\text{~~K~~} \cdot \text{~~mol~~}} \times 298 \text{ ~~K~~}}{0.980 \text{ ~~atm~~}} = \underline{3.47 \text{ L}}$

10-92 (a) g C_4H_{10} $\longrightarrow$ mol C_4H_{10} $\longrightarrow$ mol CO_2 $\longrightarrow$ Vol CO_2

$$85.0 \text{ g } C_4H_{10} \times \frac{1 \text{ mol } C_4H_{10}}{58.12 \text{ g } C_4H_{10}} \times \frac{8 \text{ mol } CO_2}{2 \text{ mol } C_4H_{10}} \times \frac{22.4 \text{ L}}{\text{mol } CO_2} = \underline{131 \text{ L}}$$

(b) g C_4H_{10} $\longrightarrow$ mol C_4H_{10} $\longrightarrow$ mol O_2 $\longrightarrow$ Vol O_2

$$85.0 \text{ g } C_4H_{10} \times \frac{1 \text{ mol } C_4H_{10}}{58.12 \text{ g } C_4H_{10}} \times \frac{13 \text{ mol } O_2}{2 \text{ mol } C_4H_{10}} = 9.51 \text{ mol O}$$

$$V = \frac{nRT}{P} = \frac{9.51 \text{ mol} \times 0.0821 \frac{L \cdot atm}{K \cdot mol} \times 400 \text{ K}}{3.25 \text{ atm}} = \underline{96.1 \text{ L } O_2}$$

(c) Vol C_4H_{10} $\longrightarrow$ mol C_4H_{10} $\longrightarrow$ mol CO_2 $\longrightarrow$ Vol CO_2

$$C_4H_{10}: \quad n = \frac{PV}{RT} = \frac{0.750 \text{ atm} \times 45.0 \text{ L}}{0.0821 \frac{L \cdot atm}{K \cdot mol} \times 298 \text{ K}} = 1.38 \text{ mol}$$

$$1.38 \text{ mol } C_4H_{10} \times \frac{8 \text{ mol } CO_2}{2 \text{ mol } C_4H_{10}} \times \frac{22.4 \text{ L}}{\text{mol } CO_2} = \underline{124 \text{ L } CO_2}$$

10-93 $$n(H_2) = \frac{PV}{RT} = \frac{2.80 \times 10^4 \text{ L} \times 70.0 \text{ atm}}{0.0821 \frac{L \cdot atm}{K \cdot mol} \times 523 \text{ K}} = 4.56 \times 10^4 \text{ mol } H_2$$

mol H_2 $\longrightarrow$ mol Zr $\longrightarrow$ g Zr $\longrightarrow$ kg Zr

$$4.56 \times 10^4 \text{ mol } H_2 \times \frac{1 \text{ mol Zr}}{2 \text{ mol } H_2} \times \frac{91.22 \text{ g Zr}}{\text{mol Zr}} \times \frac{1 \text{ kg Zr}}{10^3 \text{ g Zr}} = 2080 \text{ kg Zr}$$

$$2080 \text{ kg} \times \frac{2.205 \text{ lb}}{\text{kg}} \times \frac{1 \text{ ton}}{2000 \text{ lb}} = \underline{2.29 \text{ tons Zr}}$$

10-95 Vol O_2 $\longrightarrow$ mol O_2 $\longrightarrow$ mol CO_2 $\longrightarrow$ Vol CO_2

$$P = \frac{825 \text{ torr}}{760 \text{ torr/atm}} = 1.09 \text{ atm} \quad n(O_2) = \frac{PV}{RT} = \frac{27.5 \text{ L} \times 1.09 \text{ atm}}{0.0821 \frac{L \cdot atm}{K \cdot mol} \times 250 \text{ K}} = 1.46 \text{ mol } O_2$$

$$1.46 \text{ mol } O_2 \times \frac{1 \text{ mol } CO_2}{2 \text{ mol } O_2} = 0.730 \text{ mol } CO_2$$

$$V = \frac{nRT}{P} = \frac{0.730 \text{ mol} \times 0.0821 \frac{L \cdot atm}{K \cdot mol} \times 300 \text{ K}}{1.50 \text{ atm}} = \underline{12.0 \text{ L } CO_2}$$

10-96 Force = 12.0 cm^2 x 15.0 cm x $\dfrac{13.6 \text{ g}}{\text{cm}^3}$ = 2450 g

$P = \dfrac{2450 \text{ g}}{12.0 \text{ cm}^2}$ = 204 g/cm^2 1 atm = 76.0 cm x $\dfrac{13.6 \text{ g}}{\text{cm}^3}$ = 1030 g/cm^2

204 g/cm^2 x $\dfrac{1 \text{ atm}}{1030 \text{ g/cm}^2}$ = $\underline{0.198 \text{ atm}}$

10-98 $n = \dfrac{1.00 \text{ L} \times 1.45 \text{ atm}}{0.0821 \dfrac{\text{L} \cdot \text{atm}}{\text{K} \cdot \text{mol}} \times 308 \text{ K}}$ = 0.0573 mol $\dfrac{8.37 \text{ g}}{0.0573 \text{ mol}}$ = $\underline{146 \text{ g/mol}}$

10-99 (1) Using the ideal gas law, find the molar mass of the compound.

$n = \dfrac{PV}{RT}$ = $\dfrac{4.50 \text{ L} \times 1.00 \text{ atm}}{350 \text{ K} \times 0.0821 \dfrac{\text{L} \cdot \text{atm}}{\text{K} \cdot \text{mol}}}$ = 0.157 mol

Molar mass = $\dfrac{6.58 \text{ g}}{0.157 \text{ mol}}$ = $\underline{41.9 \text{ g/mol}}$

(2) Using the percent composition, find the empirical formula.

In 100 g of compound there are 85.7 g of C and 14.3 g of H.

85.7 g C x $\dfrac{1 \text{ mol C}}{12.01 \text{ g C}}$ = 7.14 mol C 14.3 g H x $\dfrac{1 \text{ mol H}}{1.008 \text{ g H}}$ = 14.2 mol H

C: $\dfrac{7.14}{7.14}$ = 1.0   H: $\dfrac{14.2}{7.14}$ = 2.0   Emp. formula = CH_2

(3) Using the empirical formula, molar mass and empirical mass, find the molecular formula.

Emp. mass. = 12.01 + (2 x 1.008) = 14.03 g/emp. unit

$\dfrac{41.9 \text{ g/mol}}{14.03 \text{ g/emp. unit}}$ = 3 emp. units/mol   Molecular formula = $\underline{C_3H_6}$

10-101 n_T = 0.265 mol O_2 + 0.353 mol N_2 + 0.160 mol CO_2 = 0.778 mol of gas   V = 6.92 L

$P(O_2) = \dfrac{0.265}{0.778}$ x 2.86 atm = 0.974 atm; $P(N_2)$ = 1.30 atm; $P(CO_2)$ = 0.59 atm

10-102 $3 \times 10^4 \text{ molecules}$ x $\dfrac{1 \text{ mol}}{6.022 \times 10^{23} \text{ molecules}}$ = 5×10^{-20} mol

$P = \dfrac{nRT}{V}$ = $\dfrac{5 \times 10^{-20} \text{ mol} \times 0.0821 \dfrac{\text{L} \cdot \text{atm}}{\text{K} \cdot \text{mol}} \times 10 \text{ K}}{10^{-3} \text{ L}}$ = 4×10^{-17} atm

10-103 $\dfrac{V}{n}$ = $\dfrac{RT}{P}$ = $\dfrac{0.0821 \dfrac{\text{L} \cdot \text{atm}}{\text{K} \cdot \text{mol}} \times 298 \text{ K}}{1.25 \text{ atm}}$ = 19.6 L/mol

$\dfrac{44.01 \text{ g/mol}}{19.6 \text{ L/mol}}$ = = $\underline{2.25 \text{ g/L (Density)}}$

10-104 Density $= \dfrac{\text{mass}}{V} = \dfrac{P \times MM}{RT} = \dfrac{1.00 \text{ atm} \times 29.0 \text{ g/mol}}{0.0821 \frac{L \cdot atm}{K \cdot mol} \times 673 K} = \underline{0.525 \text{ g/L (hot)}}$

Density at STP $= 1.29$ g/L

$0.525/1.29 = 0.41$ (Hot air is less than half as dense as air at STP.)

10-106 $H_3BCO + 3H_2O(l) \longrightarrow B(OH)_3(aq) + CO(g) + 3H_2(g)$

$P = \dfrac{565 \text{ torr}}{760 \text{ torr/atm}} = 0.743$ atm

$n(H_3BCO) = \dfrac{PV}{RT} = \dfrac{0.743 \text{ atm} \times 0.425 \text{ L}}{0.0821 \frac{L \cdot atm}{K \cdot mol} \times 373 K} = 0.0103$ mol of H_3BCO

$0.0103 \text{ mol } H_3BCO \times \dfrac{4 \text{ mol gas}}{1 \text{ mol } H_3BCO} = 0.0412$ mol gas

$V = \dfrac{nRT}{P} = \dfrac{0.0412 \text{ mol} \times 0.0821 \frac{L \cdot atm}{K \cdot mol} \times 298 K}{0.900 \text{ atm}} = \underline{1.12 \text{ L}}$

10-108 $2Al(s) + 3F_2(g) \longrightarrow 2AlF_3(s)$ $P = \dfrac{725 \text{ torr}}{760 \text{ torr/atm}} = 0.954$ atm

(original F_2) $= \dfrac{0.954 \text{ atm} \times 8.23 \text{ L}}{0.0821 \frac{L \cdot atm}{K \cdot mol} \times 308 K} = 0.310$ mol F_2

(left over F_2) $= \dfrac{3.50 \text{ g}}{38.00 \text{ g/mol}} = 0.0921$ mol $0.310 - 0.092 = 0.218$ mol F_2 reacts

$0.218 \text{ mol } F_2 \times \dfrac{2 \text{ mol } AlF_3}{3 \text{ mol } F_2} \times \dfrac{83.98 \text{ g } AlF_3}{\text{mol } AlF_3} = \underline{12.2 \text{ g } AlF_3}$

10-110 $2.54 \times 10^{24} \text{ molecules} \times \dfrac{1 \text{ mol } N_2O_3}{6.022 \times 10^{23} \text{ molecules}} \times \dfrac{2 \text{ mol gas}}{\text{mol } N_2O_3} = 8.44$ mol gas

$V = \dfrac{nRT}{PV} = \dfrac{8.44 \text{ mol} \times 0.0821 \frac{L \cdot atm}{K \cdot mol} \times 308 K}{1.58 \text{ atm}} = \underline{135 \text{ L}}$

10-112 $15.0 \text{ mL} \times \dfrac{0.917 \text{ g}}{\text{mL}} \times \dfrac{1 \text{ mol}}{18.02 \text{ g } H_2O} = 0.763$ mol H_2O

$V = \dfrac{nRT}{P} = \dfrac{0.763 \text{ mol} \times 0.0821 \frac{L \cdot atm}{K \cdot mol} \times 298 K}{\dfrac{22 \text{ torr}}{760 \text{ torr/atm}}} = \underline{645 \text{ L}}$

10-113 $\quad n = \dfrac{PV}{RT} = \dfrac{1.20 \text{ atm} \times 0.0200 \text{ L}}{0.0821 \dfrac{\text{L} \cdot \text{atm}}{\text{K} \cdot \text{mol}} \times 298 \text{ K}} = 9.81 \times 10^{-4} \text{ mol } NH_3$

$9.81 \times 10^{-4} \text{ mol } NH_3 \times \dfrac{1 \text{ mol } N_2H_4}{2 \text{ mol } NH_3} \times \dfrac{22.4 \text{ L } N_2H_4}{\text{mol } N_2H_4} = 0.0110 \text{ L} = \underline{11.0 \text{ mL } N_2H_4}$

10-115 (1) Find empirical formula of reactant compound.

N: $30.4 \text{ g N} \times \dfrac{1 \text{ mol N}}{14.01 \text{ g N}} = 2.17 \text{ mol N}$

O: $69.6 \text{ g O} \times \dfrac{1 \text{ mol O}}{16.00 \text{ g O}} = 4.35 \text{ mol O}$

N: $\dfrac{2.17}{2.17} = 1.0$ $\qquad$ O: $\dfrac{4.35}{2.17} = 2.0$ $\qquad$ Empirical formula = NO_2

(2) Find molar mass of product compound.

$n = \dfrac{PV}{RT} = \dfrac{\dfrac{715 \text{ torr}}{760 \text{ torr/atm}} \times 1.05 \text{ L}}{0.0821 \dfrac{\text{L} \cdot \text{atm}}{\text{K} \cdot \text{mol}} \times 273 \text{ K}} = 0.0441 \text{ mol}$ $\quad$ Molar mass $= \dfrac{2.03 \text{ g}}{0.0441 \text{ mol}} = \underline{46.0 \text{ g/mol}}$

(3) Since one compound decomposes to one other compound, the reactant compound must have the same empirical formula as the product compound. Since the empirical mass of NO_2 = 46.01 g/emp unit, then the product must be NO_2(MM = 46.0 g/mol). Since 0.0220 mol of reactant form 0.0441 mol of product (1:2 ratio), the reaction must be

$$N_2O_4(l) \longrightarrow 2NO_2(g)$$

10-117 $\quad 6Li(s) + N_2(g) \longrightarrow 2Li_3N(s) \quad 3Mg(s) + N_2(g) \longrightarrow Mg_3N_2(s)$

$n = \dfrac{PV}{RT} = \dfrac{\dfrac{985 \text{ torr}}{760 \text{ torr/atm}} \times 256 \text{ L}}{0.0821 \dfrac{\text{L} \cdot \text{atm}}{\text{K} \cdot \text{mol}} \times 373 \text{ K}} = 10.8 \text{ mol } N_2$

$10.8 \text{ mol } N_2 \times \dfrac{6 \text{ mol Li}}{\text{mol } N_2} \times \dfrac{6.941 \text{ g Li}}{\text{mol Li}} = \underline{450 \text{ g Li}}$

$10.8 \text{ mol } N_2 \times \dfrac{3 \text{ mol Mg}}{\text{mol } N_2} \times \dfrac{24.31 \text{ g Mg}}{\text{mol Mg}} = \underline{788 \text{ g Mg}}$

11

The Solid and Liquid States

Review of Part A *The Properties of Condensed States and the Forces Involved*

Review of Part B *The Liquid State and Changes in State*

Chapter Summary Assessment

Answers to Assessments of Objectives

Answers and Solutions to Green Text Problems

Review of Part A *The Properties of Condensed States and the Forces Involved*

OBJECTIVES AND DETAILED TABLE OF CONTENTS

11-1 Properties of the Solid and Liquid States

OBJECTIVE *Use kinetic molecular theory to distinguish the physical properties of solids and liquids from gases.*

11-2 Intermolecular Forces and Physical State

OBJECTIVE *Describe the types of intermolecular forces that can occur between two molecules and their relative strengths.*

11-3 The Solid State: Melting Point

OBJECTIVE *Classify a solid as ionic, molecular, network, or metallic based on its physical properties.*

SUMMARY OF PART A

The solid and liquid states are known as condensed states primarily because the ions or molecules are close together, unlike the gaseous state. This fact accounts for four common properties. Specifically, they have high densities, are incompressible, expand little when heated, and have a definite volume.

In solids and liquids, the basic particles obviously "stick together." It is therefore important to describe the **intermolecular forces** that cause them to coalesce. All molecules have inherent forces of attraction between them called **London forces**. London forces are weakest for small, low molar mass compounds but become stronger for larger molecules. For nonpolar molecules, London forces are the only forces of attraction.

Polar molecules can align themselves so that the negative end of one molecule is attracted to the positive end of another. Thus, in addition to London forces, polar molecules have a second attractive force known as a **dipole-dipole force**. Given two compounds of similar molar mass, one polar and one nonpolar, the polar compound will have the stronger intermolecular forces of attraction and is more likely to exist in a condensed state at a specific temperature. It is more difficult to compare the forces of a polar, low-molar mass compound with those of a nonpolar, high-molar mass compound.

A third but very important interaction is known as **hydrogen bonding**. Hydrogen bonding is a considerably stronger interaction than normal dipole-dipole interactions and it is restricted to an electropositive hydrogen on one molecule interacting with the unshared pair of electrons on a nitrogen, oxygen, or fluorine on another molecule. It accounts for the fact that water is a liquid at room temperature rather than a gas. When one looks more closely at the three-dimensional structure of a water molecule, it becomes clearer how hydrogen bonding interactions occur.

We first look at the solid state and then the liquid state. Solids may be either **amorphous** or **crystalline**. In crystalline solids, the basic particles (atoms, molecules, or ions) exist in an orderly arrangement known as a **crystal lattice**. The temperature at which the lattice breaks down (the melting point) reflects the degree of attraction between the basic particles. **Ionic solids** have high melting points; those of **molecular solids** are generally lower because of weaker forces between basic particles. **Network solids,** such as the two main allotropes of carbon (*diamond* and *graphite*), generally have high melting points. **Metallic solids** have a wide range of melting points.

ASSESSMENT OF OBJECTIVES

A-1 Matching

Which state or states of matter (liquid, solid, gas) can be described by the following statements?

(a)   The most compressible   _____

(b)   The state where molecules mix most slowly   _____

(c)   The state where molecules have the most motion   _____

(d)   The condensed state or states of matter   _____

(e)   The state where molecules move the farthest between collisions   _____

(f)   The state or states where molecules are free to move
      past one another   _____

(g)   The state where molecules are in fixed positions   _____

(h)   The state or states where water molecules do not
      form hydrogen bonds   _____

(i)   The most likely state for ionic compounds at room temperature   _____

(j)   The most likely state for nonpolar compounds with
      a low molar mass at room temperature   _____

A-2 Multiple Choice

_____ 1. Which of the following bonds is the most polar?

(a) N-N   (b) B-F   (c) H-Cl   (d) N-Cl

_____ 2. Which of the following molecules is predicted to be nonpolar?

(a) S-S-O

(b) S-O

(c) H-Be-H

(d) $O \diagup^{S}\diagdown O$

(e) $H \diagup^{Se}\diagdown H$

_____ 3. Which of the following compounds could form hydrogen bonds?

(a) CH_3OH (b) PH_3 (c) KCN   (d) CH_3CN (e) H_2S

_____ 4. How many hydrogen bonds can form to the oxygen in a H_2O molecule?

(a) 4   (b) 3   (c) 2   (d) 1   (e) 0

_____ 5. How many hydrogen bonds can form to the nitrogen in a NH_3 molecule?

(a) 4   (b) 3   (c) 2   (d) 1   (e) 0

_____ 6. A compound is ionic. Which of the following temperatures is the most likely melting point of the compound?

(a) $658^\circ C$ (b) 325 K   (c) $-112^\circ C$ (d) $15^\circ C$

_____ 7. Diamond is an example of which kind of solid?

(a) ionic   (b) molecular   (c) network   (d) metallic

_____ 8. In what type of solid do only cations occupy lattice positions?

(a) ionic   (b) molecular   (c) network   (d) metallic

_____ 9. The bonds of a certain molecule with a low molar mass are polar but the molecule itself is nonpolar. Which of the following temperatures would most likely represent its boiling point?

(a) 378 K   (b) $-20^\circ C$ (c) $343^\circ C$ (d) $1200^\circ C$

_____ 10. Which of the following molecular compounds should have the highest melting point?

(a) BF_3 (polar bonds - nonpolar molecule)
(b) NH_2OH (polar bonds - polar molecule)
(c) SCl_2 (polar bonds - polar molecule)
(d) HI (essentially nonpolar bond)

_____ 11. The compound $PbCl_2$ melts at $501^\circ C$ and $PbCl_4$ melts at $-15^\circ C$. Which of the following statements is most likely true?

(a) $PbCl_2$ is ionic and $PbCl_4$ is molecular.
(b) $PbCl_2$ is molecular and $PbCl_4$ is ionic.
(c) Both compounds are probably ionic.
(d) Both compounds are probably molecular.
(e) No conclusions can be drawn.

_____ 12. Although SF_4 is a polar molecule and SF_6 is nonpolar, SF_6 melts at a higher temperature than SF_4. Which of the following conclusions can be made?

(a) Nonpolar compounds melt at higher temperatures than polar compounds.
(b) SF_6 is actually ionic.
(c) London forces (which are proportional to the molar mass) can be more important than dipole-dipole forces.
(d) The S-F bond is nonpolar.

Review of Part B *The Liquid State and Changes in State*

OBJECTIVES AND DETAILED TABLE OF CONTENTS

11-4 The Liquid State: Surface Tension, Viscosity, and Boiling Point

OBJECTIVE *Describe the relationship between intermolecular forces and the physical properties of the liquid state.*

11-5 Energy and Changes in State

OBJECTIVE *Given the appropriate heats of fusion and vaporization, calculate the energy required to melt and vaporize a given compound.*

11-6 The Heating Curve of Water

OBJECTIVE *Calculate the energy absorbed or released as a substance progresses along a heating curve.*

SUMMARY OF PART B

In the liquid state, the basic particles are still close together but are not in fixed positions. Thus, they can move past one another. This attraction creates a **surface tension** in the liquid and produces **viscosity**.

Molecules of a liquid can escape to the gaseous phase in a process known as **vaporization**. In a closed container, an equilibrium is established between molecules in the vapor and in the liquid. As the temperature increases, the **vapor pressure** of a liquid increases since a higher fraction of molecules have enough kinetic energy to escape to the gaseous phase. The temperature at which the vapor pressure of a liquid equals the restraining pressure is the **boiling point**. The **normal boiling point** is the temperature at which the vapor pressure is exactly one atmosphere. If a liquid is allowed to **evaporate**, the average kinetic energy of the molecules in the remaining liquid decreases which means its temperature decreases. Other important phase changes are known as **condensation** and **sublimation**.

In the last section, we discussed how intermolecular forces affected the temperature at which compounds undergo a change of state (melt or boil). In addition to the temperature at which changes of state occur, the amount of heat energy required to effect these changes varies for different compounds. The amount of heat energy required to cause melting of a specified mass of solid is known as the **heat of fusion**. The amount of heat energy required to cause vaporization of a specified mass of liquid is the **heat of vaporization**. The magnitude of these quantities, like the melting and boiling points, depends once again on the strength of the interactions between the basic particles of the compound. For a small, covalent molecule, water has unusually large values for all of these because of the extent and strength of its hydrogen bonding.

During the processes of melting or boiling a pure substance, heat supplied to a substance is transferred into potential energy, which means the temperature remains constant. Water is used as a model compound in a **heating curve** to illustrate what happens on the molecular level as it is heated from the solid state at -10°C to the vapor state above 100°C.

ASSESSMENT OF OBJECTIVES

B-1 Multiple Choice

_____ 1. Which of the following types of compounds has the highest heat of fusion?

    (a) nonpolar molecular   (c) hydrogen-bonded molecular
    (b) polar molecular   (d) ionic

_____ 2. A compound boils at -78°C. It is likely that the compound

    (a) has a high melting point.   (c) has a high heat of fusion.
    (b) has a low heat of vaporization.   (d) is ionic.

_____ 3. What is the vaporization of a solid called?

    (a) condensation   (c) fusion
    (b) evaporization   (d) sublimation

_____ 4. Which of the following would cool the fastest if allowed to evaporate under the same conditions? (Refer to Fig. 11-15 in the text.)

    (a) water at 95°C   (c) water at 25°C
    (b) ethyl ether at 0°C   (d) water ice at 0°C

_____ 5. Of the liquids discussed in Figure 11-15 in the text, which would exist as a gas at 25°C and 300 torr?

    (a) all three liquids   (d) alcohol and water
    (b) alcohol and ether   (e) alcohol
    (c) ether

_____ 6. The heat of vaporization of PCl_3 is 217 J/g. For BCl_3 it is 160 J/g. Which of the following statements is a possible explanation for the order of these values?

    (a) Both compounds are ionic.
    (b) PCl_3 is ionic and BCl_3 is molecular.
    (c) BCl_3 is polar and PCl_3 is nonpolar.
    (d) PCl_3 has a greater molar mass than BCl_3.

_____ 7. Carbon tetrachloride has a vapor pressure of 680 torr at 70°C. Which temperature would most likely represent its normal boiling point?

      (a) 50°C   (b) 70°C   (c) 135°C   (d) 76°C

B-2 Problems

(For the following problems, use these values: heat of fusion, 105 J/g at the melting point of ethyl alcohol of -114°C; heat of vaporization, 854 J/g at the boiling point of 78°C. The specific heat of liquid ethyl alcohol is 2.26 J/(g · C).

1. How many joules are required to melt 285 g of ethyl alcohol?

2. What mass of ethyl alcohol can be vaporized at the boiling point by 10.7 kJ?

3. If 3.70 kJ of heat energy is added to 150 g of liquid ethyl alcohol, how many Celsius degrees does the temperature rise?

4. How many kilojoules are required to change 15.0 g of solid ethyl alcohol at -114°C to vapor at 78°C?

Chapter Summary Assessment

S-1 Problems

1. Fill in the following values from those listed below.

	Melting point	Heat of fusion	Boiling point	Heat of vaporization
CH_4	_____	_____	_____	_____
NH_3	_____	_____	_____	_____
KF	_____	_____	_____	_____

    melting points: -78°C, 880°C, -183°C

    boiling points: 1500°C, -3°C, -156°C

    heats of fusion: 452 J/g, 350 J/g, 61 J/g

    heats of vaporization: 577 J/g, 15.1 kJ/g, 1.36 kJ/g

2. The normal boiling point of carbon disulfide is 46°C. Circle the correct answer in the following statements.

(a) The heat of vaporization of CS_2 is (higher, lower) than H_2O.

(b) The attractions between CS_2 molecules are (stronger, weaker) than between H_2O molecules in the liquid state.

(c) At room temperature, CS_2 is (more, less) volatile than H_2O.

(d) The freezing point of CS_2 is probably (higher, lower) than H_2O.

3. In Chapter 9, we learned that metals and nonmetals usually combine to form ionic compounds. A compound formed between titanium and chlorine ($TiCl_4$) has a heat of fusion of 49.4 J/g and a melting point of -25°C.

(a) What does this information tell you about the nature of the Ti-Cl bond?

(b) When a 500 g quantity of $TiCl_4$ melts, how many kJ are released? How does this compare to the kJ released when the same mass of water is allowed to melt?

Answers to Assessments of Objectives

A-1 Matching

(a) gas

(b) solid

(c) all the same at the same temperature

(d) solid and liquid

(e) gas

(f) liquid and gas

(g) solid

(h) gas

(i) solid

(j) gas

A-2 Multiple Choice

1.   **b** B-F (These two atoms have the largest difference in electronegativity.)

2.   **c** H-Be-H (A linear molecule.)

3.   **a** CH_3OH (The S-H bond in CH_3CN is nearly nonpolar.)

4.   **c** (There are two unshared pairs of electrons on an oxygen.)

5.   **d** 1 (There is one lone pair of electrons on the nitrogen.)

6.   **a** 658°C

7. **c** network

8. **d** metallic

9. **b** -20°C

10. **b** NH_2OH (Has hydrogen bonds.)

11. **a** $PbCl_2$ is ionic and $PbCl_4$ is molecular.

12. **c** London forces can be more important than dipole-dipole forces.

B-1 Multiple Choice

1. **d** ionic

2. **b** The compound likely has a low heat of vaporization.

3. **d** sublimation

4. **a** Water at 95°C has the highest vapor pressure so it cools the fastest.

5. **c** ether

6. **d** PCl_3 has a greater molar mass than BCl_3. (PCl_3 is also polar and BCl_3 is nonpolar.)

7. **d** 76°C

B-2 Problems

1. $285 \text{ g} \times 105 \text{ J/g} = 2.99 \times 10^4 \text{ J}$

2. $10.7 \text{ kJ} \times \dfrac{10^3 \text{ J}}{\text{kJ}} \times \dfrac{1 \text{ g}}{854 \text{ J}} = \underline{12.5 \text{ g}}$

3. $3.70 \text{ kJ} \times \dfrac{10^3 \text{ J}}{\text{kJ}} = 150 \text{ g} \times \dfrac{2.26 \text{ J}}{\text{g} \cdot {}^\circ\text{C}}$ °C = $\underline{11°\text{C rise}}$

4. melt the solid: $15.0 \text{ g} \times 10^5 \text{ J/g} = 1580 \text{ J} = 1.58 \text{ kJ}$

   heat the liquid: $15.0 \text{ g} \times 192°\text{C} \times \dfrac{2.26 \text{ J}}{\text{g} \cdot °\text{C}} = 6,510 \text{ J} = 6.51 \text{ kJ}$

   vaporize the liquid: $15.0 \text{ g} \times 854 \text{ J/g} = 12,800 \text{ J} = 12.8 \text{ kJ}$

   Total heat: $1.58 \text{ kJ} + 6.51 \text{ kJ} + 12.8 \text{ kJ} = \underline{20.9 \text{ kJ}}$

S-1 Problems

1. CH_4 is a nonpolar molecule with a small molar mass. Therefore, it has the smallest values of the given properties. NH_3 is polar covalent with hydrogen bonding. It has intermediate properties.

   KF is ionic and has large values for these properties.

   CH_4: melting point, -183°C; heat of fusion, 61 J/g; boiling point -156°C; heat of vaporization, 577 J/g.

   NH_3: melting point, -78°C; heat of fusion, 350 J/g; boiling point, -33°C; heat of vaporization, 1.36 kJ/g.

   KF: melting point, 880°C; heat of fusion, 452 J/g; boiling point, 1500°C; heat of vaporization, 15.1 kJ/g.

2. (a) lower   (b) weaker   (c) more   (d) lower

3. (a) Since the melting point and the heat of fusion of $TiCl_4$ are typical of molecular compounds, we can conclude that the Ti-Cl bond is primarily covalent rather than ionic.

   (b) 500 g x   49.4 J/g = 24,700 J = <u>24.7 kJ</u>

   For H_2O: 500 g x   334 J/g = 167,000 J = <u>167 kJ</u>

Answers and Solutions to Green Text Problems

11-1 Since gas molecules are far apart they move a comparatively great distance between collisions. Liquid molecules, on the other hand, are close together so do not move far between collisions. The farther molecules move, the faster they mix.

11-3 The liquid molecules are in motion as well as the food coloring molecules. Through constant motion and collisions the food coloring molecules will eventually become dispersed.

11-5 Generally, solids have greater densities than liquids. (Ice and water are notable exceptions.) Since the molecules of a solid are held in fixed positions, more of them usually fit into the same volume compared to the liquid state. This is similar to being able to get more people into a room if they are standing still than if they are moving around.

11-6 $CH_4 < CCl_4 < GeCl_4$

11-8 All are nonpolar molecules with only London forces between molecules. The higher the molar mass, the greater the London forces and the more likely the compound is a solid. I_2 is the heaviest and is a solid; Cl_2 is the lightest and is a gas.

11-10 If H_2O were linear, the two equal bond dipoles would be exactly opposite and would therefore cancel. Hydrogen bonding can occur only when the molecule is polar.

11-12

NH$_3$ molecules interact with hydrogen bonding.

11-14 (a) HBr, (b) SO$_2$ (V-shaped), and (f) CO

11-17 (a) HF, (c) H$_2$NCl (d) H$_2$O, (f) HCOOH

11-19 (a) ion-ion   (b) hydrogen bonding plus London   (c) dipole-dipole plus London (d) London only

11-20 F$_2$ < HCl < HF < KF

11-23 CO$_2$ is a nonpolar molecular compound. SiO$_2$ is a network solid.

11-25 PbCl$_2$ is most likely ionic (Pb^{2+}, 2Cl$^-$) while the melting point of PbCl$_4$ indicates that it is a molecular compound.

11-26 Motor oil is composed of large molecule, which increase viscosity. Motor oil also has a higher surface tension.

11-28 Water evaporates quickly on a hot day, which lowers the air temperature.

11-30 The comparatively low boiling point of ethyl chloride indicates that it has a high vapor pressure and even boils at room temperature (25°C). The rapid boiling means that it will cool rapidly even to temperatures below the freezing point of water.

11-32 The liquid is ethyl ether. The higher the vapor pressure, the faster the liquid evaporates and the liquid cools.

11-33 Equilibrium refers to a state where opposing forces are balanced. In the case of a liquid in "equilibrium" with its vapor, it means that a molecule escaping to the vapor is replaced by one condensing to the liquid.

11-35 The substance is a gas at one atmosphere and at 75°C. It would boil at a temperature below 75°C.

11-38 Ethyl alcohol boils at about 52°C at that altitude. At 10°C ethyl ether is a gas at that altitude.

11-39 Death Valley is below sea level. Its atmospheric pressure is more than one atmosphere, so water boils above its normal boiling point.

11-41 Both exist as atoms with only London forces, but the higher atomic mass of argon accounts for its higher boiling point.

11-43 Figure 11-15 is hard to read but at 10°C the actual vapor pressure of water is 9.2 torr. Liquid water could theoretically exist but it would rapidly evaporate and change to ice due to the cooling effect.

11-44 Hexane (75 torr) has the lower boiling point. It also probably has the lower heat of vaporization.

11-46 Molecular O_2 is heavier than Ne atoms so has higher intermolecular forces (London). CH_3OH has hydrogen bonding which would indicate a much higher heat of vaporization than the other two.

11-47 This is a comparatively high heat of fusion, so the melting point is probably also comparatively high.

11-49 This is a comparatively high boiling point, so the compound probably also has a high melting point.

11-50 (H_2O) (H_2S) (H_2Se) (H_2Te)

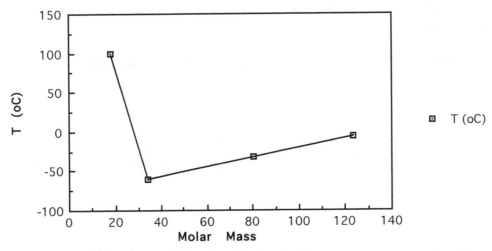

The boiling point of H_2O would be about -75°C without hydrogen bonding.

11-53 $18.0 \text{ g} \times \dfrac{393 \text{ J}}{\text{g}} = 7.07 \times 10^3 \text{ J}$

11-55 $850 \text{ J} \times \dfrac{1.00 \text{ g}}{334 \text{ J}} = 2.54 \text{ g } H_2O$ $850 \text{ J} \times \dfrac{1.00 \text{ g}}{519 \text{ J}} = \underline{1.64 \text{ g NaCl}}$

$850 \text{ J} \times \dfrac{1.00 \text{ g}}{127 \text{ J}} = \underline{6.69 \text{ g benzene}}$

11-57 $25.0 \text{ g} \times \dfrac{22.2 \text{ cal}}{\text{g}} = \underline{555 \text{ cal (ether)}}$ $25.0 \text{ g} \times \dfrac{79.8 \text{ cal}}{\text{g}} = \underline{2000 \text{ cal } (H_2O)}$

Water would be more effective.

$125 \text{ g} \times \dfrac{104 \text{ J}}{\text{g}} = \underline{1.30 \times 10^4 \text{ J } (13.0 \text{ kJ})}$

11-59 NH_3: $450 \text{ g} \times \dfrac{1.36 \text{ kJ}}{\text{g}} \times \dfrac{10^3 \text{ J}}{\text{kJ}} = 6.12 \times 10^5 \text{ J}$

Freon: $450 \text{ g} \times \dfrac{161 \text{ J}}{\text{g}} = 7.25 \times 10^4 \text{ J}$

On the basis of mass, ammonia is the more effective refrigerant.

11-61 Condensation.: $275 \text{ g} \times \dfrac{2260 \text{ J}}{\text{g}} = 62.2 \times 10^4 \text{ J}$

Cooling: $275 \text{ g} \times 75.0 \text{ °C} \times \dfrac{4.184 \text{ J}}{\text{g} \cdot \text{°C}} = 8.63 \times 10^4 \text{ J}$

Total $= (62.2 \times 10^4) + (8.63 \times 10^4) = 70.8 \times 10^4 \text{ J} = \underline{7.08 \times 10^5 \text{ J} \ (708 \text{ kJ})}$

11-63 $120 \text{ g} \times 53.0 \text{ °C} \times \dfrac{0.590 \text{ cal}}{\text{g} \cdot \text{°C}} = 3750 \text{ cal}$

$120 \text{ g} \times \dfrac{204 \text{ cal}}{\text{g}} = 24{,}500 \text{ cal}$

Total $= 28{,}300 \text{ cal} = \underline{2.83 \times 10^4 \text{ cal}}$

11-65 heat ice: $132 \text{ g} \times 20.0 \text{ °C} \times \dfrac{0.492 \text{ cal}}{\text{g} \cdot \text{°C}} = 1300 \text{ cal}$

melt ice: $132 \text{ g} \times \dfrac{79.8 \text{ cal}}{\text{g}} = 10{,}500 \text{ cal}$

heat H_2O: $132 \text{ g} \times 100.0 \text{ °C} \times \dfrac{1.00 \text{ cal}}{\text{g} \cdot \text{°C}} = 13{,}200 \text{ cal}$

vap. H_2O: $132 \text{ g} \times \dfrac{540 \text{ cal}}{\text{g}} = 71{,}300 \text{ cal}$

Total $= \underline{96{,}300 \text{ cal} \ (96.3 \text{ kcal})}$

11-67 Let Y = the mass of the sample in grams. Then

$(\dfrac{2260 \text{ J}}{\text{g}} \times Y) + (25.0 \text{ °C} \times Y \times \dfrac{4.184 \text{ J}}{\text{g} \cdot \text{°C}}) = 28{,}400 \text{ J} \qquad Y = 12.0 \text{ g}$

11-69 Find the heat released by condensation: $10.0 \text{ g} \times \dfrac{393 \text{ J}}{\text{g}} = 3930 \text{ J}$

The remainder of the 5000 J is released by the benzene as the liquid cools.

$5000 - 3930 = 1070 \text{ J} \qquad$ Let Y = temperature change in °C, then

$10.0 \text{ g} \times Y \text{ °C} \times \dfrac{1.72 \text{ J}}{\text{g} \cdot \text{°C}} = 1070 \text{ J}$

$Y = 62 \text{ °C}$ change $\qquad$ Final temp. $= 80 - 62 = \underline{18 \text{ °C}}$

11-70 (b) melting and (c) boiling

11-71 The average kinetic energy of all molecules of water at the same temperature is the same regardless of the physical state.

11-73 Because H_2O molecules have an attraction for each other, moving them apart increases the potential energy. Since the molecules in a gas at $100^{\circ}C$ are farther apart than in a liquid at the same temperature, the potential energy of the gas molecules is greater.

11-74 The water does not become hotter but it will boil faster.

11-76

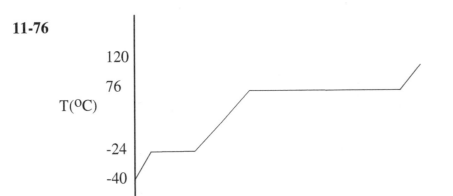

The time of boiling is over ten times longer than the time of melting.

11-78 $C_2H_5NH_2$: $17^{\circ}C$ - hydrogen bonding; CH_3OCH_3: $-25^{\circ}C$ - dipole-dipole (polar)
CO_2: $-78^{\circ}C$ - London forces only (nonpolar)

11-79 SF_6 is a molecular compound and SnO is an ionic compound (i.e., Sn^{2+}, O^{2-}). All ionic compounds are found as solids at room temperature because of the strong ion-ion forces.

11-81 There is hydrogen bonding in CH_3OH. Hydrogen bonding is a considerably stronger interaction than ordinary dipole-dipole forces. The stronger the interactions, the higher the boiling point.

11-83 SiH_4 is nonpolar. PH_3 and H_2S are both polar but H_2S is more polar with more and stronger dipole-dipole interactions.

11-85 Carbon monoxide is polar whereas nitrogen is not. The added dipole-dipole interaction in CO may account for the slightly higher boiling point.

11-87 $2000\,\cancel{lb} \times \dfrac{453.6\,g}{\cancel{lb}} \times \dfrac{266\,J}{g} = 2.41 \times 10^8\,J$

Heating 1 g of H_2O from $25.0^{\circ}C$ to $100.0^{\circ}C$ and the vaporizing the water requires

$(75.0\,\cancel{^{\circ}C} \times 4.184\,J/\cancel{^{\circ}C}) + 2,260\,J = 2.57 \times 10^3\,J/g\ H_2O$

$\dfrac{2.41 \times 10^8\,\cancel{J}}{2.57 \times 10^3\,\cancel{J}/g\ H_2O} = 9.38 \times 10^4\,g\ H_2O = \underline{93.8\ kg\ H_2O}$

11-88 $V = 100$ L, $T = 34 + 273 = 307$ K, $P(H_2O) = 0.700 \times 39.0$ torr $= 27.3$ torr

Use the ideal gas law to find moles of water.

$$n = \frac{PV}{RT} = \frac{\frac{27.3 \text{ torr}}{760 \text{ torr/atm}} \times 100 \text{ L}}{0.0821 \frac{\text{L} \cdot \text{atm}}{\text{K} \cdot \text{mol}} \times 307 \text{ K}} = 0.143 \text{ mol}$$

$$0.143 \text{ mol H}_2\text{O} \times \frac{18.02 \text{ g H}_2\text{O}}{\text{mol H}_2\text{O}} = \underline{2.58 \text{ g H}_2}$$

11-90 First calculate the heat required to heat the water from 25°C to 100°C then to vaporize the water.

$$1.00 \text{ kg} \times \frac{10^3 \text{ g}}{\text{kg}} \times [(100 - 25)°C \times \frac{4.184 \text{ J}}{\text{g} \cdot °C}] = 3.10 \times 10^5 \text{ J (to heat water)}$$

$$1.00 \text{ kg} \times \frac{10^3 \text{ g}}{\text{kg}} \times 2{,}260 \frac{\text{J}}{\text{g}} = 2.260 \times 10^6 \text{ J (to vaporize water)}$$

$$[3.10 \times 10^5 \text{ J} \times \frac{1 \text{ kJ}}{10^3 \text{ J}}] + [2.260 \times 10^6 \text{ J} \times \frac{1 \text{ kJ}}{10^3 \text{ J}}] = 2570 \text{ kJ}$$

Cr: $\dfrac{2570 \text{ kJ}}{21.0 \text{ kJ/mol}} \times \dfrac{52.00 \text{ g Cr}}{\text{mol Cr}} \times \dfrac{1 \text{ kg}}{10^3 \text{ g}} = \underline{6.36 \text{ kg Cr}}$

Mo: $\dfrac{2570 \text{ kJ}}{28.0 \text{ kJ/mol}} \times \dfrac{95.94 \text{ g Mo}}{\text{mol Mo}} \times \dfrac{1 \text{ kg}}{10^3 \text{ g}} = \underline{8.81 \text{ kg Mo}}$

W: $\dfrac{2570 \text{ kJ}}{35.0 \text{ kJ/mol}} \times \dfrac{183.9 \text{ g W}}{\text{mol W}} \times \dfrac{1 \text{ kg}}{10^3 \text{ g}} = \underline{13.5 \text{ kg W}}$

11-91 The element must be mercury (Hg) since it is a liquid at room temperature and must be a metal since it forms a +2 ion.

The 15.0 kJ represents the heat required to melt X g of Hg, heat X g from -39°C to 357°C (i.e., 396°C), and vaporize X g of Hg.

$$(11.5 \text{ J/g} \times X) + [396°C \times 0.139 \frac{\text{J}}{\text{g} °C} \times X] + (29.5 \text{ J/g} \times X) = 15.00 \text{ kJ} \times \frac{10^3 \text{ k}}{\text{kJ}}$$

$$11.5 \text{ J/g } X + 55.0 \text{ J/g } X + 29.5 \text{ J/g } X = 15{,}000 \text{ J}$$

$$X = \underline{156 \text{ g Hg}}$$

11-93 Calculate the mass of water in the vapor state in the room at -5°C (268 K) using the ideal gas law.

$$n = \frac{PV}{RT} = \frac{\frac{2.50 \text{ torr}}{760 \text{ torr/atm}} \times 20000 \text{ L}}{0.0821 \frac{\text{L} \cdot \text{atm}}{\text{K} \cdot \text{mol}} \times 268 \text{ K}} = 2.99 \text{ mol H}_2\text{O (in vapor)}$$

$$2.99 \text{ mol H}_2\text{O} \times \frac{18.02 \text{ g H}_2\text{O}}{\text{mol H}_2\text{O}} = 53.9 \text{ g H}_2\text{O (capacity of the room)}$$

Eventually, all of the 50.0 g should sublime since it is less than the room could contain.

12

Aqueous Solutions

Review of Part A *Solutions and the Quantities Involved*

OBJECTIVES AND DETAILED TABLE OF CONTENTS

12-1 The Nature of Aqueous Solutions

OBJECTIVE *Describe the forces that interact during the formation of aqueous solutions of ionic compounds, strong acids, and polar molecules.*

12-1.1 Mixtures of Two Liquids
12-1.2 The Formation of Aqueous Solutions of Ionic Compounds
12-1.3 Formation of Aqueous Solutions of Molecular Compounds

12-2   The Effects of Temperature and Pressure on Solubility

OBJECTIVE *Describe the effects of temperature and pressure on the solubility of solids and gases in a liquid.*

12-3 Concentration: Percent by Mass

OBJECTIVE *Perform calculations of concentration involving percent by mass, ppm, and ppb.*

12-4 Concentration: Molarity

OBJECTIVE *Perform calculations involving molarity and the dilution of solutions.*

12-5 Stoichiometry Involving Solutions

OBJECTIVE *Perform calculations involving titrations and other solution stoichiometry.*

SUMMARY OF PART A

If chemists were to design the ideal **solvent** in which to study chemical reactions, the solvent would probably have the following properties: (a) it would be inexpensive and plentiful, (b) it would be nontoxic, (c) it would be a liquid at room temperature with a long temperature range in the liquid state. And (d), it would dissolve a large number of **solutes** to form **solutions**. Fortunately such a solvent exists, and it is just ordinary water. Chemists use many other solvents, but none is quite so useful and versatile as H_2O.

When two liquids dissolve in each other to form a solution, the liquids are said to be **miscible**. If they do not mix, they are said to be **immiscible**.

As we learned in Chapter 11, water molecules are attracted to each other in the liquid state by electrostatic interactions called hydrogen bonds. The electrostatic forces that hold ions together in ionic compounds are known as **ion-ion forces**. The sum of these forces is known as the **lattice energy**. When an ionic compound dissolves in water, there are electrostatic interactions between solvent and solute called **ion-dipole forces**. Ionic compounds dissolve in water when the ion-dipole forces between water and ion overcome the ion-ion forces between oppositely charged ions in the crystal. The solution process of an ionic compound in water can be represented by an equation. The ions present in the original compound are now present in solution as hydrated ions [indicated by (aq)].

$$Na_2SO_4(s) \xrightarrow{\;H_2O\;} 2Na^+(aq) + SO_4{}^{2-}(aq)$$

Certain polar covalent compounds also dissolve in water. In some cases (e.g., HCl), the compound undergoes **ionization** in solution. In other cases (e.g., CH_3OH), the compound is dissolved without ion formation. Nonpolar compounds do not generally dissolve in polar solvents such as water, however. Nonpolar solutes do dissolve in nonpolar solvents.

We now turn our attention to the quantitative aspects of solutions. In laboratory situations it is often necessary to know the amount of solute in a certain mass or volume of solution, which is known as the **concentration** of the solute. Concentration can be expressed in several ways with each way being useful for a certain purpose. The **percent by mass** of solute is given by the following relationship that was previously introduced in Chapter 3.

$$\frac{\text{mass of solute}}{\text{mass of solution}} \times 100\% = \text{percent by mass}$$

Even smaller units of concentration are obtained by using **parts per million (ppm)** or even **parts per billion (ppb)**. In ppm, the ratio of mass of solute to solution above is multiplied by 10^6 and in ppb, the ratio is multiplied by 10^9 rather than 10^2 as in percent.

Molarity (M) is the most commonly used unit of concentration and is defined as:

$$M = \frac{\text{number of moles of solute (n)}}{\text{liters of solution (V)}}$$

Knowledge of two of the three variables of molarity allows calculation of the third, as illustrated by the following example.

Example A-1 Conversion of Molarity to Mass

What mass of K_2CO_3 is dissolved in 350 mL of a 0.455 M solution?

PROCEDURE

From V and M we can calculate the moles of K_2CO_3. With the molar mass, we can convert moles to mass.

SOLUTION

$$M = \frac{n}{V} \qquad n = M \times V = 0.455 \ \text{mol/}\cancel{\text{L}} \times 0.350 \ \cancel{\text{L}} = 0.159 \ \text{mol} \ K_2CO_3$$

$$0.159 \ \cancel{\text{mol } K_2CO_3} \times \frac{138.2 \ \text{g} \ K_2CO_3}{\cancel{\text{mol } K_2CO_3}} = \underline{22.0 \ \text{g} \ K_2CO_3}$$

A common laboratory procedure is the **dilution** of a concentrated solution. The addition of solvent lowers the molarity of a solution according to the following equation:

$$M_{con} \times V_{con} = M_{dil} \times V_{dil}$$

An example of a dilution problem follows.

Example A-2   Molarity of a Diluted Solution

If 625 mL of a 0.837 M solution of HCl is diluted to 2.75 L, what is the molarity of the dilute solution?

PROCEDURE

Solve the dilution equation for the molarity of the dilute solution.

$$M_{dil} = \frac{M_{con} \times V_{con}}{V_{dil}}$$ Convert 625 mL to 0.625 L

SOLUTION

$$M_{dil} = \frac{0.837 \text{ mol/L} \times 0.625 \text{ L}}{2.75 \text{ L}} = \underline{0.190 \text{ mol/L}}$$

Since molarity and volume relate to moles, the general procedure for stoichiometry problems discussed in Chapter 7 and 10 can be expanded to include solutions.

GIVEN　　　　　　　　**REQUESTED**

VOLUME (Solution)　　　　　　　VOLUME (Solution)

$(n = M \times V)$　　(Mole Ratio)　　$(V = n/V)$

MOLE　　⟶　　MOLE

A sample problem involving stoichiometry and solutions is as follows.

Example A-3 Stoichiometry and Molarity

Given the following balanced equation:

$$2AgNO_3(aq) + Na_2CrO_4(aq) \longrightarrow Ag_2CrO_4(s) + 2NaNO_3(aq)$$

If 214 mL of a 0.182 M $AgNO_3$ solution is added to a solution containing excess sodium chromate, what mass of silver chromate precipitates?

PROCEDURE

1. Find the moles of $AgNO_3$ from M and V.

2. Convert moles of $AgNO_3$ to moles of Ag_2CrO_4.

3. Convert moles of Ag_2CrO_4 to mass of Ag_2CrO_4.

$$0.214 \text{ L} \times \frac{0.182 \text{ mol AgNO}_3}{\text{L}} \times \frac{1 \text{ mol Ag}_2\text{CrO}_4}{2 \text{ mol AgNO}_3} \times \frac{331.8 \text{ g Ag}_2\text{CrO}_4}{\text{mol Ag}_2\text{CrO}_4}$$

$$= 6.46 \text{ g Ag}_2\text{CrO}_4$$

ASSESSMENT OF OBJECTIVES

A-1 Multiple Choice

_____ 1. What are the electrostatic forces between solute and solvent when an ionic compound dissolves in water?

(a) ion-ion   (c) dipole-dipole
(b) ion-dipole   (d) none of these

_____ 2. Which of the following statements is false?

(a) Ion-ion forces refer to the forces holding an ionic compound together.
(b) Only ionic compounds dissolve in water.
(c) Most compounds are more soluble in water at a higher temperature.
(d) An ion in aqueous solution is hydrated.

_____ 3. When one mole of potassium sulfate dissolves in water, which of the following are present?

(a) 2 mol K^+ (c) 2 mol SO_4^{2-}

(b) 1 mol K^{2+} (d) 1 mol K^+ (e) 2 mol K^{2+}

_____ 4. When one liquid dissolves in another, we say that the two liquids

(a) form a supersaturated solution. (d) form an unsaturated solution.
(b) are immiscible.   (e) are miscible.
(c) are both solvents.

_____ 5. An ionic compound has a solubility of 20 g/100 g H_2O. If 12 g of solute is present in 67 g of water, the solution is

(a) saturated.   (c) unsaturated.
(b) supersaturated.   (d) heterogeneous.

A-2 Problems

1. A solution of K_2SO_3 contains 0.0441 g of K_2SO_3 dissolved in 1.00 g of H_2O. What is the percent by mass of K_2SO_3?

2. A solution is 12.0% by mass alcohol (C_2H_6O). What mass of alcohol is dissolved in each gram of water?

3. Calculate the molarity of the following:

   (a) 0.117 mol of HNO_3 in 0.644 L of solution

   (b) 126 g of H_3PO_4 in 1.45 L of solution

   (c) 8.79 x 10^{22} molecules of H_2SO_3 in 450 mL of solution

4. What volume of 0.335 M HNO_3 is needed to provide 18.0 g of HNO_3?

5. A solution of NaOH has a density of 1.23 g/mL and is 10.0% by mass NaOH. What is the molarity of the solution?

6. If 500 mL of a 0.330 M solution of NaOH is required, what volume of 6.00 M NaOH is needed to dilute with water?

7. When 10.0 mL of 2.00 M HCl is diluted to 115 mL, what is the molarity of the dilute solution?

8. Given the following balanced equation:
   $$2Na_3PO_4(aq) + 3Ca(NO_3)_2(aq) \longrightarrow Ca_3(PO_4)_2(s) + 6NaNO_3(aq)$$

   (a) What volume of 0.662 M Na_3PO_4 completely reacts with 450 mL of 0.752 M $Ca(NO_3)_2$?

   (b) What is the molarity of a $Ca(NO_3)_2$ solution if 8500 mL of the solution produced 230 g of $Ca_3(PO_4)_2$?

9. Given the balanced equation:
   $$CaCO_3(s) + 2HCl(aq) \longrightarrow CO_2(g) + CaCl_2(aq) + H_2O$$

   What volume of CO_2 gas measured at 25°C and 0.945 atm is produced from the complete reaction of 1.27 L of 0.125 M HCl?

230

Review of Part B   *The Effects of Solutes on the Properties of Water*

OBJECTIVES AND DETAILED TABLE OF CONTENTS

12-6 Electrical Properties of Solutions

OBJECTIVE *Explain the differences between nonelectrolytes, strong electrolytes, and weak electrolytes.*

12-7 Colligative Properties of Solutions

OBJECTIVE *Calculate the boiling and melting points of aqueous solutions of electrolytes and nonelectrolytes.*

SUMMARY OF PART B

How can a clear solution be distinguished from the pure solvent? In many cases they appear identical to the naked eye. The answer is that the solution has many physical properties that distinguish it from the solvent. One difference involves the conduction of electricity. Although water itself is a **nonconductor** of electricity, a solution may or may not be a **conductor** depending on the nature of the solute in the aqueous solution. It is the presence of ions in solution that allows water to become a conductor. Polar covalent compounds that dissolve in water without ion formation are known as **nonelectrolytes**. Compounds that produce ions are known as **electrolytes**. Ionic compounds and some polar covalent compounds (i.e., strong acids) are almost completely dissociated into ions in solution and are known as **strong electrolytes**. Some polar covalent compounds produce only a limited concentration of ions, and these are known as **weak electrolytes**.

The properties of a solution differ from a solvent in other ways. The presence of a nonvolatile solute causes **vapor pressure lowering**, **boiling point elevation**, and **freezing point lowering** of the solution from the levels of the pure solvent. A solute also produces an effect known as **osmotic pressure**. The magnitude of these effects depends on the amount of solute present in a given amount of solvent and not on the identity of the solute particles. Such a property is known as a **colligative property**.

The magnitude of the boiling point elevation (ΔT_b) and freezing point lowering (ΔT_f) are given by the equations

$$\Delta T_b = K_b m \qquad \qquad \Delta T_f = K_f m$$

where K_b and K_f are constants characteristic of the solvent and "m" is the **molality**. Molality is a unit of concentration defined as

$$m = \frac{\text{moles of solute}}{\text{kg of solvent}}$$

Freezing point lowering and boiling point elevation are often used in chemistry laboratories to determine the molar mass of an unknown pure compound. This procedure is illustrated by the following example.

Example B-1 Molar Mass by Freezing Point Lowering

A pure compound dissolves in water and is found to be a nonelectrolyte. When 50.0 g of this compound is dissolved in 473 g of water, the solution freezes at -2.13°C. What is the molar mass of the compound? For H_2O, $K_f = 1.86$°C · kg/mol.

PROCEDURE

Calculate the value of T and solve the equation for molality to obtain the molar mass (M.M.):

$$m = \frac{\text{mol Solute}}{\text{kg solvent}} = \frac{\frac{\text{g solute}}{\text{M.M.}}}{\frac{\text{g solvent}}{1000 \text{ g/kg}}}$$

Solving for M.M. we get $\text{M.M.} = \frac{1000 \text{ g/kg x g solute}}{\text{g solvent x } m}$

SOLUTION

Since the freezing point of pure water is 0.00°C,

$$T_f = 0.00 - (-2.13) = 2.13\text{°C degrees}$$

$$T_f = K_f m \qquad m = \frac{T}{K_f} = \frac{2.13 \text{ °C}}{1.86 \frac{\text{°C} \cdot \text{kg}}{\text{mol}}} = 1.15 \text{ mol/kg}$$

$$\text{M.M.} = \frac{1000 \text{ g/kg x 50.0 g}}{473 \text{ g x 1.15 mol/kg}} = \underline{91.9 \text{ g/mol}}$$

Osmotic pressure, another colligative property, is important in many life processes. The process of **osmosis** concerns the unequal passage of solvent molecules through a semipermeable membrane separating solutions of different concentrations. We also discussed the effect of electrolytes on colligative properties. Since one mole of a solute such as NaCl produces two moles of particles (ions), the effect on colligative properties is about twice as much as the effect of one mole of a nonelectrolyte.

ASSESSMENT OF OBJECTIVES

B-1 Multiple Choice

_____ 1. Which of the following properties of a solvent is not necessarily changed by the presence of a solute?

(a) boiling point   (c) conduction of electricity
(b) freezing point   (d) vapor pressure

_____ 2. Given the mass of a certain solute, what other quantity is needed to calculate the molality?

(a) mass of solvent   (c) volume of solution
(b) mass of solution   (d) molar mass of solvent

_____ 3. If the freezing point of an aqueous solution is -1.86°C, what is the normal boiling point? (For H_2O, $K_b = 0.512$°C · kg/mol.)

(a) 98.14°C   (b) 101.8°C   (c) 100.512°C   (d) 99.488°C

_____ 4. Which of the following can be accomplished by reverse osmosis?

(a) dilute a concentrated solution
(b) concentrate a dilute solution
(c) increase the freezing point of a solution
(d) decrease the boiling point of a solution

_____ 5. Which of the following solutes is the most effective (per mole) in raising the osmotic pressure of water?

(a) KCl   (c) HBr
(b) $C_3H_8O_3$ (a nonelectrolyte)   (d) Li_2CO_3

B-2 Problems

1. What is the molality of a solution made by dissolving 6.50 g of glycerol in 76.0 g of water? Glycerol is a nonvolatile nonelectrolyte with the formula $C_3H_8O_3$.

2. What are the freezing point and boiling point of the solution in problem 1 above?

Chapter Summary Assessment

S-1 Problems

Two hypothetical ionic compounds with the formulas AX_2 and B_2Y are soluble in water. When solutions of these two compounds are mixed in stoichiometric amounts, however, a precipitate forms. The solution after filtration of the precipitate is found to contain 40.0 g of solute dissolved in 500 g of water. The freezing point of this solution was found to be -2.98°C. The atomic masses of the elements are as follows: A = 20, B = 30, X = 70, and Y = 40.

1.   What is the precipitate, AY or BX? (Remember that the solution contains an electrolyte consisting of two ions. Therefore, the calculated molar mass must be multiplied by two to get the molar mass of the compound.)

2.   The charge on the A ion is +2 and on the Y ion is -2. Write the molecular, total ionic, and net ionic equation illustrating the reaction that occurred.

3.   What is the molarity of the solution containing the dissolved compound after the precipitate is removed? The density of the solution is 1.06 g/mL.

4.   If 325 mL of an aqueous solution of AX_2 was originally mixed with a solution of B_2Y, what was the molarity of the original B_2Y solution? (Use the volume of the solution after mixing that was calculated in problem 3. Assume that the two solutions are mixed in exactly stoichiometric amounts.)

Answers to Assessments of Objectives

A-1 Multiple Choice

1.   **b** ion-dipole

2.   **b** Only ionic compounds dissolve in water.

3.   **a** 2 mol K^+ (and 1 mol SO_4^{2-})

4.   **e** are miscible

5.   **c** 20.0 g solute/100 g H_2O x 67 g H_2O = 13.4 g solute. If 12 g is present, the solution is unsaturated.

A-2 Problems

1.   Mass of solution = mass of solute + mass of solvent = 0.0441 + 1.00 = 1.04 g

$$\frac{0.044 \text{ g}}{1.04 \text{ g}} \times 100\% = \underline{4.24\% \text{ solute}}$$

2.   Let X = mass of solute: therefore the mass of solution is 1.00 + X.

$$\frac{X}{1.00 + X} \times 100\% = 12.0\% \qquad X = \underline{0.136 \text{ g}}$$

3.   (a) $M = \dfrac{n}{V} = \dfrac{0.117 \text{ mol}}{0.664 \text{ L}} = \underline{0.182 \text{ mol/L}}$

    (b) $126 \text{ g } H_3PO_4 \times \dfrac{1 \text{ mol}}{97.99 \text{ g } H_3PO_4} = 1.29 \text{ mol} \quad M = \dfrac{1.29 \text{ mol}}{1.45 \text{ L}} = \underline{0.890 \text{ mol/L}}$

    (c) $8.79 \times 10^{22} \text{ molecules} \times \dfrac{1 \text{ mol}}{6.022 \times 10^{23} \text{ molecules}} = 0.146 \text{ mol}$

$$M = \frac{0.146 \text{ mol}}{0.450 \text{ L}} = \underline{0.324 \text{ mol/L}}$$

4.   First find moles of HNO_3: $18.0 \text{ g } HNO_3 \times \dfrac{1 \text{ mol}}{63.02 \text{ g } HNO_3} = 0.286 \text{ mol}$

    Find requested volume: $M = \dfrac{n}{V} \quad V = \dfrac{n}{M} = \dfrac{0.286 \text{ mol}}{0.335 \text{ mol/L}} = \underline{0.854 \text{ L}}$

5.   Assume exactly one liter (1000 mL) of solution.

    mass of solution: $1000 \text{ mL} \times 1.23 \text{ g/mL} = 1230 \text{ g}$

    mass of NaOH in solution: $1230 \text{ g} \times 0.100 = 123 \text{ g}$

    mol NaOH in solution: $123 \text{ g NaOH} \times \dfrac{1 \text{ mol}}{40.00 \text{ g NaOH}} = 3.08 \text{ mol}$

    $M = \dfrac{n}{V} = \dfrac{3.08 \text{ mol}}{1.00 \text{ L}} = \underline{3.08 \text{ mol/L}}$

6. $V_{con} = \dfrac{M_{dil} \times V_{dil}}{M_{con}} = \dfrac{500 \text{ mL} \times 0.330 \text{ M}}{6.00 \text{ M}} = \underline{27.5 \text{ mL}}$

7. $M_{dil} = \dfrac{M_{con} \times V_{con}}{V_{dil}} = \dfrac{2.00 \text{ M} \times 10.0 \text{ mL}}{115 \text{ mL}} = \underline{0.174 \text{ mol/L}}$

8.   (a)   (1) Convert M and V of $Ca(NO_3)_2$ to moles $Ca(NO_3)_2$.
                (2) Convert moles of $Ca(NO_3)_2$ to moles of Na_3PO_4.
                     (1)   (2)

$$0.450 \text{ L} \times \frac{0.752 \text{ mol } Ca(NO_3)_2}{\text{L}} \times \frac{2 \text{ mol } Na_3PO_4}{3 \text{ mol } Ca(NO_3)_2} = 0.226 \text{ mol } Na_3PO_4$$

(3)

$$V = \frac{n}{M} = \frac{0.226 \text{ mol}}{0.662 \text{ mol/L}} = \underline{0.341 \text{ L}}$$

(b)   (1) Convert mass of $Ca_3(PO_4)_2$ to moles of $Ca_3(PO_4)_2$.
      (2) Convert moles of $Ca_3(PO_4)_2$ to moles of $Ca(NO_3)_2$.
      (3) Convert n and V of $Ca(NO_3)_2$ to M of $Ca(NO_3)_2$.

$$\qquad\qquad (1) \qquad\qquad\qquad (2)$$

$$230 \text{ g } Ca_3(PO_4)_2 \times \frac{1 \text{ mol } Ca_3(PO_4)_2}{310.2 \text{ g } Ca_3(PO_4)_2} \times \frac{3 \text{ mol } Ca(NO_3)_2}{1 \text{ mol } Ca_3(PO_4)_2} = 2.22 \text{ mol } Ca(NO_3)_2$$

$$(3)$$
$$M = \frac{n}{V} = \frac{2.22 \text{ mol}}{8.50 \text{ L}} = \underline{0.261 \text{ mol/L}}$$

9.   (1) Convert V and M of HCl to mol of HCl.
     (2) Convert mol of HCl to mol of CO_2.
     (3) Convert n, T, and P of CO_2 to V of CO_2 using Ideal Gas Law.

$$\qquad\qquad (1) \qquad\qquad\qquad (2)$$

$$1.27 \text{ L } \times \frac{0.125 \text{ mol HCl}}{L} \times \frac{1 \text{ mol } CO_2}{2 \text{ mol HCl}} = 0.0794 \text{ mol } CO_2.$$

$$(3)$$

$$V = \frac{nRT}{P} = \frac{0.0794 \text{ mol} \times 0.0821 \frac{L \cdot atm}{K \cdot mol} \times 298 \text{ K}}{0.945 \text{ atm}} = \underline{2.06 \text{ L}}$$

B-1 Multiple Choice

1.   **c** conduction of electricity (solute may be a nonelectrolyte)

2.   **a** mass of solvent

3.   **c** $100.512\,^{\circ}C$

4.   **b** concentrating a dilute solution (Choices **c** and **d** would result when a solution is diluted.)

5.   **d** Li_2CO_3 (produces three moles of ions)

B-2 Problems

1. $m = \dfrac{\text{mol solute}}{\text{kg solvent}}$ $\quad \text{mol solute} = \dfrac{6.50 \text{ g}}{92.0 \text{ g/mol}} = 0.707 \text{ mol}$

$$\text{kg solvent} = \frac{76.0 \text{ g}}{1000 \text{ g/kg}} = 0.0760 \text{ kg}$$

$$m = 0.0707 \text{ mol}/0.760 \text{ kg} = \underline{0.930 \text{ mol/kg}}$$

236

2. $\Delta T_f = K_f m = 1.86^{\circ}C \cdot kg/mol \times 0.930\ mol/kg = 1.73^{\circ}C$

       $FP = 0.00^{\circ}C - 1.73^{\circ}C = -1.73^{\circ}C$

       $\Delta T_b = K_b m = 0.512^{\circ}C \cdot \cancel{kg/mol} \times 0.930\ \cancel{mol/kg} = 0.476^{\circ}C$

       $BP = 100.000^{\circ}C + 0.476^{\circ}C = 100.476^{\circ}C$

S-1 Problems

1. $\Delta T_f = K_f m \quad m = \dfrac{\Delta T}{K_f} = \dfrac{2.98\ ^{\circ}C}{1.86\ ^{\circ}C \cdot kg/mol} = 1.60\ mol/kg$

       Use the equation from the sample problem.

       $M.M. = \dfrac{1000\ g/kg \ \times\ g\ solute}{g\ solvent \times m} = \dfrac{1000\ g/kg \ \times\ 40.0\ g}{500\ g\ \times\ 1.60\ \cancel{mol/kg}} = 50.0\ g/mol$

       Since there are two ions per mole, the measured molar mass is the average of the ions. Twice this value is therefore the molar mass of the compound in solution. A molar mass of 100 g/mol corresponds to the compound BX. Therefore, <u>AY precipitates</u>.

2.   Molecular: $AX_2(aq) + B_2Y(aq) \longrightarrow AY(s) + 2BX(aq)$

       Total ionic: $A^{2+}(aq) + 2X^{-}(aq) + 2B^{+}(aq) + Y^{2-}(aq) \longrightarrow$

                                       $AY(s) + 2B^{+}(aq) + 2X^{-}(aq)$

       Net ionic: $A^{2+}(aq) + Y^{2-}(aq) \longrightarrow AY(s$

3.   Convert the total mass of the solution (500 g + 40.0 g = 540 g) into the equivalent volume.

       $540\ g \ \times\ \dfrac{1\ mL}{1.06\ g} = 509\ mL = 0.509\ L$ moles $BX = \dfrac{40.0\ g}{100\ \cancel{g/mol}} = 0.400\ mol$

       $M = n/V = 0.400\ mol/0.509\ L = \underline{0.786\ mol/L}$

4.   Use the balanced molecular equation from (2) to calculate moles of B_2Y.

       $g\ BX \longrightarrow mol\ BX \longrightarrow mol\ B_2Y$

       $40.0\ \cancel{g\ BX} \ \times\ \dfrac{1\ \cancel{mol\ BX}}{100\ \cancel{g\ BX}} \ \times\ \dfrac{1\ mol\ B_2Y}{2\ \cancel{mol\ BX}} = 0.200\ mol\ B_2Y$

       Volume of the B_2Y solution = 509 mL - 325 mL = 184 mL = 0.184 L

       $M = n/V = 0.200\ mol/0.184\ L = \underline{1.09\ mol/L}$

Answers and Solutions to Green Text Problems

12-1 The ion-ion forces in the crystal hold the crystal together and resist the ion-dipole forces between water and the ions attached in the crystal. The ion-dipole forces remove the ions from the crystal.

12-3 Calcium bromide is soluble, lead(II) bromide is insoluble, benzene and water are immiscible, and alcohol and water are miscible.

12-4 (a) $LiF \longrightarrow Li^+(aq) + F^-(aq)$

(b) $(NH_4)_3PO_4 \longrightarrow 3NH_4^+(aq) + PO_4^{3-}(aq)$

(c) $Na_2CO_3 \longrightarrow 2Na^+(aq) + CO_3^{2-}(aq)$

(d) $Ca(C_2H_3O_2)_2 \longrightarrow Ca^{2+}(aq) + 2C_2H_3O_2^-(aq)$

12-6

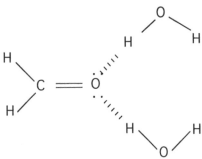

There is hydrogen bonding between solute and solvent.

12-8 At 10^oC, Li_2SO_4 is the most soluble; at 70^oC, KCl is the most soluble.

12-10 (a) unsaturated   (b) supersaturated
(c) saturated   (d) unsaturated

12-12 A specific amount of Li_2SO_4 will precipitate unless the solution becomes supersaturated.

$$(500 \text{ g } H_2O \text{ x } \frac{35 \text{ g}}{100 \text{ g } H_2O}) - (500 \text{ g } H_2O \text{ x } \frac{28 \text{ g}}{100 \text{ g } H_2O}) = \underline{35 \text{ g } Li_2SO_4 \text{ (precipitate)}}$$

12-13 Mass of solution = 650 g + 9.85 g = 660 g

$$\frac{9.85 \text{ g}}{660 \text{ g}} \text{ x } 100\% = \underline{1.49\%}$$

12-15 $150 \text{ g solution} \text{ x } \frac{10.0 \text{ g NaOH}}{100 \text{ g solution}} = 15.0 \text{ g NaOH}$

$$15.0 \text{ g NaOH} \text{ x } \frac{1 \text{ mol NaOH}}{40.00 \text{ g NaOH}} = \underline{0.375 \text{ mol NaOH}}$$

12-17 Solution is 23.2% KNO_3 and 76.8% H_2O
Let Y = Mass of the solution, then $0.768 \text{ x } Y = 100 \text{ g}$ Y = 130 g
Mass of solute = 130 g – 100 g = $\underline{30 \text{ g } KNO_3}$

12-18 M.M. $(C_2H_5OH) = 46.07$ g/mol   M.M. NaOH = 40.00 g/mol
Mass of solution = 40.00 g + (9 x 46.07 g) = 454.6 g

$$\frac{40.00 \text{ g}}{454.6 \text{ g}} \text{ x } 100\% = \underline{8.799\% \text{ NaOH}}$$

12-19 $\dfrac{10 \times 10^{-3}\ g}{100\ g} \times 10^6\ ppm = \underline{100\ ppm}$

12-21 $\dfrac{1.00\ g\ gold}{x\ g\ sea\ water} \times 10^9\ ppb = 1.2 \times 10^{-2}\ ppb$

$x = 8.3 \times 10^{10}\ \cancel{g\ sea\ water} \times \dfrac{1\ \cancel{mL}}{1.0\ \cancel{g}} \times \dfrac{1\ L}{10^3\ \cancel{mL}} = \underline{8.3 \times 10^7\ L}$

12-23 $\dfrac{2.44\ mol}{4.50\ L} = \underline{0.542\ M}$

12-24 (a) $\dfrac{n}{V} = \dfrac{2.40\ mol}{2.75\ L} = \underline{0.873M}$

(b) $26.5\ \cancel{g} \times \dfrac{1\ mol}{46.07\ \cancel{g}} = 0.575\ mol \qquad \dfrac{n}{V} = \dfrac{0.575\ mol}{0.410\ L} = \underline{1.40\ M}$

(c) $V = \dfrac{n}{M} = \dfrac{3.15\ mol}{0.255\ mol/L} = \underline{12.4\ L}$

(d) $n = M \times V = 0.625\ mol/\cancel{L} \times 1.25\ \cancel{L} = 0.781\ mol$

$0.781\ \cancel{mol} \times \dfrac{52.95\ g}{\cancel{mol}} = \underline{41.4\ g}$

(e) $M = \dfrac{n}{V} = \dfrac{0.250\ mol}{0.850\ L} = \underline{0.294\ M}$

(f) $n = M \times V = 0.054\ mol/\cancel{L} \times 0.45\ \cancel{L} = \underline{0.024\ mol}$

(g) $14.7\ \cancel{g} \times \dfrac{1\ mol}{138.2\ \cancel{g}} = 0.106\ mol; \qquad V = \dfrac{0.106\ \cancel{mol}}{0.345\ \cancel{mol}/L} = 0.310\ L = \underline{307\ mL}$

(h) $n = 1.24\ mol/\cancel{L} \times 1.65\ \cancel{L} = 2.05\ mol \qquad 2.05\ \cancel{mol} \times \dfrac{23.95\ g}{\cancel{mol}} = \underline{49.1\ g}$

(i) $0.178\ \cancel{g} \times \dfrac{1\ mol}{98.09\ \cancel{g}} = 1.81 \times 10^{-3}\ mol$

$V = \dfrac{n}{M} = \dfrac{1.81 \times 10^{-3}\ \cancel{mol}}{0.905\ \cancel{mol}/L} = 0.00200\ L = \underline{2.00\ mL}$

12-28 $2.50 \times 10^{-4}\ \cancel{g\ NaHCO_3} \times \dfrac{1\ mol\ NaHCO_3}{84.01\ \cancel{g\ NaHCO_3}} = 2.98 \times 10^{-6}\ mol$

$\dfrac{2.98 \times 10^{-6}\ mol}{2.54 \times 10^{-3}\ L} = 1.17 \times 10^{-3}\ M$

12-29 $13.5\ \cancel{g\ Ba(OH)_2} \times \dfrac{1\ mol\ Ba(OH)_2}{171.3\ \cancel{g\ Ba(OH)_2}} = 0.0788\ mol\ Ba(OH)_2$

$\dfrac{0.0788\ mol}{0.475\ L} = 0.166\ M\ [Ba(OH)_2]$

$Ba(OH)_2 \longrightarrow Ba^{2+}(aq) + 2OH^-(aq)$

$0.166\ \cancel{M\ Ba(OH)_2} \times \dfrac{1\ M\ Ba^{2+}}{1\ \cancel{M\ Ba(OH)_2}} = \underline{0.166\ M\ Ba^{2+}}$

$$0.166 \; \cancel{\text{M Ba(OH)}_2} \times \frac{2 \text{ M OH}^-}{1 \; \cancel{\text{M Ba(OH)}_2}} = \underline{0.332 \text{ M OH}^-}$$

12-31 Assume one L (1000 mL) of solution $\qquad$ 1000 mL x 1.21 g/mL = 1210 g solution

$$1210 \; \cancel{\text{g}} \times \frac{25 \text{ g solute}}{100 \; \cancel{\text{g solution}}} = 302.5 \text{ g solute} \qquad 302.5 \; \cancel{\text{g}} \times \frac{1 \text{ mol}}{164.1 \; \cancel{\text{g}}} = 1.84 \text{ mol}$$

$$\frac{n}{V} = \frac{1.84 \text{ mol}}{1.00 \text{ L}} = \underline{1.84 \text{ M}}$$

12-33 Assume one L (1000 mL) of solution

$$n = M \times V = 14.7 \text{ mol/}\cancel{\text{L}} \times 1.00 \; \cancel{\text{L}} = 14.7 \text{ mol solute} \qquad 14.7 \; \cancel{\text{mol}} \times \frac{63.02 \text{ g}}{\cancel{\text{mol}}} = 926 \text{ g solute}$$

Since 70% of the mass of the solution is solute,

mass of solution x $\dfrac{70 \text{ g solute}}{100 \text{ g solution}} = 926 \text{ g solute}$ $\qquad$ mass of solution = 1320 g

Density = $\dfrac{1320 \text{ g}}{1000 \text{ mL}} = \underline{1.32 \text{ g/mL}}$

12-34 $M_d \times V_d = M_c \times V_c \qquad V_c = \dfrac{1.50 \; \cancel{\text{M}} \times 2.50 \text{ L}}{4.50 \; \cancel{\text{M}}} = \underline{0.833 \text{ L}}$

12-36 First, find the volume of concentrated NaOH needed.

$$V_c = \frac{1.00 \text{ L} \times 0.250 \; \cancel{\text{mol/L}}}{0.800 \; \cancel{\text{mol/L}}} = 0.313 \text{ L (313 mL)}$$

Slowly add 313 mL of the 0.800 M NaOH to about 500 mL of water in a 1-L volumetric flask. Dilute to the 1-L mark with water.

12-38 $M_d = \dfrac{0.200 \text{ M} \times 3.50 \; \cancel{\text{L}}}{5.00 \; \cancel{\text{L}}} = \underline{0.140 \text{ M}}$

12-39 $V_d = \dfrac{M_c \times V_c}{M_d} = \dfrac{0.860 \; \cancel{\text{M}} \times 1.25 \text{ L}}{0.545 \; \cancel{\text{M}}} = 1.97 \text{ L} \qquad 1.97 \text{ L} - 1.25 \text{ L} = 0.72 \text{ L} = \underline{720 \text{ mL}}$

12-41 $\qquad$ L $\longrightarrow$ mol $\longrightarrow$ g $\longrightarrow$ mL

$$0.250 \; \cancel{\text{L}} \times \frac{0.200 \; \cancel{\text{mol}}}{\cancel{\text{L}}} \times \frac{60.05 \; \cancel{\text{g}}}{\cancel{\text{mol}}} \times \frac{1.00 \text{ mL}}{1.05 \; \cancel{\text{g}}} = \underline{2.86 \text{ mL}}$$

12-42 Find the total moles and the total volume:

$n = V \times M = 0.150 \; \cancel{\text{L}} \times 0.250 \text{ mol/}\cancel{\text{L}} = 0.0375 \text{ mol (solution 1)}$

$n = V \times M = 0.450 \; \cancel{\text{L}} \times 0.375 \text{ mol/}\cancel{\text{L}} = 0.169 \text{ mol (solution 2)}$
$V_T = 0.600 \text{ L}; \; n_T = 0.206 \text{ mol}$

$\dfrac{n}{V} = \dfrac{0.206 \text{ mol}}{0.600 \text{ L}} = \underline{0.343 \text{ M}}$

12-43 $\qquad$ **Vol. KOH** $\longrightarrow$ **mol KOH** $\longrightarrow$ **mol Cr(OH)$_3$** $\longrightarrow$ **g Cr(OH)$_3$**

$$0.500 \; \cancel{\text{L}} \times \frac{0.250 \; \cancel{\text{mol KOH}}}{\cancel{\text{L}}} \times \frac{1 \; \cancel{\text{mol Cr(OH)}_3}}{3 \; \cancel{\text{mol KOH}}} \times \frac{103.0 \text{ g Cr(OH)}_3}{\cancel{\text{mol Cr(OH)}_3}} = \underline{4.29 \text{ g Cr(OH)}_3}$$

12-45 Vol. $Al_2(SO_4)_3$ $\longrightarrow$ mol $Al_2(SO_4)_3$ $\longrightarrow$ mol $BaSO_4$ $\longrightarrow$ g $BaSO_4$

$$0.650\ \cancel{L}\ \times\ \frac{0.320\ \cancel{mol\ Al_2(SO_4)_3}}{\cancel{L}}\ \times\ \frac{3\ \cancel{mol\ BaSO_4}}{1\ \cancel{mol\ Al_2(SO_4)_3}}\ \times\ \frac{233.4\ g\ BaSO_4}{\cancel{mol\ BaSO_4}}\ =\ \underline{146\ g\ BaSO_4}$$

12-46 g $Al(OH)_3$ $\longrightarrow$ mol $Al(OH)_3$ $\longrightarrow$ mol $Ba(OH)_2$ $\longrightarrow$ Vol. $Ba(OH)_2$

$$265\ \cancel{g\ Al(OH)_3}\ \times\ \frac{1\ \cancel{mol\ Al(OH)_3}}{78.00\ \cancel{g\ Al(OH)_3}}\ \times\ \frac{3\ mol\ Ba(OH)_2}{2\ \cancel{mol\ Al(OH)_3}}\ =\ 5.10\ mol\ Ba(OH)_2$$

$$V = \frac{n}{M} = \frac{5.10\ mol}{1.25\ mol/L} = \underline{4.08\ L}$$

12-48 Vol. $Ca(ClO_3)$ $\longrightarrow$ mol $Ca(ClO_3)_2$ $\longrightarrow$ mol Na_3PO_4 $\longrightarrow$ Vol. Na_3PO_4

$$0.580\ \cancel{L}\ \times\ \frac{3.75\ \cancel{mol\ Ca(ClO_3)_2}}{\cancel{L}}\ \times\ \frac{2\ mol\ Na_3PO_4}{3\ \cancel{mol\ Ca(ClO_3)_2}}\ =\ 1.45\ mol\ Na_3PO_4$$

$$V = \frac{n}{M} = \frac{1.45\ \cancel{mol}}{2.22\ \cancel{mol}/L} = \underline{0.653\ L}$$

12-50 Vol. NaOH $\longrightarrow$ mol NaOH $\longrightarrow$ mol $HC_2H_3O_3$ $\longrightarrow$ M $HC_2H_3O_3$

$$0.0288\ \cancel{L\ NaOH}\ \times\ \frac{0.300\ \cancel{mol\ NaOH}}{\cancel{L\ NaOH}}\ \times\ \frac{1\ mol\ HC_2H_3O_2}{1\ \cancel{mol\ NaOH}}\ =\ 0.00864\ mol\ HC_2H_3O_2$$

$$0.00864\ mol / 0.0100\ L = \underline{0.864\ M}$$

12-52 Vol. H_2SO_4 $\longrightarrow$ mol H_2SO_4 $\longrightarrow$ mol NH_3 $\longrightarrow$ M NH_3

$$0.0226\ \cancel{L\ H_2SO_4}\ \times\ \frac{0.220\ \cancel{mol\ H_2SO_4}}{\cancel{L\ H_2SO_4}}\ \times\ \frac{2\ mol\ NH_3}{1\ \cancel{mol\ H_2SO_4}}\ =\ 0.00994\ mol\ NH_3$$

$$\frac{0.00994\ mol}{0.0100\ L} = \underline{0.994\ M}$$

12-54 $n(NaOH) = M \times V = 0.250\ \cancel{L}\ \times\ 0.240\ mol/\cancel{L} = 0.0600\ mol\ NaOH$

$n(MgCl_2) = 0.400\ \cancel{L}\ \times\ 0.100\ mol/\cancel{L} = 0.0400\ mol\ Mg(OH)_2$

NaOH: $0.0600\ \cancel{mol\ NaOH}\ \times\ \dfrac{1\ mol\ Mg(OH)_2}{2\ \cancel{mol\ NaOH}}\ =\ 0.0300\ mol\ Mg(OH)_2$

$MgCl_2$: $0.0400\ \cancel{mol\ MgCl_2}\ \times\ \dfrac{1\ mol\ Mg(OH)_2}{1\ \cancel{mol\ MgCl_2}}\ =\ 0.0400\ mol\ Mg(OH)_2$

Therefore, NaOH is the limiting reactant.

$$0.0300\ \cancel{mol\ Mg(OH)_2}\ \times\ \frac{58.33\ g\ Mg(OH)_2}{\cancel{mol\ Mg(OH)_2}}\ =\ \underline{1.75\ g\ Mg(OH)_2}$$

12-56 An aqueous solution of AB is a good conductor of electricity. AB is dissociated into ions such as A^+ and B^-. A solution of AC is a weak conductor of electricity, which means that AC is only partially dissociated into ions:

i.e., $AC \rightleftharpoons A^+ + C^-$

A solution of AD is a nonconductor of electricity because it is present as undissociated molecules in solution.

12-58 1 mole of NaBr has a mass of 102.9 g, 1 molar NaBr is a solution containing 102.9 g of NaBr per liter of solution, and 1 molal NaBr is a solution containing 102.9 g of NaBr per kg of solvent.

12-59 $n(NaOH) = 25.0 \text{ g NaOH} \times \dfrac{1 \text{ mol NaOH}}{40.00 \text{ g NaOH}} = 0.625 \text{ mol NaOH}$

$250 \text{ g} \times \dfrac{1 \text{ kg}}{10^3 \text{ g}} = 0.250 \text{ kg} \qquad \dfrac{0.625 \text{ mol}}{0.250 \text{ kg}} = \underline{2.50 \text{ m}}$

The molality is the same in both solvents since the mass of solvent is the same.

12-61 $m = \dfrac{n \text{ (solute)}}{\text{kg (solvent)}} \qquad n = m \times \text{kg (solvent)}$

$0.550 \text{ kg} \times \dfrac{0.720 \text{ mol NaOH}}{\text{kg}} \times \dfrac{40.00 \text{ g NaOH}}{\text{mol NaOH}} = \underline{15.8 \text{ g NaOH}}$

12-63 $\Delta T_f = K_f m = 1.86 \times 0.20 = 0.37^\circ C \qquad F.P. = 0.00^\circ C - 0.37^\circ C = \underline{-0.37^\circ C}$

12-65 The salty water removes water from the cells of the skin by osmosis. After a prolonged period one would dehydrate and become thirsty.

12-68 One can concentrate a dilute solution by boiling away some solvent if the solute is not volatile. Reverse osmosis can also be used to concentrate a solution if pressure greater than the osmotic pressure is applied on the concentrated solution separated from the solvent by a semipermeable membrane. As the solution becomes more concentrated, the osmotic pressure becomes greater and the corresponding pressure that is applied must be increased.

12-69 Assume 1000 g of solution. There are then 100 g of $CaCl_2$ and 900 g of H_2O in the solution.

$100 \text{ g CaCl}_2 \times \dfrac{1 \text{ mol CaCl}_2}{111.0 \text{ g CaCl}_2} = 0.901 \text{ mol CaCl}_2 \qquad \dfrac{0.901 \text{ mol}}{0.900 \text{ kg}} = \underline{1.00 \text{ m}}$

12-71 $m = \Delta T_f / K_f = 5.0/1.86 = 2.7 \quad m = \dfrac{n \text{ (solute)}}{\text{kg (solvent)}} = 2.7 \quad \dfrac{n}{5.00 \text{ kg}} = 2.7$

$n = 14 \text{ mol} \quad 14 \text{ mol} \times \dfrac{62.07 \text{ g}}{\text{mol}} = \underline{870 \text{ g glycol}}$

12-72 $\Delta T_b = K_b m = 0.512 \times 2.7 = 1.4^\circ C \qquad B.P. = 100.0^\circ C + 1.4^\circ C = \underline{101.4^\circ C}$

12-74 $101.5^\circ C - 100.0^\circ C = 1.5^\circ C = T_b \qquad m = \Delta T_b / K_b = 1.5/0.512 = \underline{2.9 \text{ m}}$

12-77 $100 \text{ g } \cancel{CH_3OH} \times \dfrac{1 \text{ mol } CH_3OH}{32.04 \text{ g } \cancel{CH_3OH}} = 3.12 \text{ mol } CH_3OH$

$\Delta T_f = K_f m = 5.12 \times \dfrac{3.12 \text{ mol}}{0.800 \text{ kg}} = 20.0^{\circ}C$ 　　　　　 F.P. $= 5.5^{\circ}C - 20.0^{\circ}C = -14.5^{\circ}C$

12-79 HCl is a strong electrolyte that produces two particles (ions) for each mole of HCl that dissolves. HF is a weak electrolyte that is essentially present as un-ionized molecules in solution.

12-82 (a) $\Delta T_f = K_f m = 1.86 \times \dfrac{\frac{10.0 \text{ g}}{32.04 \text{ g/mol}}}{0.100 \text{ kg solvent}} = 5.8^{\circ}C$ 　 F.P. $= 0.0^{\circ}C - 5.8^{\circ}C = \underline{-5.8^{\circ}C}$

(b) $\Delta T_f = K_f m = 1.86 \times \dfrac{\frac{10.0 \text{ g}}{58.44 \text{ g/mol}}}{0.100 \text{ kg solvent}} = \underline{3.2^{\circ}C}$

$NaCl \longrightarrow Na^+(aq) + Cl^-(aq)$

The two ions per mole of NaCl cause the melting point to be lowered twice as much as one mole of a nonelectrolyte.

$2 \times 3.2^{\circ}C = 6.4^{\circ}C$ 　　　　　　　 F.P. $= 0.0^{\circ}C - 6.4^{\circ}C = \underline{-6.4^{\circ}C}$

(c) $\Delta T_f = K_f m = 1.86 \times \dfrac{\frac{10.0 \text{ g}}{111.0 \text{ g/mol}}}{0.100 \text{ g solvent}} = 1.68^{\circ}C$

$CaCl_2 \longrightarrow Ca^{2+}(aq) + 2Cl^-(aq)$

The three ions per mole of $CaCl_2$ cause the melting point to be lowered three times as much as one mole of a nonelectrolyte.

$3 \times 1.68^{\circ}C = 5.0^{\circ}C$ 　　　　　　 F.P. $= 0.0^{\circ}C - 5.0^{\circ}C = \underline{-5.0^{\circ}C}$

12-83 Dissolve the mixture in 100 g of H_2O and heat to over $45^{\circ}C$. Cool to $0^{\circ}C$ where the solution is saturated with about 10 g of KNO_3 and about 25 g of KCl. 50 – 25 = 25 g of KCl precipitates.

12-85 $150 \text{ } \cancel{mL} \times \dfrac{1.00 \text{ g } H_2O}{\cancel{mL}} = 150 \text{ g } H_2O$ 　 mass of solution = 150 g + 10.0 g + 5.0 g = 165 g

$\dfrac{150 \text{ g}}{165 \text{ g}} \times 100\% = 90.9\% \text{ } H_2O$; 　$\dfrac{10.0 \text{ g}}{165 \text{ g}} \times 100\% = 6.06 \% \text{ sugar}$

$100\% - (90.9 + 6.06)\% = 3.0\%$ salt

12-87 $0.500 \text{ L} \times 0.20 \text{ mol/L} = 0.10 \text{ mol } Ag^+$ 　　　 $0.500 \text{ L} \times 0.30 \text{ mol/L} = 0.15 \text{ mol } Cl^-$

0.10 mol of Ag^+ reacts with 0.10 mol of Cl^- to form AgCl leaving 0.05 mol of Cl^- in 1.00 L of solution. $[Cl^-] = 0.05 \text{ mol/1L} = 0.05 \text{ M}$

12-89 $0.225 \text{ } \cancel{L} \times \dfrac{0.196 \text{ } \cancel{\text{mol HCl}}}{\cancel{L}} \times \dfrac{1 \text{ mol M}}{2 \text{ } \cancel{\text{mol HCl}}} = 0.0221 \text{ mol M}$

$\dfrac{1.44 \text{ g}}{0.0221 \text{ mol}} = \underline{65.2 \text{ g/mol (Zn)}}$

12-91 $\Delta T = 5.67 - 2.22 = 3.45^\circ C$ molality $= \dfrac{\Delta T}{K_f} = \dfrac{3.45}{8.10} = 0.426 \ m$

Molar mass $= \dfrac{1000 \times \text{mass solute}}{m \times \text{mass solution}} = \dfrac{1000 \times 3.07}{0.426 \times 120} = 60.1 \ g/mol$

C: $40.0 \ \cancel{g \ C} \times \dfrac{1 \ mol \ C}{12.01 \ \cancel{g \ C}} = 3.33 \ mol \ C$ H: $13.3 \ \cancel{g \ H} \times \dfrac{1 \ mol \ H}{1.008 \ \cancel{g \ H}} = 13.2 \ mol \ H$

N: $46.7 \ \cancel{g \ N} \times \dfrac{1 \ mol \ N}{14.01 \ \cancel{g \ N}} = 3.33 \ mol \ N$

H: $\dfrac{13.2}{3.33} = 4.0$ C & N: $\dfrac{3.33}{3.33} = 1.0$ empirical formula $= CH_4N$;

Emp mass = 30 g/emp unit $\dfrac{30.05 \ \cancel{g/\text{emp unit}}}{60.1 \ \cancel{g/\text{mol}}} = 2 \ \text{emp unit/mol}$

Molecular formula $= C_2H_8N_2$

12-93 0.30 m sugar < 0.12 m KCl (0.24 m ions) < 0.05 m $CrCl_3$ (0.20 m ions)
< 0.05 m K_2CO_3 (0.15 m ions) < pure water

12-95 n(hydroxide) $= M \times V = 0.120 \ mol/\cancel{L} \times 0.487 \ \cancel{L} = 0.0584 \ mol$

$n(H_2) = \dfrac{PV}{RT} = \dfrac{0.650 \ \cancel{atm} \times 1.10 \ \cancel{L}}{0.0821 \dfrac{\cancel{L} \cdot \cancel{atm}}{\cancel{K} \cdot mol} \times 298 \cancel{K}} = 0.0292 \ mol \ H_2$

$\dfrac{0.0584}{0.0292} = 2$ (2 mol hydroxide: 1 mol H_2)

Balanced Equations:

$2Na(s) + 2H_2O(l) \longrightarrow 2NaOH(aq) + H_2(g)$ (2 mol sodium: 1 mol H_2)

$Ca(s) + 2H_2O(l) \longrightarrow Ca(OH)_2(aq) + H_2(g)$ (1 mol calcium: 1 mol H_2)

The answer is sodium since it produces the correct ratio of hydrogen.

12-96 $NH_3(g) \longrightarrow NH_3(aq)$

n(aq) $= 0.450 \ mol/\cancel{L} \times 0.250 \ \cancel{L} = 0.113 \ mol$

$V(gas) = \dfrac{nRT}{P} = \dfrac{0.113 \ \cancel{mol} \times 0.0821 \dfrac{L \cdot \cancel{atm}}{\cancel{K} \cdot \cancel{mol}} \times 298 \ \cancel{K}}{0.951 \ \cancel{atm}} = \underline{2.91 \ L}$

12-98 $NaHCO_3(aq) + HCl(aq) \longrightarrow NaCl(aq) + H_2O + CO_2(g)$

n $= M \times V = 0.340 \ mol/\cancel{L} \times 1.00 \ \cancel{L} = 0.340 \ mol \ NaHCO_3$

$0.340 \ \cancel{mol \ NaHCO_3} \times \dfrac{1 \ mol \ CO_2}{1 \ \cancel{mol \ NaHCO_3}} = 0.340 \ mol \ CO_2$

$$V(CO_2) = \frac{nRT}{P} = \frac{0.340 \text{ mol} \times 0.0821 \frac{L \cdot atm}{K \cdot mol} \times 308 \text{ K}}{1.00 \text{ atm}} = \underline{8.60 \text{ L}}$$

12-100 The formula must be PCl_3 with the Lewis structure

$$:\overset{..}{Cl}-\overset{..}{\underset{|}{P}}-\overset{..}{Cl}: \quad \text{which would be trigonal pyramidal.}$$
$$\quad :\overset{..}{\underset{..}{Cl}}:$$

The reaction is $PCl_3(g) + 3H_2O \longrightarrow H_3PO_3(aq) + 3HCl(aq)$

$$0.750 \text{ L} \times \frac{1 \text{ mol } PCl_3}{22.4 \text{ L}} = 0.0335 \text{ mol } PCl_3$$

$$0.0335 \text{ mol } PCl_3 \times \frac{3 \text{ mol } HCl}{1 \text{ mol } PCl_3} = 0.101 \text{ mol } HCl$$

$$M = n/V = 0.101 \text{ mol}/0.250 \text{ L} = \underline{0.404 \text{ M } HCl}$$

12-102 KCl (K^+Cl^-) $10.0 \text{ g } KCl \times \frac{1 \text{ mol } KCl}{74.55 \text{ g } KCl} = 0.134 \text{ mol } KCl$

$$\Delta T = 1.86 \times \frac{\frac{0.134 \text{ mol}}{100 \text{ g}}}{1000 \text{ g/kg}} \times 2 \text{ mol ions/mol } KCl = 4.98^oC \text{ change}$$

Na_2S ($2Na^+S^{2-}$) $10.0 \text{ g } Na_2S \times \frac{1 \text{ mol } Na_2S}{78.05 \text{ g } Na_2S} = 0.128 \text{ mol } Na_2S$

$$\Delta T = 1.86 \times \frac{\frac{0.128 \text{ mol}}{100 \text{ g}}}{1000 \text{ g/kg}} \times 3 \text{ mol ions/mol } Na_2S = 7.14^oC \text{ change}$$

$CaCl_2$ ($Ca^{2+}2Cl^-$) $10.0 \text{ g } CaCl_2 \times \frac{1 \text{ mol } CaCl_2}{111.0 \text{ g } CaCl_2} = 0.0901 \text{ mol } CaCl_2$

$$\Delta T = 1.86 \times \frac{\frac{0.0901 \text{ mol}}{100 \text{ g}}}{1000 \text{ g/kg}} \times 3 \text{ mol ions/mol } CaCl_2 = 5.03^oC \text{ change}$$

The answer is $\underline{Na_2S}$ since it would have a freezing point of $\underline{-7.14^oC}$.

12-104 $\Delta T = 1.50^oC \quad m = \Delta T/K_f = 1.50/1.86 = 0.806$

$n/1.00 \text{ kg} = 0.806 \quad n = 0.806 \text{ mol} \quad 1 \text{ L} = 1000 \text{ mL} \times \frac{1.00 \text{ g}}{mL} = 1000 \text{ g (1 kg) } H_2O$

For dilute solution, Let X = kg of added water, then

$$\Delta T = K_f m \quad 1.15^oC = 1.86 \times \frac{0.806}{1.00 + X}$$

Solving for X, X = 0.30 kg $\quad$ 0.30 kg = 300 g = $\underline{300 \text{ mL of water added.}}$

13

Acids, Bases, and Salts

Review of Part A *Acids, Bases, and the Formation of Salts*

OBJECTIVES AND DETAILED TABLE OF CONTENTS

13-1 Properties of Acids and Bases

OBJECTIVE *List the general properties of acids and bases.*

13-1.1 Arrhenius Acids and Bases
13-1.2 Strong Acids in Water
13-1.3 Strong Bases in Water

13-2 Bronsted-Lowry Acids and Bases

OBJECTIVE *Identify Bronsted acids and bases and conjugate acid-base pairs in a proton exchange reaction.*

13-2.1 Conjugate Acid-Base Pairs
13-2.2 Amphiprotic Ions

13-3 Strengths of Acids and Bases

OBJECTIVE *Calculate the hydronium ion concentration in a solution of a strong acid and a weak acid given the initial concentration of the acid and the percent ionization of the weak acid.*

13-3.1 The Strength of Acids
13-3.2 The Strength of Bases

13-4 Neutralization and the Formation of Salts

OBJECTIVE *Write the molecular, total ionic, and net ionic equations for neutralization reactions.*

13-4.1 Neutralization of a Strong Acid with a Strong Base
13-4.2 Neutralization of a Weak Acid with a Strong Base
13-4.3 Neutralization of a Polyprotic Acid with a Strong Base

SUMMARY OF PART A

Three ancient yet still important classes of compounds are known as **acids**, **bases**, and **salts**. This chapter deals exclusively with these compounds and the nature of their reactions in water. Some quantitative considerations of acids and bases will be presented in Chapter 15.

Although these types of compounds have been classified since antiquity, an early but still commonly used definition was advanced by Arrhenius in 1884. Acids are defined as substances that produce $H^+(aq)$ ions in water. Thus the acid nature of hydrobromic acid is illustrated as follows:

$$HBr(g) + H_2O \longrightarrow H_3O^+(aq) + Br^-(aq)$$

Acids are molecular compounds when pure. But, in the presence of polar water molecules, the acid molecules are **ionized**. It is the formation of the **hydronium ion** [H_3O^+ or $H^+(aq)$] that gives acids their unique character.

Bases obtain their unique character from the production of hydroxide (OH^-) ions in aqueous solution. The common strong bases are ionic solids that dissolve in water to form ions.

$$NaOH(s) \xrightarrow{\;H_2O\;} Na^+(aq) + OH^-(aq)$$

To expand upon our concept of acids and bases in water and to extend this behavior to substances such as ionic compounds, it is helpful to employ a somewhat broader definition. In the **Bronsted-Lowry** definition, **acids** are identified as proton (H^+) donors and **bases** as proton acceptors. In this approach, the list of acids and bases is much more extensive than in the

Arrhenius definition. An acid or a base is identified by the role (or potential role), it plays in a proton exchange reaction.

The reaction of a Bronsted-Lowry acid produces what is known as its **conjugate base**. The reaction of a base produces its **conjugate acid**. Conjugate acid-base pairs can be identified by the gain or loss of an H^+. An **amphiprotic** substance is one that has both a conjugate acid and a conjugate base. The relationships between acids and their conjugate bases are illustrated as follows:

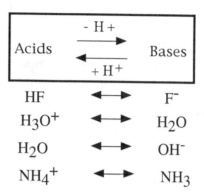

The strengths of acids vary according to the percentage of the acid molecules that ionize to produce hydronium ions. **Strong acids** are 100% ionized in water and thus all have the same acid strength. **Weak acids**, on the other hand, are only partially ionized, which means that the following reaction goes to the right to a limited extent.

$$HNO_2(aq) + H_2O \rightleftharpoons H_3O^+(aq) + NO_2^-(aq)$$

The double arrows ($\rightleftharpoons$) indicate an incomplete reaction that reaches a state of dynamic equilibrium. In a reaction that reaches a state of equilibrium, the forward reaction ($\longrightarrow$) and the reverse reaction ($\longleftarrow$) are both occurring, but at the same rate, so that reactants and products coexist in the same solution in definite proportions. Thus weak acids (which are also weak electrolytes) produce small concentrations of H_3O^+ compared to the strong acids.

In a weak acid, most of the acid molecules are present as neutral molecules as shown on the left side of the equation. Likewise, **strong bases** are completely ionized, but **weak bases** such as ammonia produce a limited concentration of OH^- ions.

$$NH_3(aq) + H_2O \rightleftharpoons NH_4^+(aq) + OH^-(aq)$$

When solutions of acids and bases are mixed, a reaction known as neutralization occurs. A **neutralization reaction** is a type of double replacement reaction similar to the precipitation reaction in which the cation and anion combine to form a solid precipitate. In neutralization, the cation (of the acid) and the anion (of the base) combine to form the molecular compound water. The neutralization reaction between a strong acid and base was first illustrated in Chapter 6; an example of the balanced molecular, total ionic, and net ionic equations of such a reaction involving a **monoprotic acid** follows.

Molecular: $HNO_3(aq) + NaOH(aq) \longrightarrow NaNO_3(aq) + H_2O(l)$

Total ionic: $H^+(aq) + NO_3^-(aq) + Na^+(aq) + OH^-(aq) \longrightarrow Na^+(aq) +$

$NO_3^-(aq) + H_2O(l)$

Net ionic: $H^+(aq) + OH^-(aq) \longrightarrow H_2O$

The molecular, total ionic, and net ionic equations for the neutralization of a weak monoprotic acid and a strong base are shown below. In this case, the acid is not represented as ions in the ionic equations because most of the acid is present in solution as molecules.

Molecular: $HNO_2(aq) + KOH(aq) \longrightarrow KNO_2(aq) + H_2O(l)$

Total ionic: $HNO_2(aq) + K^+(aq) + OH^-(aq) \longrightarrow K^+(aq) + NO_2^-(aq) + H_2O(l)$

Net ionic: $HNO_2(aq) + OH^-(aq) \longrightarrow NO_2^-(aq) + H_2O(l)$

Polyprotic acids have more than one replaceable hydrogen per molecule. Sulfurous acid, H_2SO_3, has two and is referred to as a **diprotic acid**. Phosphoric acid, H_3PO_4, has three and is an example of a **triprotic acid**. Polyprotic acids can react with bases in steps. For example, consider the reaction of one mole of sulfurous acid with one mole of KOH. The three appropriate equations representing the reaction are as follows:

Molecular: $H_2SO_3(aq) + KOH(aq) \longrightarrow KHSO_3(aq) + H_2O(l)$

Total ionic: $H_2SO_3(aq) + K^+(aq) + OH^-(aq) \longrightarrow K^+(aq) + HSO_3^-(aq) + H_2O(l)$

Net ionic: $H_2SO_3(aq) + OH^-(aq) \longrightarrow HSO_3^-(aq) + H_2O(l)$

The potassium hydrogen sulfite ($KHSO_3$) is known as an **acid salt**. It is a salt whose anion contains at least one acid proton that can react with another mole of base. Acid salts are formed from the partial neutralization of polyprotic acids.

ASSESSMENT OF OBJECTIVES

A-1 Multiple Choice

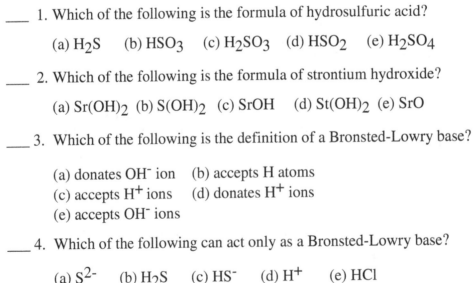

___ 1. Which of the following is the formula of hydrosulfuric acid?

    (a) H_2S (b) HSO_3 (c) H_2SO_3 (d) HSO_2 (e) H_2SO_4

___ 2. Which of the following is the formula of strontium hydroxide?

    (a) $Sr(OH)_2$ (b) $S(OH)_2$ (c) $SrOH$ (d) $St(OH)_2$ (e) SrO

___ 3. Which of the following is the definition of a Bronsted-Lowry base?

    (a) donates OH^- ion   (b) accepts H atoms
    (c) accepts H^+ ions   (d) donates H^+ ions
    (e) accepts OH^- ions

___ 4. Which of the following can act only as a Bronsted-Lowry base?

    (a) S^{2-} (b) H_2S (c) HS^- (d) H^+ (e) HCl

___ 5. Which of the following can react as either a Bronsted acid or a Bronsted base?

(a) HS^- (b) Cl^- (c) NH_4^+ (d) H_2S (e) S^{2-}

___ 6. Which of the following is the conjugate base of $H_2PO_4^-$?

(a) HPO_4^{2-} (b) H_3PO_4 (c) HPO_4^- (d) PO_4^{3-} (e) OH^-

___ 7. Which of the following is a weak acid?

(a) HBr (b) H_2SO_3 (c) HNO_3 (d) HCl (e) HI

___ 8. Which of the following compounds is a strong base?

(a) NH_3 (b) $B(OH)_3$ (c) $LiOH$ (d) HNO_2 (e) $CaCl_2$

___ 9. A 0.10 M solution of a substance has a H_3O^+ concentration of 0.01 M. The substance is:

(a) a weak base.   (b) a strong base.
(c) a strong acid.   (d) a weak acid.

___ 10. Which of the following is a salt formed by the complete neutralization of barium hydroxide with nitrous acid?

(a) $BaHNO_2$ (b) $Ba(NO_3)_2$
(c) $Ba(OH)_2$ (d) $BaNO_2$ (e) $Ba(NO_2)_2$

___ 11. Which of the following is an acid salt?

(a) NH_4Cl (b) NH_4NO_3
(c) $BaHPO_4$ (d) $CaSO_3$ (e) H_2SO_3

___ 12. Which of the following is a product when one mole of H_3AsO_4 reacts with one mole of $Ca(OH)_2$?

(a) $Ca(H_2AsO_4)_2$ (b) $Ca(HAsO_4)$
(c) $Ca(AsO_4)_2$ (d) $Ca(HAsO_4)_2$ (e) CaH_3AsO_4

___ 13. Which of the following is a salt formed from the neutralization of a strong acid and a strong base?

(a) CaI_2 (b) $Ca(NO_2)_2$
(c) NH_4Br (d) NH_4NO_2 (e) KF

A-2 Problems

1. Identify each of the following as an acid or base when dissolved in water:

    (a) HClO _____   (b) $Mn(OH)_2$ _____

    (c) H_2Se _____   (d) CH_3NH_2 _____

    (e) H_2SO_3 _____   (f) CuOH _____

2. Complete the following equations illustrating Bronsted-Lowry acid-base reactions:

    Acid₁ +   *Base₂* ⟶   *Acid₂*   +   *Base₁*

    (a) $HClO_3$ + H_2O ⟶

    (b) HSO_4^- + NH_3 ⟶

    (c) H_2O + HCO_3^- ⟶

    (d) HCO_3^- + H_2O ⟶

    (e) H_2O + NH_2^- ⟶

    (f) H_2O + CH_3NH_2 ⟶

    (g) H_2S + H_2O ⟶

    (h) $H_2PO_4^-$ + CH_3O^- ⟶

3. Complete and balance the following equations illustrating <u>complete</u> neutralizations:

    (a) HNO_3 + $Ca(OH)_2$ ⟶ $Ca(NO_3)_2$ + _____

    (b) $HC_2H_3O_2$ + NaOH ⟶ _____ + _____

    (c) _____ + KOH ⟶ K_2CO_3 + H_2O

    (d) H_2S + $Mg(OH)_2$ ⟶ _____ + H_2O

    (e) HClO + NH_3 ⟶ _____

    (f) _____ + _____ ⟶ $Fe_2(SO_4)_3$ + _____

4. Write the total ionic and the net ionic equation for 3(a), (b), and (c).

5. Write equations representing the preparation of the following salts or acid salts produced by the reaction of the appropriate acids and bases:

   (a) RbHS

   (b) $Ca(H_2PO_4)_2$

   (c) NiS

6. Write the net ionic equations for 5 (a) and (b).

Review of Part B   *The Measurement of Acid Strength*

OBJECTIVES AND DETAILED TABLE OF CONTENTS

13-5 Equilibrium of Water

OBJECTIVE *Using the ion product of water, relate the hydroxide ion and the hydronium ion concentrations.*

13-6 The pH Scale

OBJECTIVE *Given the hydronium or hydroxide concentrations, calculate the pH and pOH and vice versa.*

SUMMARY OF PART B

We are now ready to expand our view of acids and bases in water based on the fact that there is a small but important equilibrium concentration of H_3O^+ and OH^- present in pure water. This occurs because of the **autoionization** of water, illustrated as follows:

$$2H_2O \rightleftharpoons H_3O^+ + OH^-$$

The product of the molar concentrations of H_3O^+ and OH^- is known as the **ion product** of water and has a constant value of 1.0×10^{-14} at $25^\circ C$. Thus in pure water

$$K_w = [H_3O^+][OH^-] = 1.0 \times 10^{-14} \qquad [H_3O^+] = [OH^-] = 1.0 \times 10^{-7}$$

In this light, we can redefine an acid as a substance that increases $[H_3O^+]$. Since there is an inverse relation between $[H_3O^+]$ and $[OH^-]$, an increase in $[H_3O^+]$ is the same as a decrease in $[OH^-]$. (As an analogy, pushing one side of a see saw down is the same as pushing the other up.) A base does the opposite; it increases $[OH^-]$ while decreasing $[H_3O^+]$.

The $[H_3O^+]$ in a solution is more conveniently expressed on a logarithmic scale rather than in scientific notation. This scale, known as **pH**, is defined as follows:

$$pH = -\log [H_3O^+] \quad (\text{or } \mathbf{pOH} = -\log [OH^-]) \quad \text{and} \quad pH + pOH = 14.00$$

Some example calculations follow.

Example B-1 Conversion of $[H_3O^+]$ to pH

What is the pH of a solution if $[H_3O^+] = 8.3 \times 10^{-4}$?

PROCEDURE

$pH = -\log [H_3O^+]$ Take the log of 8.3×10^{-4}.

Using a calculator, enter 8.3, EXP, 4, +/-, and take the log.
The answer on the calculator is changed to a positive number.

SOLUTION

$pH = -\log (8.3 \times 10^{-4}) = \underline{3.08}$

Example B-2 Conversion of pH to $[H_3O^+]$

What is $[H_3O^+]$ if pH = 10.43?

PROCEDURE

$-\log [H_3O^+] = 10.43 \qquad \log [H_3O^+] = -10.43$
Using a calculator, enter 10.43, +/-, and INV (or shift), and log.

Find the antilog of 0.57 which is then multiplied by 10^{-11}.

(Review the discussion of logarithms in Appendix C in the text if necessary.)

SOLUTION

Inverse log (antilog) of $-10.43 = [H_3O^+] = \underline{3.7 \times 10^{-11}}$

Acidic, basic, and neutral solutions can be defined in terms of $[H_3O^+]$, $[OH^-]$, pH, and pOH as follows:

	$[H_3O^+]$	$[OH^-]$	pH	pOH
Neutral	1.0×10^{-7}	1.0×10^{-7}	7.00	7.00
Acidic	$> 1.0 \times 10^{-7}$	$< 1.0 \times 10^{-7}$	< 7.00	> 7.00
Basic	$< 1.0 \times 10^{-7}$	$> 1.0 \times 10^{-7}$	> 7.00	< 7.00

ASSESSMENT OF OBJECTIVES

B-1 Multiple Choice

____ 1. Which of the following is an acidic solution?

(a) $[H_3O^+] = 10^{-1}$ (b) $[OH^-] = 10^{-7}$

(c) $[OH^-] = 10^{-4}$ (d) $[OH^-] = 10^{-14}$

____ 2. When $[H_3O^+] = 10^{-3}$, what is $[OH^-]$?

(a) $[OH^-] = 10^{-11}$ (b) $[OH^-] = 10^{-3}$

(c) $[OH^-] = 10^{-7}$ (d) $[OH^-] = 10^{-9}$

____ 3. If there is 10^{-4} mol of H_3O^+ in 10 L of water, what is $[OH^-]$?

(a) 10^{-5} (b) 10^{11} (c) 10^{-4} (d) 10^{-9} (e) 10^{-3}

____ 4. What is the pH of the solution in the preceding problem?

(a) 4   (b) 5   (c) 9   (d) 11   (e) 3

____ 5. When $[H_3O^+] = 1.0 \times 10^{-9}$, pOH equals which of the following?

(a) 5.0   (b) 7.0   (c) 4.0   (d) 9.0   (e) 10.0

____ 6. Which of the following is the most acidic solution?

(a) pH = 7.0   (b) pOH = 1.2
(c) pOH = 13.8   (d) pH = 4.3   (e) pH = 11.2

B-2 Problems

1. What is the $[H_3O^+]$ in the following solutions?

(a) $[OH^-] = 1.0 \times 10^{-12}$

(b) $[OH^-] = 5.9 \times 10^{-4}$

(c) pH = 9.0

(d) pH = 7.85

(e) pOH = 4.18

2. What is the pH of the following solutions?

(a) pOH = 4.51

(b) $[OH^-] = 1.0 \times 10^{-7}$

(c) $[H_3O^+] = 3.6 \times 10^{-4}$

(d) $[H_3O^+] = 48 \times 10^{-10}$

Review of Part C *Salts and Oxides as Acids and Bases*

OBJECTIVES AND DETAILED TABLE OF CONTENTS

13-7 The Effect of Salts on pH – Hydrolysis

OBJECTIVE *Write a hydrolysis reaction, if one occurs, for salt solutions, to determine whether they are acidic, basic, or neutral.*

13-8 Control of pH – Buffer Solutions

OBJECTIVE *Write equations illustrating how a buffer solution can absorb either added acid or base.*

13-9 Oxides as Acids and Bases

OBJECTIVE *Determine, if possible, whether a specific oxide is acidic or basic.*

SUMMARY OF PART C

Besides molecular compounds, solutions of certain salts may be acidic or basic as well as neutral. Cations of strong bases (i.e., Na^+ and Ca^{2+}) do not affect the pH of water. Anions of strong acids (Cl^-, Br^-, I^-, NO_3^-, ClO_4^- but not HSO_4^-) also do not affect the pH of water. Salts containing any

256

combination of such cations and anions form neutral aqueous solutions (pH = 7.0). Anions of weak acids behave as weak bases in water and cations of weak acids behave as weak acids in water. Such reactions are known as **hydrolysis** reactions. The reaction of an anion as a base is illustrated as follows:

$$X^- + H_2O \rightleftharpoons HX + OH^-$$

Thus solutions of salts such as $KC_2H_3O_2$, $Sr(NO_2)_2$, and $NaClO$ are weakly basic. Cations of weak bases also hydrolyze in water to form acidic solutions illustrated by the following equation:

$$MH^+ + H_2O \rightleftharpoons M + H_3O^+$$

Thus solutions of salts such as NH_4Br and $N_2H_5^+Cl^-$ are weakly acidic.

The partial ionization of a weak acid can be utilized in the formation of a **buffer** solution. A buffer solution is one that can absorb limited quantities of added H_3O^+ or OH^- so that the pH of a solution remains relatively constant. Obviously, such a solution contains two substances: one that is a potential base to remove added H_3O^+ and one that is a potential acid to remove added OH^-. A solution of a weak acid and a salt of its conjugate base (or a weak base and a salt of its conjugate acid) serve this purpose. A solution of acetic acid ($HC_2H_3O_2$) containing sodium acetate ($NaC_2H_3O_2$) acts as a buffer as shown below. Note that the Na^+ ion is a spectator ion and is not involved in the process.

$$\overset{0.25 \text{ M}}{\mathbf{HC_2H_3O_2}} + H_2O \rightleftharpoons H_3O^+ + \overset{0.25 \text{ M}}{\mathbf{C_2H_3O_2^-}}$$

Thus a solution that is 0.25 M in both $HC_2H_3O_2$ and $C_2H_3O_2^-$ has a reservoir of species that are available to react with either H_3O^+ or OH^- added to the solution, as illustrated by the following two equations:

$$HC_2H_3O_2 + OH^- \longrightarrow C_2H_3O_2^- + H_2O$$
$$C_2H_3O_2^- + H_3O^+ \longrightarrow HC_2H_3O_2 + H_2O$$

Buffers are important in many chemical and life processes to control pH within strict limits.

The final topic concerns an important environmental issue. This is the role of oxides as acids or bases. Generally, nonmetal oxides react with water to form acids and are thus called **acid anhydrides**.

$$SO_2(g) + H_2O(l) \longrightarrow H_2SO_3(aq)$$
$$Cl_2O_5(aq) + H_2O(l) \longrightarrow 2HClO_3(aq)$$

Many metal oxides (generally, ionic metal oxides) are **base anhydrides** and react with water to form a hydroxide.

$$K_2O(s) + H_2O \longrightarrow 2KOH(aq)$$

$$BaO(s) + H_2O \longrightarrow Ba(OH)_2(aq)$$

ASSESSMENT OF OBJECTIVES

C-1 Multiple Choice

___ 1. Which of the following anions does not hydrolyze?

(a) I^- (b) F^- (c) ClO_2^- (d) CHO_2^- (e) NO_2^-

___ 2. A solution of which of the following salts produces a basic solution?

(a) NH_4Cl (b) NH_4NO_3

(c) KBr (d) $Mg(ClO_4)_2$ (e) Rb_2S

___ 3. Which of the following solutions is a buffer solution?

(a) 0.20 M HNO_2 and 0.20 M KNO_3
(b) 0.20 M HNO_3 and 0.20 M KNO_2
(c) 0.20 M HNO_3 and 0.20 M KNO_3
(d) 0.20 M HNO_2 and 0.20 M KNO_2
(e) 0.20 M HNO_2 and 0.20 M HNO_3

___ 4. Given one mole of NaOH in 10 L of water, which of the following makes a buffer solution when added to this solution?

(a) one mole of HCl   (b) two moles of $HC_2H_3O_2$
(c) one mole of $HC_2H_3O_2$ (d) one mole of $NaC_2H_3O_2$
(e) two moles of HCl

___ 5. Which of the following is the acid formed when P_4O_{10} reacts with water?

(a) H_3PO_3 (b) HPO_2 (c) H_3PO_4 (d) $H_2P_4O_{11}$ (e) H_2SO_4

___ 6. Which of the following is the anhydride of $Mn(OH)_3$?

(a) Mn_2O_7 (b) MnO (c) MnO_2 (d) Mn_2O_3 (e) MnO_2H

C-2 Problems

1. Indicate whether solutions of the following salts in water are acidic, basic, or neutral. If a solution is acidic or basic, write the equation illustrating this behavior.

   (a) $MgCl_2$

   (b) NH_4I

   (c) K_2CO_3

2. Complete the following equations concerning anhydrides:

   (a) SO_2 + H_2O $\longrightarrow$ _____

   (b) $2HClO_4$ $\longrightarrow$ _____ + H_2O

   (c) $2Fe(OH)_3$ $\longrightarrow$ _____ + $3 H_2O$

Chapter Summary Assessment

S-1 Matching I

____ A strong acid

____ A weak base

____ An acid salt

____ A salt formed from a strong acid and a strong base

____ The conjugate acid of CH_3NH_2

____ The anhydride of $HClO_3$

____ The anhydride of KOH

(a) $NaNO_3$

(b) $HClO_2$

(c) NH_4Br

(d) Cl_2O_3

(e) $CH_3NH_3^+$

(f) KO

(g) $KHSO_3$

(h) $HClO_4$

(i) Cl_2O_5

(j) K_2O

(k) CH_3NH^-

(l) CH_3NH_2

S-2 Matching II

For each of the following solutions pick a pH and any acid, base, salt, or oxide listed whose solution could produce the appropriate acidity. (Assume that a significant amount of each compound is in solution.)

weakly acidic _____   strongly acidic _____

weakly basic _____   strongly basic _____

neutral _____

pH	Acid or base	Salt	Soluble oxide
1.0	NH_3	K_2SO_3	SO_3
5.0	HNO_3	NH_4NO_3	SO_2
7.0	H_2SO_3	KI	BaO
9.0	KOH		
13.0			

S-3 Problems

1. Write equations illustrating reactions of the following with water. Where no reaction occurs write "NR."

(a) $NH_3 + H_2O$

(b) $HClO + H_2O$

(c) $ClO^- + H_2O$

(d) $I^- + H_2O$

(e) $N_2H_5^+ + H_2O$

(f) $CN^- + H_2O$

(g) $H_2S + H_2O$

(h) $Ca^{2+} + H_2O$

(i) $HCN + H_2$

Answers to Assessments of Objectives

A-1 Multiple Choice

1. **a** H_2S

2. **a** $Sr(OH)_2$

3. **c** accepts H^+ ions

4. **a** S^{2-}

5. **a** HS^- [It has a conjugate base (S^{2-}) and a conjugate acid (H_2S).]

6. **a** HPO_4^{2-}

7. **b** H_2SO_3

8. **c** LiOH

9. **d** a weak acid (It is 10% ionized.)

10. **e** $Ba(NO_2)_2$ [$2HNO_2 + Ba(OH)_2 \longrightarrow Ba(NO_2)_2 + 2H_2O$]

11. **c** $BaHPO_4$

12. **b** $CaHAsO_4$ [$H_3AsO_4 + Ca(OH)_2 \longrightarrow CaHAsO_4 + 2H_2O$]

13. **a** CaI_2 [formed from $Ca(OH)_2$ and HI]

A-2 Problems

1. (a) acid   (b) base   (c) acid   (d) base   (e) acid   (f) base

2.

Acid₁	+	Base₂	→	Acid₂	+	Base₁
(a) $HClO_3$	+	H_2O	→	H_3O^+	+	ClO_3^-
(b) HSO_4^-	+	NH_3	→	NH_4^+	+	SO_4^{2-}
(c) H_2O	+	HCO_3^-	→	H_2CO_3	+	OH^-
(d) HCO_3^-	+	H_2O	→	H_3O^+	+	CO_3^{2-}
(e) H_2O	+	NH_2^-	→	NH_3	+	OH^-
(f) H_2O	+	CH_3NH_2	→	$CH_3NH_3^+$	+	OH^-
(g) H_2S	+	H_2O	→	H_3O^+	+	HS^-
(h) $H_2PO_4^-$	+	CH_3O^-	→	CH_3OH	+	HPO_4^{2-}

3.   (a) $2HNO_3 + Ca(OH)_2 \longrightarrow Ca(NO_3)_2 + \mathbf{2H_2O}$

       (b) $HC_2H_3O_2 + NaOH \longrightarrow \mathbf{NaC_2H_3O_2 + H_2O}$

       (c) $\mathbf{H_2CO_3} + 2KOH \longrightarrow K_2CO_3 + 2H_2O$

       (d) $H_2S + Mg(OH)_2 \longrightarrow \mathbf{MgS} + 2H_2O$

       (e) $HClO + NH_3 \longrightarrow \mathbf{NH_4ClO}$

       (f) $\mathbf{3H_2SO_4 + 2Fe(OH)_3} \longrightarrow Fe_2(SO_4)_3 + \mathbf{6H_2O}$

4.   (a) $2H^+(aq) + 2NO_3^-(aq) + Ca^{2+}(aq) + 2OH^-(aq) \longrightarrow Ca^{2+}(aq) + 2NO_3^-(aq)$
                                           $+ 2H_2O(l)$

           $H^+(aq) + OH^-(aq) \longrightarrow H_2O(l)$

       (b) $HC_2H_3O_2 + Na^+(aq) + OH^-(aq) \longrightarrow Na^+(aq) + C_2H_3O_2^-(aq) + H_2O(l)$

           $HC_2H_3O_2(aq) + OH^-(aq) \longrightarrow C_2H_3O_2^-(aq) + H_2O(l)$

       (c) $H_2CO_3(aq) + 2K^+(aq) + 2OH^-(aq) \longrightarrow 2K^+(aq) + CO_3^{2-}(aq) + 2H_2O(l)$

           $H_2CO_3(aq) + 2OH^-(aq) \longrightarrow CO_3^{2-}(aq) + 2H_2O(l)$

5.   (a) $H_2S + RbOH \longrightarrow RbHS + H_2O$

       (b) $2H_3PO_4 + Ca(OH)_2 \longrightarrow Ca(H_2PO_4)_2 + 2H_2O$

       (c) $H_2S + Ni(OH)_2 \longrightarrow NiS + 2H_2O$

6.   (a) $H_2S(aq) + OH^-(aq) \longrightarrow HS^-(aq) + H_2O(l)$

       (b) $H_3PO_4(aq) + OH^-(aq) \longrightarrow H_2PO_4^-(aq) + H_2O(l)$

B-1 Multiple Choice

    1.   **d** $[OH^-] = 10^{-14}$

    2.   **a** $[OH^-] = 10^{-11}$

    3.   **d** $[H_3O^+] = \dfrac{10^{-4} \text{ mol}}{10 \text{ L}} = 10^{-5}$ M, $[OH^-] = \dfrac{K_w}{[H_3O^+]} = \dfrac{10^{-14}}{10^{-5}} = 10^{-9}$

    4.   **b** $pH = -\log [H_3O^+] = -\log (10^{-5}) = 5$

    5.   **a** $pOH = 5$ $(pOH = 14 - pH = 14.0 - 9.0 = 5.0)$

    6.   **c** $pOH = 13.8$ $(pH = 14.0 - pOH = 14.0 - 13.8 = 0.2)$

B-2 Problems

1. (a) $[H_3O^+] = \underline{1.0 \times 10^{-2}}$ (b) $[H_3O^+] = \underline{1.7 \times 10^{-11}}$

     (c) $[H_3O^+] = \underline{1.0 \times 10^{-9}}$ (d) $[H_3O^+] = \underline{1.4 \times 10^{-8}}$

     (e) pOH = 4.18; pH = 14 - 4.18 = $\underline{9.82}$

       $-\log [H_3O^+] = \underline{9.82}$ $[H_3O^+] = \underline{1.5 \times 10^{-10}}$

2. (a) pH = $\underline{9.49}$

     (b) $[H_3O^+] = 1.0 \times 10^{-8}$; pH = $\underline{8.00}$

     (c) pH = 4.00 - log 3.6 = 4.00 - 0.56 = $\underline{3.44}$

     (d) $[H_3O^+] = 48 \times 10^{-10} = 4.8 \times 10^{-9}$; pH = $\underline{8.32}$

C-1 Multiple Choice

1. **a** I^- (I^- is the anion or conjugate base of the strong acid HI.)

2. **e** Rb_2S (The S^{2-} ion hydrolyzes according to the equation

       $S^{2-} + H_2O \rightleftharpoons HS^- + OH^-$.)

3. **d**   0.20 M HNO_2 and 0.20 M KNO_2. (This is the only solution shown of a weak acid and a salt containing its conjugate base.)

4. **b**   two moles of $HC_2H_3O_2$ (One mole of NaOH neutralizes one mole of $HC_2H_3O_2$ to produce one mole of $NaC_2H_3O_2$,

       i.e., $NaOH + HC_2H_3O_2 \longrightarrow NaC_2H_3O_2 + H_2O$.
The solution now contains one remaining mole of $HC_2H_3O_2$ and one mole of $NaC_2H_3O_2$ and is thus a buffer solution.)

5. **c** H_3PO_4 ($P_4O_{10} + 6H_2O \longrightarrow 4H_3PO_4$)

6. **d** Mn_2O_3 ($2Mn(OH)_3 \longrightarrow Mn_2O_3 + 3H_2O$)

C-2 Problem

1. (a) neutral (Neither ion hydrolyzes.)

     (b) acidic (cation hydrolysis) $NH_4^+ + H_2O \rightleftharpoons NH_3 + H_3O^+$

     (c) basic (anion hydrolysis) $CO_3^{2-} + H_2O \rightleftharpoons HCO_3^- + OH^-$

2. (a) $SO_2 + H_2O \longrightarrow H_2SO_3$

     (b) $2HClO_4 \longrightarrow Cl_2O_7 + H_2O$

     (c) $2Fe(OH)_3 \longrightarrow Fe_2O_3 + 3H_2O$

S-1 Matching I

h $HClO_4$ is a strong acid.

g $KHSO_3$ is an acid salt.

e $CH_3NH_3^+$ is the conjugate acid of CH_3NH_2.

i Cl_2O_5 is the anhydride of $HClO_3$.

l CH_3NH_2 is a weak base.

a $NaNO_3$ is formed from a strong acid (HNO_3) and a strong base ($NaOH$).

j K_2O is the anhydride of KOH.

S-2 Matching II

weakly acidic	pH = 5.0, H_2SO_3, NH_4NO_3, SO_2 (forms H_2SO_3 in water)
strongly acidic	pH = 1.0, HNO_3, SO_3 (forms H_2SO_4 in water)
neutral	pH = 7.0, KI
weakly basic	pH = 9.0, NH_3, K_2SO_3
strongly basic	pH = 13.0, KOH, BaO

S-3 Problems

(a) $NH_3 + H_2O \rightleftharpoons NH_4^+ + OH^-$

(b) $HClO + H_2O \rightleftharpoons H_3O^+ + ClO^-$

(c) $ClO^- + H_2O \rightleftharpoons HClO + OH^-$

(d) $I^- + H_2O \longrightarrow NR$

(e) $N_2H_5^+ + H_2O \rightleftharpoons N_2H_4 + H_3O^+$

(f) $CN^- + H_2O \rightleftharpoons HCN + OH^-$

(g) $H_2S + H_2O \rightleftharpoons H_3O^+ + HS^-$

$HS^- + H_2O \rightleftharpoons H_3O^+ + S^{2-}$

(h) $Ca^{2+} + H_2O \longrightarrow NR$

(i) $HCN + H_2O \rightleftharpoons H_3O^+ + CN^-$

Answers and Solutions to Green Text Problems

13-1 (a) HNO_3, nitric acid (b) HNO_2, nitrous acid
(c) $HClO_3$, chloric acid (d) H_2SO_3, sulfurous acid

13-3 (a) $CsOH$, cesium hydroxide (b) $Sr(OH)_2$, strontium hydroxide
(c) $Al(OH)_3$, aluminum hydroxide (d) $Mn(OH)_3$, manganese(III) hydroxide

13-5 (a) $HNO_3 + H_2O \longrightarrow H_3O^+ + NO_3^-$

(b) $CsOH \longrightarrow Cs^+ + OH^-$

(c) $Ba(OH)_2 \longrightarrow Ba^{2+} + 2OH^-$

(d) $HBr + H_2O \longrightarrow H_3O^+ + Br^-$

13-7

Acid	Conjugate base		Acid	Conjugate base
(a) HNO_3	NO_3^-		(d) CH_4	CH_3^-
(b) H_2SO_4	HSO_4^-		(e) H_2O	OH^-
(c) HPO_4^{2-}	PO_4^{3-}		(f) NH_3	NH_2^-

13-9 (a) $HClO_4$, ClO_4^- and H_2O, OH^-

(b) HSO_4^-, SO_4^{2-} and $HClO$ and ClO^-

(c) H_2O, OH^- and NH_3, NH_2^-

(d) NH_4^+, NH_3 and H_3O^+, H_2O

13-12 (a) $NH_3 \;+\; H_2O \longrightarrow NH_4^+ \;+\; OH^-$
$\quad\quad\;\; B_1 \quad\quad\quad\quad\quad A_1 \quad\quad\quad B_2$
$\quad\quad\quad\;\; A_2 \quad\quad\quad\quad\quad\quad\quad B_2$

(b) $N_2H_4 \;+\; H_2O \longrightarrow N_2H_5^+ \;+\; OH^-$
$\quad\quad\;\; B_1 \quad\quad\quad\quad\quad A_1 \quad\quad\quad B_2$
$\quad\quad\quad\;\; A_2 \quad\quad\quad\quad\quad\quad\quad B_2$

(c) $HS^- \;+\; H_2O \longrightarrow H_2S \;+\; OH^-$
$\quad\quad\;\; B_1 \quad\quad\quad\quad\quad A_1 \quad\quad\quad B_2$
$\quad\quad\quad\;\; A_2 \quad\quad\quad\quad\quad\quad\quad B_2$

(d) $H^- \;+\; H_2O \longrightarrow H_2 \;+\; OH^-$
$\quad\quad\;\; B_1 \quad\quad\quad\quad\quad A_1 \quad\quad\quad B_2$
$\quad\quad\quad\;\; A_2 \quad\quad\quad\quad\quad\quad\quad B_2$

(e) F^- + H_2O $\longrightarrow$ HF + OH^-

$\underset{B_1}{\rule{0pt}{0pt}}$ $\underset{A_2}{\rule{0pt}{0pt}}$ $\underset{A_1}{\rule{0pt}{0pt}}$ $\underset{B_2}{\rule{0pt}{0pt}}$

13-13 Base: HS^- + H_3O^+ $\rightleftharpoons$ H_2S + H_2O

Acid: HS^- + OH^- $\rightleftharpoons$ S^{2-} + H_2O

13-16 The information indicates that HBr is a strong acid. The Br^- ion, however, would be a very weak base in water.

13-18 (a) HNO_3 + H_2O $\longrightarrow$ H_3O^+ + NO_3^-

(b) HNO_2 + H_2O $\rightleftharpoons$ H_3O^+ + NO_2^-

(c) $HClO_3$ + H_2O $\rightleftharpoons$ H_3O^+ + ClO_3^-

(d) H_2SO_3 + H_2O $\rightleftharpoons$ H_3O^+ + HSO_3^-

13-20 $(CH_3)_2NH$ + H_2O $\rightleftharpoons$ $(CH_3)_2NH_2^+$ + OH^-

13-22 HX is a weak acid   HX + H_2O $\rightleftharpoons$ H_3O^+ + X^-

The concentration of H_3O^+ must equal the concentration of the HX that is ionized.

$\dfrac{0.010}{0.100}$ x 100% = 10% ionized

13-24 Since $HClO_4$ is one of the strong acids, it is completely dissociated.

Thus, $[H_3O^+]$ = 0.55 M

13-25 3.0% of the original HX concentration is ionized to form H_3O^+.

Thus, $[H_3O^+]$ = 0.030 x 0.55 = 0.017 M

13-27   100%

H_2SO_4 + H_2O ------> H_3O^+ + HSO_4^-

From the first ionization: $[H_3O^+]$ = 0.354 M

HSO_4^- + H_2O $\rightleftharpoons$ H_3O^+ + SO_4^{2-} (25% to the right)

The concentration of HSO_4^- from the first ionization is equal to 0.354 M.
Of that, 25% dissociates.

0.25 x 0.354 = 0.089 M $[H_3O^+]$ from the second ionization.

The total $[H_3O^+]$ = 0.354 + 0.089 = 0.443 M

13-28 The concentration of H_3O^+ is equal to the concentration of the ionized acid.

Therefore, $\dfrac{0.050\ M}{1.0\ M}$ x 100% = 5.0% ionized

13-30 The conjugate acid of HSO_4^- is H_2SO_4. This is a strong acid so it completely ionizes in water. Thus it cannot exist in the molecular form in water.

13-31 (a) acid   (b) salt   (c) acid   (d) acid salt
       (e) salt   (f) base

13-33 (a) $HNO_3 + NaOH \longrightarrow NaNO_3 + H_2O$

(b) $2HI + Ca(OH)_2 \longrightarrow CaI_2 + 2H_2O$

(c) $HClO_2 + KOH \longrightarrow KClO_2 + H_2O$

13-35 (a) $H^+(aq) + NO_3^-(aq) + Na^+(aq) + OH^-(aq) \longrightarrow Na^+(aq) + NO_3^-(aq) + H_2O(l)$

$H^+(aq) + OH^-(aq) \longrightarrow H_2O$ (net ionic)

(b) $2H^+(aq) + 2I^-(aq) + Ca^{2+}(aq) + 2OH^-(aq) \longrightarrow Ca^{2+}(aq) + 2I^-(aq) + 2H_2O(l)$

$H^+(aq) + OH^-(aq) \longrightarrow H_2O$ (net ionic)

(c) $HClO_2(aq) + K^+(aq) + OH^-(aq) \longrightarrow K^+(aq) + ClO_2^-(aq) + H_2O$

$HClO_2(aq) + OH^-(aq) \longrightarrow ClO_2^-(aq) + H_2O(l)$ (net ionic)

13-37 $H_2C_2O_4(aq) + 2NH_3(aq) \longrightarrow (NH_4)_2C_2O_4(aq)$

$H_2C_2O_4(aq) + 2NH_3(aq) \longrightarrow 2NH_4^+(aq) + C_2O_4^{2-}(aq)$
(The net ionic equation is the same as the total ionic equation.)

13-39 (a) KOH and $HClO_3$ (b) $Al(OH)_3$ and H_2SO_3
       (c) $Ba(OH)_2$ and HNO_2 (d) NH_3 and HNO_3

13-41 (a) $2HBr + Ca(OH)_2 \longrightarrow CaBr_2 + 2H_2O$

(b) $2HClO_2 + Sr(OH)_2 \longrightarrow Sr(ClO_2)_2 + 2H_2O$

(c) $2H_2S + Ba(OH)_2 \longrightarrow Ba(HS)_2 + 2H_2O$

(d) $H_2S + 2LiOH \longrightarrow Li_2S + 2H_2O$

13-43 $LiOH + H_2S \longrightarrow LiHS + H_2O$

$LiOH + LiHS \longrightarrow Li_2S + H_2O$

13-44 $OH^-(aq) + H_2S(aq) \longrightarrow HS^-(aq) + H_2O(l)$

$OH^-(aq) + HS^-(aq) \longrightarrow S^{2-}(aq) + H_2O(l)$

13-47 $H_2S + NaOH \longrightarrow NaHS + H_2O$

13-50 The system would not be at equilibrium if $[H_3O^+] = [OH^-] = 10^{-2}$ M.

Therefore, H_3O^+ reacts with OH^- until the concentration of each is reduced to 10^{-7} M. This is a neutralization reaction.

i.e., $H_3O^+ + OH^- \longrightarrow 2H_2O$

13-51 (a) $[H_3O^+] = \dfrac{K_w}{[OH^-]} = \dfrac{10^{-14}}{10^{-12}} = \underline{10^{-2} \text{ M}}$ (b) $[H_3O^+] = \dfrac{10^{-14}}{10} = \underline{10^{-15} \text{ M}}$

(c) $[OH^-] = \dfrac{K_w}{[H_3O^+]} = \dfrac{1.0 \times 10^{-14}}{2.0 \times 10^{-5}} = \underline{5.0 \times 10^{-10} \text{ M}}$

13-53 $HClO_4 + H_2O \longrightarrow H_3O^+ + ClO_4^-$ (100% to the right)

Since all $HClO_4$ ionizes, $[H_3O^+] = 0.250$ mol/10.0 L $= \underline{0.0250 \text{ M}}$

$[OH^-] = \dfrac{K_w}{[H_3O^+]} = \dfrac{1.0 \times 10^{-14}}{2.50 \times 10^{-2}} = \underline{4.0 \times 10^{-13} \text{ M}}$

13-55 (a) acidic   (b) basic   (c) acidic

13-57 (a) acidic   (b) basic   (c) acidic   (d) basic

13-59 (a) $-\log(1.0 \times 10^{-6}) = \underline{6.00}$ (b) $-\log(1.0 \times 10^{-9}) = \underline{9.00}$
(c) pOH $= -\log(1.0 \times 10^{-2}) = 2.00$ pH $= 14.00 -$ pOH $= \underline{12.00}$
(d) pOH $= -\log(2.5 \times 10^{-5}) = 4.60$ pH $= 14.00 - 4.60 = \underline{9.40}$
(e) pH $= -\log(6.5 \times 10^{-11}) = \underline{10.19}$

13-61 (a) $0.0001 = 10^{-4}$ $\underline{\text{pH} = 4.0}$, $\underline{\text{pOH} = 10.0}$ (b) $0.00001 = 10^{-5}$ $\underline{\text{pOH} = 5.0}$, $\underline{\text{pH} = 9.0}$
(c) $0.020 = 2.0 \times 10^{-2}$ $\underline{\text{pH} = 1.70}$, $\underline{\text{pOH} = 12.30}$
(d) $0.000320 = 3.20 \times 10^{-4}$ $\underline{\text{pOH} = 3.495}$, $\underline{\text{pH} = 10.505}$

13-63 (a) $-\log[H_3O^+] = 3.00$ $[H_3O^+] = \underline{1.0 \times 10^{-3} \text{ M}}$
(b) $-\log[H_3O^+] = 3.54$ $[H_3O^+] =$ antilog (-3.54) $[H_3O^+] = \underline{2.9 \times 10^{-4} \text{ M}}$
(c) $-\log[OH^-] = 8.00$ $[OH^-] = 1.0 \times 10^{-8}$ M $[H_3O^+] = \underline{1.0 \times 10^{-6} \text{ M}}$
(d) pH $= 14.00 - 6.38 = 7.62$ $-\log[H_3O^+] = 7.62$ $[H_3O^+] = \underline{2.4 \times 10^{-8} \text{ M}}$
(e) $-\log[H_3O^+] = 12.70$ $[H_3O^+] = \underline{2.0 \times 10^{-13} \text{ M}}$

13-65 For Problem 13-59: (a) acidic   (b) basic   (c) basic   (d) basic   (e) basic
For Problem 13-63: (a) acidic   (b) acidic   (c) acidic   (d) basic   (e) basic

13-67 If pH $= 3.0$, $[H_3O^+] = 10^{-3}$ $\dfrac{10^{-3}}{100} = 10^{-5}$ $\underline{\text{pH} = 5.0}$ (less acidic)

$10^{-3} \times 10 = 10^{-2}$ $\underline{\text{pH} = 2.0}$ (more acidic)

13-69 $HNO_3 + H_2O \longrightarrow H_3O^+ + NO_3^-$ (100% to the right)

Since this is a strong acid $[H_3O^+] = 0.075$ M $-\log(7.5 \times 10^{-2}) = \underline{1.12}$

13-71 $Ca(OH)_2$ is a strong base producing 2 mol of OH^- per mole of $Ca(OH)_2$.

$[OH^-] = 2 \times 0.018 = 0.036 = 3.6 \times 10^{-2}$ M

$pOH = -\log(3.6 \times 10^{-2}) = 1.44$ $pH = 14.00 - 1.44 = \underline{12.56}$

13-72 $HX \rightleftharpoons H_3O^+ + X^-$ $[H_3O^+] = 10\%$ of $[HX]$

$[H_3O^+] = 0.100 \times 0.10 = 0.010 = 1.0 \times 10^{-2}$ M $-\log(1.0 \times 10^{-2}) = \underline{2.00}$

13-74 (a) $pH = 1.5$ strongly acidic   (b) $pOH = 13.0$ strongly acidic
(c) $pH = 5.8$ weakly acidic   (d) $pH = 13.0$ strongly basic
(e) $pOH = 7.0$ neutral   (f) $pH = 8.5$ weakly basic
(g) $pOH = 7.5$ weakly acidic   (h) $pH = -1.00$ strongly acidic

13-75 Ammonia ($pH = 11.4$), eggs ($pH = 7.8$), rain water ($[H_3O^+] = 2.0 \times 10^{-6}$ M, $pH = 5.70$), grape juice($[OH^-] = 1.0 \times 10^{-10}$ M, $pH = 4.0$), vinegar ($[H_3O^+] = 2.5 \times 10^{-3}$ M, $pH = 2.60$), sulfuric acid ($pOH = 13.6$, $pH = 0.4$)

13-78 (b) $C_2H_3O_2^- + H_2O \rightleftharpoons HC_2H_3O_2 + OH^-$

(c) $NH_3 + H_2O \rightleftharpoons NH_4^+ + OH^-$

13-80 (a) K^+ The cation of the strong base KOH.
(d) NO_3^- The conjugate base of the strong acid HNO_3.
(e) O_2 Dissolves in water without formation of ions.

13-81 (a) $S^{2-} + H_2O \rightleftharpoons HS^- + OH^-$

(b) $N_2H_5^+ + H_2O \rightleftharpoons N_2H_4 + H_3O^+$

(c) $HPO_4^{2-} + H_2O \rightleftharpoons H_2PO_4^- + OH^-$

(d) $(CH_3)_2NH_2^+ + H_2O \rightleftharpoons (CH_3)_2NH + H_3O^+$

13-83 (a) $F^- + H_2O \rightleftharpoons HF + OH^-$

(b) $SO_3^{2-} + H_2O \rightleftharpoons HSO_3^- + OH^-$

(c) $(CH_3)_2NH_2^+ + H_2O \rightleftharpoons (CH_3)_2NH + H_3O^+$

(d) $HPO_4^{2-} + H_2O \rightleftharpoons H_2PO_4^- + OH^-$

(e) $CN^- + H_2O \rightleftharpoons HCN + OH^-$
(f) No hydroylsis (cation of a strong base)

13-85 $Ca(ClO)_2 \rightarrow Ca^{2+}(aq) + 2ClO^-(aq)$

$ClO^-(aq) + H_2O \rightleftharpoons HClO(aq) + OH^-(aq)$

13-87 (a) neutral (neither cation nor anion hydrolysis)
(b) acidic (cation hydrolysis)

$N_2H_5^+ + H_2O \rightleftharpoons N_2H_4 + H_3O^+$

(c) basic (anion hydrolysis)

$C_2H_3O_2^- + H_2O \rightleftharpoons HC_2H_3O_2 + OH^-$

(d) neutral (neither cation nor anion hydrolysis)

(e) acidic (cation hydrolysis)

$NH_4^+ + H_2O \rightleftharpoons NH_3 + H_3O^+$

(f) basic (anion hydrolysis)

$F^- + H_2O \rightleftharpoons HF + OH^-$

13-89 $CaC_2(s) + 2H_2O(l) \rightarrow C_2H_2(g) + Ca^{2+}(aq) + 2OH^-(aq)$

13-90 Cation: $NH_4^+ + H_2O \rightleftharpoons NH_3 + H_3O^+$

Anion: $CN^- + H_2O \rightleftharpoons HCN + OH^-$

Since the solution is basic, the anion hydrolysis reaction must take place to a greater extent than the cation hydrolysis.

13-92 (a), (b), (e), (h), and (i) are buffer solutions.

13-93 There is no equilibrium when HCl dissolves in water. A reservoir of un-ionized acid must be present to react with any strong base that is added. Likewise, the Cl^- ion does not exhibit base behavior in water, so it cannot react with any H_3O^+ added to the solution.

13-94 $N_2H_4(aq) + H_2O \rightleftharpoons N_2H_5^+(aq) + OH^-(aq)$

Added H_3O^+: $H_3O^+ + N_2H_4 \rightarrow N_2H_5^+ + H_2O$

Added OH^-: $OH^- + N_2H_5^+ \rightarrow N_2H_4 + H_2O$

13-96 $HPO_4^{2-}(aq) + H_2O \rightleftharpoons PO_4^{3-}(aq) + H_3O^+(aq)$

Added H_3O^+: $H_3O^+ + PO_4^{3-} \rightarrow HPO_4^{2-} + H_2O$

Added OH^-: $OH^- + HPO_4^{2-} \rightarrow PO_4^{3-} + H_2O$

13-99 (a) $SrO + H_2O \rightarrow Sr(OH)_2$ (b) $SeO_3 + H_2O \rightarrow H_2SeO_4$

(c) $P_4O_{10} + 6H_2O \rightarrow 4H_3PO_4$ (d) $Cs_2O + H_2O \rightarrow 2CsOH$

(e) $N_2O_3 + H_2O \rightarrow 2HNO_2$ (f) $Cl_2O_5 + H_2O \rightarrow 2HClO_3$

13-101 $CO_2(g) + LiOH(aq) \rightarrow LiHCO_3(aq)$

13-102 $Li_2O(s) + N_2O_5(g) \rightarrow 2LiNO_3(s)$

13-103 $Fe(s) + 2HI \rightarrow FeI_2(aq) + H_2(g)$

13-105 $2HNO_3(aq) + Na_2SO_3(aq) \rightarrow 2NaNO_3(aq) + SO_2(g) + H_2O(l)$

13-107 (a) $HCN + NH_3 \rightarrow NH_4^+ + CN^-$

(b) $NH_3 + H^- \rightarrow H_2 + NH_2^-$

(c) $HCO_3^- + NH_3 \rightarrow NH_4^+ + CO_3^{2-}$

(d) $NH_4^+Cl^- + Na^+NH_2^- \longrightarrow Na^+Cl^- + 2NH_3$

13-108 (a) $HCl + CH_3OH \longrightarrow CH_3OH_2^+ + Cl^-$

(b) $CH_3OH + NH_2^- \longrightarrow NH_3 + CH_3O^-$

13-110 (a) Acidic: $H_2S + H_2O \rightleftharpoons H_3O^+ + HS^-$ weak acid

(b) Basic: $ClO^- + H_2O \rightleftharpoons HClO + OH^-$ anion hydro.

(c) Neutral: no hydr.

(d) Basic: $NH_3 + H_2O \rightleftharpoons NH_4^+ + OH^-$ weak base

(e) Acidic: $N_2H_5^+ + H_2O \rightleftharpoons N_2H_4 + H_3O^+$ cation hydro.

(f) Basic: $Ba(OH)_2 \longrightarrow Ba^{2+} + 2OH^-$ strong base

(g) Neutral: no hydr.

(h) Basic: $NO_2^- + H_2O \rightleftharpoons HNO_2 + OH^-$ anion hydro.

(i) Acidic: $H_2SO_3 + H_2O \rightleftharpoons H_3O^+ + HSO_3^-$ weak acid

(j) Acidic: $Cl_2O_3 + H_2O \longrightarrow HClO_2$ acid anhydride

$HClO_2 + H_2O \rightleftharpoons H_3O^+ + ClO_2^-$ weak acid

13-112 LiOH: strongly basic, pH = 13.0

$SrBr_2$: neutral, pH = 7.0 (no hydrolysis)

KClO: weakly basic, pH = 10.2 (anion hydrolysis)

NH_4Cl: weakly acidic, pH = 5.2 (cation hydrolysis)

HI: strongly acidic, pH = 1.0

When pH = 1.0, $[H_3O^+]$ = 0.10 M. If HI is completely ionized, its initial concentration must be 0.10 M.

13-114 $\textbf{kg coal} \longrightarrow \textbf{kg FeS}_2 \longrightarrow \textbf{g FeS}_2 \longrightarrow \textbf{mol FeS}_2 \longrightarrow$
$\textbf{mol SO}_2 \longrightarrow \textbf{mol SO}_3 \longrightarrow \textbf{mol H}_2\textbf{SO}_4 \longrightarrow \textbf{g H}_2\textbf{SO}_4$

$$100 \text{ kg coal} \times \frac{5.0 \text{ kg FeS}_2}{100 \text{ kg coal}} \times \frac{10^3 \text{ g FeS}_2}{\text{kg FeS}_2} \times \frac{1 \text{ mol FeS}_2}{120.0 \text{ g FeS}_2} \times \frac{8 \text{ mol SO}_2}{4 \text{ mol FeS}_2}$$

$$\times \frac{2 \text{ mol SO}_3}{2 \text{ mol SO}_2} \times \frac{1 \text{ mol H}_2\text{SO}_4}{1 \text{ mol SO}_3} \times \frac{98.09 \text{ g H}_2\text{SO}_4}{\text{mol H}_2\text{SO}_4} = 8200 \text{ g} = \underline{8.2 \text{ kg of H}_2\text{SO}_4}$$

13-115 2.50 g HCl x $\dfrac{1 \text{ mol HCl}}{36.46 \text{ g HCl}}$ = 0.0686 mol HCl

$[H_3O^+] = \dfrac{0.0686 \text{ mol}}{0.245 \text{ L}}$ = 0.280 mol/L   pH = -log(0.280) = <u>0.553 (con)</u>

$[H_3O^+] = \dfrac{0.0686 \text{ mol}}{0.890 \text{ L}}$ = 0.0771 mol/L   pH = -log(0.0771) = <u>1.113 (dilute)</u>

13-117 10.0 ~~g HCl~~ x $\dfrac{1 \text{ mol HCl}}{36.46 \text{ g HCl}}$ = 0.274 mol HCl

10.0 ~~g NaOH~~ x $\dfrac{1 \text{ mol NaOH}}{40.00 \text{ g NaOH}}$ = 0.250 mol NaOH

HCl + NaOH $\longrightarrow$ NaCl + H$_2$O   (1 mol HCl reacts with 1 mol NaOH)
0.274 mol HCl – 0.250 mol NaOH = 0.024 mol HCl remaining

[HCl] = [H$_3$O$^+$] = 0.024 mol/1.00 L   <u>pH = 1.62</u>

13-119 2HNO$_3$ + Ca(OH)$_2$ $\longrightarrow$ Ca(NO$_3$)$_2$ + 2H$_2$O

0.500 ~~L~~ x 0.10 mol/~~L~~ = 0.050 mol HNO$_3$ = 0.0500 mol [H$_3$O$^+$]
0.500 ~~L~~ x 0.10 mol/~~L~~ = 0.050 mol Ca(OH)$_2$

0.050 ~~mol HNO$_3$~~ x $\dfrac{1 \text{ mol Ca(OH)}_2}{2 \text{ mol HNO}_3}$ = 0.025 mol Ca(OH)$_2$ reacts

0.050 - 0.025 = 0.025 mol Ca(OH)$_2$ remaining in 1.00 L

[OH$^-$] = 0.025 ~~mol Ca(OH)$_2$~~ x $\dfrac{2 \text{ mol OH}^-}{\text{mol Ca(OH)}_2}$ = 0.050 mol/L

pOH = 1.30   <u>pH = 12.70</u>

14

Oxidation-Reduction Reactions

Review of Part A *Redox Reactions – The Exchange of Electrons*

OBJECTIVES AND DETAILED TABLE OF CONTENTS

14-1 The Nature of Oxidation and Reduction and Oxidation States

OBJECTIVE *Using oxidation states, determine the species oxidized, the species reduced, the oxidizing agent, and the reducing agent in an electron exchange reaction.*

14-1.1 Half-Reactions
14-1.2 Redox Reactions
14-1.3 Oxidation States
14-1.4 Using Oxidation States in Redox Reactions

14-2 Balancing Redox Equations: Oxidation State Method

OBJECTIVE *Balance redox reactions by the oxidation state (bridge) method.*

14-3 Balancing Redox Equations: Ion-Electron Method

OBJECTIVE *Balance redox reactions by the ion-electron (half-reaction) method in both acidic and basic media.*

> 14-3.1 Balancing Reactions in Acidic Solution
> 14-3.2 Balancing Reactions in Basic Solution

SUMMARY OF PART A

Another important class of chemical reactions is known as **oxidation-reduction** reactions or simply **redox** reactions. In this type of reaction (unlike the double-replacement reactions studied in the previous chapter) electrons are exchanged between two reactants. Electrons are assigned to atoms in a molecule by assuming all atoms in the molecule are present as ions. This is known as the **oxidation state** (or oxidation number) of the element in a compound. It is an electron bookkeeping method similar to formal charge, which was introduced in Chapter 9. Two examples of the calculation of oxidation states follow. Redox reactions involve changes in oxidation states of at least two elements in two different reactants. One atom undergoes an increase in oxidation state, which means it has lost electrons in going from reactant to product. An atom in another reactant does the opposite: it undergoes a decrease in oxidation state, which means that it has gained electrons in going from reactant to product.

The molecule or ion containing the atom that loses electrons is said to undergo **oxidation** and is also known as the **reducing agent**. The molecule or ion containing the atom that gains electrons is said to undergo **reduction** and is also known as the **oxidizing agent**.

Example A-1 Calculation of Oxidation States

What is the oxidation state of (a) P in H_3PO_3 and (b) the S in $K_2S_2O_6$?

   (a)   **PROCEDURE**

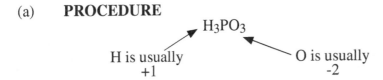

H is usually +1   O is usually -2

       SOLUTION

       Since all oxidation states in a neutral compound add to zero

$$3(H) + P + 3(O) = 0$$
$$3(+1) + P + 3(-2) = 0$$

$$P = \underline{+3}$$

(b)   **PROCEDURE**

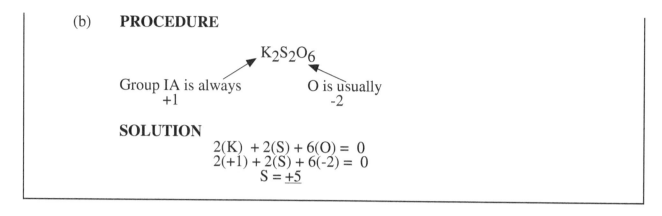

$K_2S_2O_6$

Group IA is always   O is usually
        +1   -2

SOLUTION

$$2(K) + 2(S) + 6(O) = 0$$
$$2(+1) + 2(S) + 6(-2) = 0$$
$$S = \underline{+5}$$

In order to identify oxidizing and reducing agents in a reaction containing several reactants and products, it is necessary to calculate oxidation states. For example, calculation of oxidation states of the atoms of all of the elements in the following equation allows us to identify the atoms that undergo a change.

$KMnO_4$ is reduced (the oxidizing agent)

$$\overset{+1\ \underline{+7}\ -2}{KMnO_4} + \overset{+1\ +6\ -2}{H_2SO_4} + \overset{\underline{-3}\ +1}{PH_3} \rightarrow \overset{+1\ +6\ -2}{K_2SO_4} + \overset{\underline{+2}\ +6\ -2}{MnSO_4} + \overset{+1\ -2}{H_2O} + \overset{+1\ \underline{+5}\ -2}{H_3PO_4}$$

PH_3 is oxidized (the reducing agent)

Balancing redox reactions such as the one illustrated by the previous unbalanced equation can be extremely tedious if done by inspection. Two methods are introduced in this text to balance equations. Both methods are based on the principle that "electrons gained (by the oxidizing agent) equal electrons lost (by the reducing agent)." In the **oxidation state** or **bridge method** one focuses only on the elements that change. Coefficients are introduced so as to equalize the electron exchange.

Example A-2 Balancing an Equation by the Oxidation State Method

Balance the following equation.

$$MnO_2 + PbO_2 + HNO_3 \rightarrow Pb(NO_3)_2 + HMnO_4 + H_2O$$

PROCEDURE

(a)   The two atoms that change are Mn and Pb. The Mn is oxidized from the +4 state in MnO_2 to +7 in $HMnO_4$. The Pb is reduced from the +4 state in PbO_2 to the +2 state in $Pb(NO_3)_2$.

(b)   Since MnO_2 gains three electrons and PbO_2 loses two, the MnO_2 and its product are multiplied by two and the PbO_2 and its product are multiplied by three so that exactly six electrons are exchanged.

275

(c)   Since there are six NO_3^- ions on the right, there must be six HNO_3 molecules on the left. Since there are now six Hs on the left, there must be two H_2Os on the right to balance the hydrogens. Notice that a check of the Os indicates that the equation is balanced.

SOLUTION

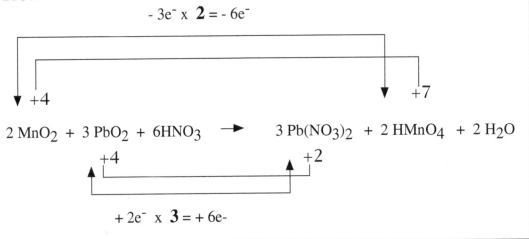

In the **ion-electron method** the focus is on the entire oxidizing or reducing agent, not just the one element that changes. It is necessary to recognize only the species that change not the actual oxidation states. In this method, the total reaction is divided into two **half-reactions** that are balanced separately. The following example refers to a reaction in acid solution.

The procedure in basic solution is modified to emphasize OH^- instead of H^+ as follows: (1) First, balance the half-reaction as in acid solution. (2) Next, convert each H^+ to H_2O by adding an OH^- ion to both sides of the equation. (The H^+ converts to H_2O by means of the neutralization reaction $H^+ + OH^- \rightarrow H_2O$.) (3) Simplify if necessary by canceling any H_2Os on both sides of the equation.

Example A-3 Balancing an Equation by the Ion-Electron Method

Balance the following by the ion-electron method.

$$MnO_4^- + NO_2 + H_2O \rightarrow Mn^{2+} + NO_3^- + H^+$$

PROCEDURE

(a)   The presence of H^+ indicates an acid solution. Notice that the MnO_4^- ion and the NO_2 molecule undergo changes in oxidation states. Balance each half-reaction separately. In each half-reaction balance the atom that changes first, oxygens second by adding H_2O in acidic solution. Next, balance the hydrogens by adding

H^+. The last step, and most important, is to balance the charge by adding electrons to the more positive side of the equation. To do this, calculate the charge on each side by multiplying the coefficient of the ion by its charge. Neutral molecules do not contribute to the charge.

(b)   The oxidation reaction is multiplied by five and then added to the reduction reaction so that each process involves five electrons. When the equations are added, the five electrons subtract out as do 4 H_2Os and 8 H^+s.

SOLUTION

(a)
$$5e^- + 8H^+ + MnO_4^- \rightarrow Mn^{2+} + 4H_2O$$
$$H_2O + NO_2 \rightarrow NO_3^- + 2H^+ + e^-$$

(b)
$$5H_2O + 5NO_2 \rightarrow 5NO_3^- + 10H^+ + 5e^-$$

Adding the oxidation and the reduction reaction, we have
$$5e^- + 8H^+ + MnO_4^- + 5H_2O + 5NO_2 \rightarrow Mn^{2+} + 4H_2O + 5NO_3^- + 10H^+ + 5e^-$$

The final balanced net ionic equation is
$$MnO_4^- + 5NO_2 + H_2O \rightarrow Mn^{2+} + 5NO_3^- + 2H^+$$

ASSESSMENT OF OBJECTIVES

A-1 Multiple Choice

_____ 1. What is the oxidation state of the Br in KBrO?

(a) +5   (b) +3   (c) -7   (d) -6   (e) none of these

_____ 2. What is the oxidation state of the Co in $Co_2(SO_4)_3$?

(a) +2   (b) +6   (c) -4   (d) +3   (e) none of these

_____ 3. Which of the following is an oxidation-reduction reaction?

(a) $K + O_2 \rightarrow KO_2$

(b) $CaO + CO_2 \rightarrow CaCO_3$

(c) $KCl + AgNO_3 \rightarrow AgCl(s) + KNO_3$

(d) $HBr + NaOH \rightarrow NaBr + H_2O$

(e) $H_2CO_3 \rightarrow H_2O + CO_2$

_____ 4. Which of the following unbalanced equations represents an impossible reaction?

(a) $Zn + 2H^+ \rightarrow Zn^{2+} + H_2$

(b) $Bi + HNO_3 \rightarrow Bi(NO_3)_3 + H_2O$

(c) $Zn^{2+} + Ni^{2+} \rightarrow Zn + Ni$

(d) $NaI + Br_2 \rightarrow NaBr + I_2$

(e) $Mg + N_2 \rightarrow Mg_3N_2$

_____ 5. In the following unbalanced equation, what is the reducing agent?

$$Fe^{2+} + H^+ + MnO_4^- \longrightarrow Fe^{3+} + Mn^{2+} + H_2O$$

(a) Fe^{2+} (b) Fe^{3+} (c) Mn^{2+} (d) H^+ (e) MnO_4^-

_____ 6. In the following equation, how many electrons have been gained $(+e^-)$ or lost $(-e^-)$ by one molecule of O_3?

$$6H^+ + O_3 \longrightarrow 3H_2O$$

(a) $+ 2e^-$ (b) $- 2e^-$ (c) $+ 6e^-$ (d) $+ 4e^-$ (e) $- 6e^-$

A-2 Problems

1. Give the oxidation states of the following:

(a) the Mn in $Ca(MnO_4)_2$ _____

(b) the Mn in K_2MnO_4 _____

(c) the Sn in $Sn(SO_4)_2$ _____

(d) the O in H_2O_2 _____

(e) the C in CH_4O _____

(f) the Cr in CrC_2O_4 _____

2. Fill in the blanks concerning the following unbalanced equations.

(a) $Cu + HNO_3 \longrightarrow Cu^{2+} + NO + H_2O$

(b) $H_2S + KMnO_4 + HCl \longrightarrow S + KCl + MnCl_2 + H_2O$

(c) $Bi(OH)_3 + K_2Sn(OH)_4 \longrightarrow Bi + K_2Sn(OH)$

reaction	(a)	(b)	(c)
species oxidized	_____	_____	_____
electrons lost	_____	_____	_____
species reduced	_____	_____	_____
electrons gained*	_____	_____	_____
oxidizing agent	_____	_____	_____
reducing agent	_____	_____	_____

*per atom or formula unit

278

3. Balance the following equations by the oxidation state (bridge) method.

    (a) $Cu + HNO_3 \longrightarrow Cu(NO_3)_2 + NO + H_2O$

    (b) $H_2S + KMnO_4 + HCl \longrightarrow S + KCl + MnCl_2 + H_2O$

    (c) $Bi(OH)_3 + K_2Sn(OH)_4 \longrightarrow Bi + K_2Sn(OH)_6$

4. Balance the following half-reactions by the ion-electron method.

    (a) $S_2O_3^{2-} \longrightarrow SO_4^{2-}$ (acid)

    (b) $CrO_2^- \longrightarrow CrO_4^{2-}$ (base)

    (c) $ClO_3^- \longrightarrow Cl^-$ (acid)

    (d) $H_2O_2 \longrightarrow O_2$ (acid)

    (e) $O_2^{2-} \longrightarrow OH^-$ (base)

    (f) $As \longrightarrow AsO_4^{3-}$ (acid)

    (g) $CN^- \longrightarrow CNO^-$ (base)

    (h) $PH_3 \longrightarrow H_3PO_4$ (acid)

5. Balance the following redox equations by the ion-electron method.

    (a) $Fe^{2+} + H_2O_2 + H^+ \longrightarrow Fe^{3+} + H_2O$

    (b) $OH^- + Cl_2 + IO_3^- \longrightarrow IO_4^- + Cl^- + H_2O$

(c) $I_2 + H^+ + NO_3^- \longrightarrow IO_3^- + NO_2 + H_2O$

(d) $H_2O_2 + ClO_2 + OH^- \longrightarrow O_2 + ClO_2^- + H_2O$

(e) $Cr_2O_7^{2-} + H_2C_2O_4 + H^+ \longrightarrow Cr^{3+} + CO_2 + H_2O$

Review of Part B   *Spontaneous and Nonspontaneous Redox Reactions*

OBJECTIVES AND DETAILED TABLE OF CONTENTS

14-4 Predicting Spontaneous Redox Reactions

OBJECTIVE *Using a table of relative strengths of oxidizing agents, determine whether a specific redox reaction is spontaneous.*

14-5 Voltaic Cells

OBJECTIVE *Describe the structure and electricity generating ability of voltaic cells and batteries.*

14-6 Electrolytic Cells

OBJECTIVE *Write the reactions that occur when a salt is electrolyzed.*

SUMMARY OF PART B

What determines the direction of a spontaneous reaction? Recall from the previous chapter how we found that a stronger acid reacts with a stronger base to produce a weaker acid and base (in the Bronsted-Lowry concept). A similar situation exists for redox reactions. That is, the stronger oxidizing agent reacts with the stronger reducing agent to produce weaker agents. For example, in the following reaction, (b) is the reverse of (a).

(a) $Zn + Cu^{2+} \longrightarrow Zn^{2+} + Cu$

(b) $Cu + Zn^{2+} \longrightarrow Cu^{2+} + Zn$

How do we determine which is the spontaneous reaction and which is the nonspontaneous reaction? The answer lies in the observations of a laboratory experiment. Since Zn forms a coating of Cu when immersed in Cu^{2+} solution, we can conclude that reaction (a) is spontaneous. Thus we conclude that Cu^{2+} is a stronger oxidizing agent than Zn^{2+}. Since there is an inverse relationship between the strength of an oxidizing agent (e.g., Cu^{2+}) and the reducing strength of its product (e.g., Cu), we can also say that Zn is a stronger reducing agent than Cu. With even more experiments, a table—such as Table 14-1 in the text—can be constructed with the strongest oxidizing agent at the top. Because of the inverse relationship, the species on the right are listed with the strongest reducing agent at the bottom. A spontaneous chemical reaction is predicted between an oxidizing agent and any reducing agent lying below it in the table. For example, the following reaction is predicted to be spontaneous.

$$Br_2 + Ni \longrightarrow Ni^{2+} + 2\,Br^-$$

Table 14-1 also implies that water can be either oxidized or reduced. Certain nonmetals (e.g., F_2 and Cl_2) oxidize water to form an acid solution and oxygen. On the other hand, water oxidizes (or, is reduced by) several metals (e.g., Na and Mg) to form a hydroxide solution and hydrogen. Other metals (e.g., Fe and Ni) are not affected by water but are oxidized by strong acid (H^+) solutions. Some metals (e.g., Cu and Ag) are not affected by either water or most strong acid solutions. (Nitric acid oxidizes copper and silver due to the oxidizing action of the nitrate ion.)

In a spontaneous reaction, the products have lower potential energy than the reactants. (In most cases, anyway.) This difference in potential energy is what accounts for the heat, light, and electricity that are liberated by the reaction. In order for the energy to be evolved as electrical energy, we need a redox reaction that can be physically separated into its two half-reactions. The electron lost in the oxidation process can then be made to travel in an external circuit (e.g., a wire) where it eventually ends up in the reduction process. As the electrons are pushed through the wire they can be made to do work such as make a light glow, a car start, or a stereo blast. A **voltaic cell** converts chemical energy directly into electrical energy. Surfaces where half-reactions take place are called **electrodes**. The electrode where oxidation occurs is called the **anode**, and the electrode where reduction occurs is called the **cathode**. One or more voltaic cells arranged in a series are known collectively as a **battery**. Examples of important batteries are the lead-acid battery used in the automobile, the dry cell, the Daniell cell, and the fuel cell. The half-reactions and the total reaction in the lead-acid battery are as follows.

anode: $\quad Pb(s) + H_2SO_4(aq) \longrightarrow PbSO_4(s) + 2H^+(aq) + 2e^-$

cathode: $\quad PbO_2(s) + 2H^+(aq) + H_2SO_4(aq) + 2e^- \longrightarrow PbSO_4(s) + 2H_2O$

total: $\quad Pb(s) + PbO_2(s) + 2H_2SO_4(aq) \longrightarrow 2PbSO_4(aq) + 2H_2O$

Some chemical reactions can be forced in the nonspontaneous direction if energy is supplied from an outside source. For example, the lead-acid battery in the automobile is especially useful because it can be forced in the reverse direction if electrical energy is supplied. After the engine starts, it supplies the energy that regenerates the battery. When electrical energy is thus converted to chemical energy, the cell is known as an **electrolytic cell**. Electrolytic cells are useful in processes such as **electrolysis** and electroplating. Also, large amounts of several elements such as aluminum and chlorine are freed from their compounds by electrolytic means.

ASSESSMENT OF OBJECTIVES

B-1 Multiple Choice

_____ 1. A strong oxidizing agent

(a) is also a strong reducing agent.
(b) produces a strong reducing agent.
(c) produces a weak reducing agent.
(d) cannot be easily reduced.

_____ 2. Which of the following elements does not react with pure water?

(a) Cl_2 (b) Na   (c) Mg   (d) Br_2

_____ 3. Which of the following metals does not react with pure water but is oxidized by strong acid solution?

(a) Zn   (b) Cu   (c) Ag   (d) Al

_____ 4. A voltaic cell

(a) converts chemical energy to electrical energy.
(b) proceeds in a nonspontaneous direction.
(c) does not utilize a redox reaction.
(d) is used to silverplate base metals.

_____ 5. In the following voltaic cell, what reaction occurs at the anode?

$$Cu^{2+} + Fe \rightarrow Fe^{2+} + Cu$$

(a) Fe^{2+} is oxidized.   (b) Cu^{2+} is reduced.
(c) Fe is reduced.   (d) Fe is oxidized.   (e) Cu is oxidized.

_____ 6. When molten $MgCl_2$ is electrolyzed, which of the following occurs?

(a) H_2 is produced at the cathode.

(b) Mg is produced at the cathode.
(c) Cl_2 is produced at the cathode.

(d) Mg is produced at the anode.
(e) O_2 is produced at the cathode.

B-2 Problems

1. A partial list from Table 14-1 is shown below. The strongest oxidizing agent (on the left) is at the top and the strongest reducing agent (on the right) is at the bottom.

$$Br_2 + 2e^- \rightleftharpoons 2Br^-$$

$$Cu^{2+} + 2e^- \rightleftharpoons Cu$$

$$Fe^{2+} + 2e^- \rightleftharpoons Fe$$

$$Mg^{2+} + 2e^- \rightleftharpoons Mg$$

(a) Write a spontaneous reaction involving Cu and Br.

(b) A piece of copper (Cu) is placed in a solution of Fe^{2+} ion. Write the spontaneous reaction, if any.

(c) A piece of iron (Fe) is placed in a solution of Cu^{2+} ion. Write the spontaneous reaction, if any.

(d) A piece of iron is placed in a solution of Br^- ion. Write the spontaneous reaction, if any.

(e) Lead metal (Pb) forms a coating of Cu when placed in a Cu^{2+} solution but does not form a coating of Fe in a Fe^{2+} solution. Where does Pb^{2+} rank in the table as an oxidizing agent?

(f) Write the spontaneous reaction between the strongest oxidizing agent and the strongest reducing agent from the partial table.

2. Write balanced equations representing:

(a) the oxidation of water by F_2

(b) the reduction of water by Al

(c) the reduction of an H^+ solution by Fe

Chapter Summary Assessment

S-1 Problems

1. The species below are arranged in a sequence such as in Table 14-1 in the text.

strongest oxidizing agent $\longrightarrow$ ClO_3^- $\rightleftarrows$ Cl_2

MnO_2 $\rightleftarrows$ Mn^{2+}

N_2O_4 $\rightleftarrows$ NO $\longleftarrow$ strongest reducing agent

(a) Balance all three of the preceding half-reactions (acid solution).

(b) Write the balanced equation representing the spontaneous reaction involving the $N_2O_4 - NO$ and the $MnO_2 - Mn^{2+}$ half-reactions.

(c) In the reaction from part (b), indicate the following:

species oxidized _____   oxidizing agent _____

species reduced _____   reducing agent _____

(d) Write balanced equations representing two other spontaneous reactions.

Answers to Assessments of Objectives

A-1 Multiple Choice

1. **e**   2. **d**   3. **a**   4. **c**   5. **a**   6. **c**

A-2 Problems

1.   (a) Mn +7   (b) Mn +6   (c) Sn +4   (d) O -1
      (e) C –2   (f) Cr +2

2.

reaction	a	b	c
species oxidized	Cu	H_2S	$K_2Sn(OH)_4$
electrons lost	2	2	2
species reduced	HNO	$KMnO_4$	$Bi(OH)_3$
electrons gained	3	5	3
oxidizing agent	HNO_3	$KMnO_4$	$Bi(OH)_3$
reducing agent	Cu	H_2S	$K_2Sn(OH)_4$

3. (a)

$$
\overset{\displaystyle -2e^- \times 3 = -6e^-}{\underset{\displaystyle +5 \qquad\qquad\qquad\qquad +2}{\underset{\displaystyle +3e^- \times 2 = +6\,e^-}{\overset{0 \qquad\qquad\qquad +2}{3Cu + 8HNO_3 \longrightarrow 3Cu(NO_3)_2 + 2NO + 4H_2O}}}}
$$

In this equation, six HNO_3 molecules are needed in addition to the two that are reduced. These provide NO_3^- ions for the Cu^{2+}.

(b)

$$-2e^- \times \mathbf{5} = -10\ e^-$$

$$\overset{-2}{5H_2S} + 2KMnO_4 + 6HCl \longrightarrow 5S + 2KCl + \overset{+2}{2MnCl_2} + 8H_2O$$

$$\overset{+7}{}$$

$$+5e^- \times \mathbf{2} = +10\ e^-$$

After balancing the oxidized and reduced species, you can balance the Ks on the right. Then from the total Cls, balance the HCl on the left. Finally, from the total Hs on the left balance the H_2Os on the right.

(c)

$$-3e^- \times \mathbf{2} = -6\ e^-$$

$$\overset{+3}{2Bi(OH)_3} + \overset{+2}{3K_2Sn(OH)_4} \longrightarrow 2Bi + \overset{+4}{3K_2Sn(OH)_6}$$

$$+2e^- \times \mathbf{3} = +6\ e^-$$

4. (a) $5H_2O + S_2O_3^{2-} \longrightarrow 2SO_4^{2-} + 10H^+ + 8e^-$

(b) $2H_2O + CrO_2^- \longrightarrow CrO_4^{2-} + 4H^+ + 3e^-$ (acid)

  $\mathbf{4OH^-} + 2H_2O + CrO_2^- \longrightarrow CrO_4^{2-} + (4H^+ + \mathbf{4OH^-}) + 3e^-$

  $4OH^- + 2H_2O + CrO_2^- \longrightarrow CrO_4^{2-} + 4H_2O + 3e^-$

  $4OH^- + CrO_2^- \longrightarrow CrO_4^{2-} + 2H_2O + 3e^-$ (base)

(c) $6e^- + 6H^+ + ClO_3^- \longrightarrow Cl^- + 3H_2O$

(d) $H_2O_2 \longrightarrow O_2 + 2H^+ + 2e^-$

(e) $2e^- + 2H_2O + O_2^{2-} \longrightarrow 4OH^-$

(f) $4H_2O + As \longrightarrow AsO_4^{3-} + 8H^+ + 5e^-$

(g) $H_2O + CN^- \longrightarrow CNO^- + 2H^+ + 2e^-$ (acid)

  $\mathbf{2OH^-} + H_2O + CN^- \longrightarrow CNO^- + (2H^+ + \mathbf{2OH^-}) + 2e^-$

  $2OH^- + H_2O + CN^- \longrightarrow CNO^- + 2H_2O + 2e^-$

  $2OH^- + CN^- \longrightarrow CNO^- + H_2O + 2e^-$ (base)

(h) $4H_2O + PH_3 \longrightarrow H_3PO_4 + 8H^+ + 8e^-$

5. (a)
$$Fe^{2+} \rightarrow Fe^{3+} + e^- \quad\Big|\ \times 2$$
$$2e^- + 2H^+ + H_2O_2 \rightarrow 2H_2O \quad\Big|\ \times 1$$
$$2Fe^{2+} + 2H^+ + H_2O_2 \rightarrow 2Fe^{3+} + 2H_2O$$

(b)
$$2OH^- + IO_3^- \rightarrow IO_4^- + H_2O + 2e^- \quad\Big|\ \times 1$$
$$2e^- + Cl_2 \rightarrow 2Cl^- \quad\Big|\ \times 1$$
$$2OH^- + Cl_2 + IO_3^- \rightarrow IO_4^- + 2Cl^- + H_2O$$

(c)
$$6H_2O + I_2 \rightarrow 2IO_3^- + 12H^+ + 10e^- \quad\Big|\ \times 1$$
$$e^- + 2H^+ + NO_3^- \rightarrow NO_2 + H_2O \quad\Big|\ \times 10$$
$$6H_2O + I_2 + 20\ H^+ + 10NO_3^- \rightarrow 2IO_3^- + 12H^+ + 10NO_2 + 10H_2O$$
$$I_2 + 8H^+ + 10NO_3^- \rightarrow 2IO_3^- + 10NO_2 + 4H_2O$$

(d)
$$2OH^- + H_2O_2 \rightarrow O_2 + 2H_2O + 2e^- \quad\Big|\ \times 1$$
$$e^- + ClO_2 \rightarrow ClO_2^- \quad\Big|\ \times 2$$
$$H_2O_2 + 2ClO_2 + 2OH^- \rightarrow 2ClO_2^- + O_2 + 2H_2O$$

(e)
$$6e^- + 14H^+ + Cr_2O_7^{2-} \rightarrow 2Cr^{3+} + 7H_2O \quad\Big|\ \times 1$$
$$H_2C_2O_4 \rightarrow 2CO_2 + 2H^+ + 2e^- \quad\Big|\ \times 3$$
$$8H^+ + Cr_2O_7^{2-} + 3H_2C_2O_4 \rightarrow 2Cr^{3+} + 7H_2O + 6CO_2$$

B-1 Multiple Choice

1. **c** 2. **d** 3. **a** 4. **a** 5. **d** 6. **b**

B-2 Problems

1. (a) $Cu + Br_2 \rightarrow Cu^{2+} + 2Br^-$
   (b) No spontaneous reaction
   (c) $Fe + Cu^{2+} \rightarrow Fe^{2+} + Cu$
   (d) No spontaneous reaction
   (e) $Pb^{2+} + 2e^- \rightarrow Pb$ half-reaction belongs between Cu^{2+} and Fe^{2+}.
   (f) $Br_2 + Mg \rightarrow Mg^{2+} + 2Br^-$

2. (a) $2F_2(g) + 2H_2O \rightarrow 4HF(aq) + O_2(g)$
   (Since HF is a weak acid, it is written in molecular form rather than as H^+ and F^-.)
   (b) $2Al(s) + 6H_2O(l) \rightarrow 2Al(OH)_3(s) + 3H_2(g)$
   [$Al(OH)_3$ is insoluble in water.]
   (c) $Fe(s) + 2H^+(aq) \rightarrow Fe^{2+}(aq) + H_2(g)$

S-1 Problems

1. (a) $10e^- + 12H^+ + 2ClO_3^- \rightleftharpoons Cl_2 + 6H_2O$

$2e^- + 4H^+ + MnO_2 \rightleftharpoons Mn^{2+} + 2H_2O$

$4e^- + 4H^+ + N_2O_4 \rightleftharpoons 2NO + 2H_2O$

(b) $4H^+ + 2MnO_2 + 2NO \rightarrow N_2O_4 + 2Mn^{2+} + 2H_2O$

(c) NO is oxidized and is also the reducing agent.
MnO$_2$ is reduced and is also the oxidizing agent.

(d) $2ClO_3^- + 5Mn^{2+} + 4H_2O \rightarrow Cl_2 + 5MnO_2 + 8H^+$

$4H^+ + 4ClO_3^- + 10NO \rightarrow 2Cl_2 + 2H_2O + 5N_2O_4$

Answers and Solutions to Green Text Problems

14-1 (a) Pb +4, O -2   (b) P +5, O -2   (c) C -1, H +1
(d) N -2, H +1   (e) Li +1, H -1   (f) B +3, Cl -1
(g) Rb +1, Se -2   (h) Bi +3, S -2

14-3 (a) Li, (e) K, and (g) Rb

14-5 +3

14-6 (a) $3(+1) + P + 4(-2) = 0$ P = +5   (b) $2(+1) + 2C + 4(-2) = 0$ C = +3
(c) $Cl + 4(-2) = -1$ Cl = +7   (d) $(+2) + 2Cr + 7(-2) = 0$ Cr = +6
(e) $S + 6(-1) = 0$ S = +6 (f) $(+1) + N + 3(-2) = 0$ N = +5
(g) $(+1) + Mn + 4(-2) = 0$ Mn = +7

14-7 (a) SO$_3$, S(+6)   (b) Co$_2$O$_3$, Co(+3)   (c) UF$_6$, U(+6)
(d) HNO$_3$, N(+5)   (e) K$_2$CrO$_4$, Cr(+6)   (f) CaMnO$_4$, Mn(+6)

14-9 The following are redox reactions

(a) $2H_2 + O_2 \rightarrow 2H_2O$

(c) $2Na + 2H_2O \rightarrow 2NaOH + H_2$

(f) $Zn + CuCl_2 \rightarrow ZnCl_2 + Cu$

14-10 (a) $Na \rightarrow Na^+ + e^-$ Oxidation

(b) $Zn^{2+} + 2e^- \rightarrow Zn$ Reduction

(c) $Fe^{2+} \rightarrow Fe^{3+} + e^-$ Oxidation

(d) $O_2 + 4H^+ + 4e^- \rightarrow 2H_2O$ Reduction

(e) $S_2O_8^{2-} + 2e^- \rightarrow 2SO_4^{2-}$ Reduction

14-12

Species Oxidized	Oxid. Product	Species Reduced	Reduc. Product	Oxidizing Agent	Reducing Agent
Br^-	Br_2	MnO_2	Mn^{2+}	MnO_2	Br^-
CH_4	CO_2	O_2	CO_2, H_2O	O_2	CH_4
Fe^{2+}	Fe^{3+}	MnO_4^-	Mn^{2+}	MnO_4^-	Fe^{2+}

14-13

Species Oxidized	Oxid. Product	Species Reduced	Reduc. Product	Oxidizing Agent	Reducing Agent
Al	AlO_2^-	H_2O	H_2	H_2O	Al
Mn^{2+}	MnO_4^-	$Cr_2O_7^{2-}$	Cr^{3+}	$Cr_2O_7^{2-}$	Mn^{2+}

14-16

(a)
$$-5e^- \times 4 = -20e^-$$

$$\underset{-3}{4NH_3} + 5O_2 \longrightarrow \underset{+2}{4NO} + \underset{-2}{6H_2O}$$

$$0$$
$$+4e^- \times 5 = +20e^-$$

(b)
$$-4e^- \times 1 = -4e^-$$

$$\underset{0}{Sn} + 4HNO_3 \longrightarrow \underset{+4}{SnO_2} + \underset{+4}{4NO_2} + 2H_2O$$

$$+5$$
$$+1e^- \times 4 = +4e^-$$

(c) Before the number of electrons lost is calculated, notice that a temporary coefficient of "2" is needed for the Na_2CrO_4 in the products since there are 2 Crs in Cr_2O_3 in the reactants.

$$+2e^- \times 3 = +6e^-$$

$$\underset{+6}{Cr_2O_3} + 2Na_2CO_3 + \underset{+5}{3KNO_3} \longrightarrow 2CO_2 + \underset{+12}{2Na_2CrO_4} + \underset{+3}{3KNO_2}$$

$$-6e^- \times 1 = -6e^-$$

(d)
$$-4e^- \times 3 = -12e^-$$

$$\underset{0}{3Se} + \underset{+5}{2BrO_3^-} + 3H_2O \longrightarrow \underset{+4}{3H_2SeO_3} + \underset{-1}{2Br^-}$$

$$+6e^- \times 2 = +12e^-$$

14-18 (a) $2H_2O + Sn^{2+} \longrightarrow SnO_2 + 4H^+ + 2e^-$

(b) $2H_2O + CH_4 \longrightarrow CO_2 + 8H^+ + 8e^-$

(c) $e^- + Fe^{3+} \longrightarrow Fe^{2+}$

(d) $6H_2O + I_2 \longrightarrow 2IO_3^- + 12H^+ + 10e^-$

(e) $e^- + 2H^+ + NO_3^- \longrightarrow NO_2 + H_2O$

14-20 (a)

$$\left.\begin{array}{l} S^{2-} \longrightarrow S + 2e^- \\ 3e^- + 4H^+ + NO_3^- \longrightarrow NO + 2H_2O \end{array}\right| \begin{array}{l} \times\ 3 \\ \times\ 2 \end{array}$$

$$3S^{2-} + 8H^+ + 2NO_3^- \longrightarrow 3S + 2NO + 4H_2O$$

(b)

$$\left.\begin{array}{l} 2S_2O_3^{2-} \longrightarrow S_4O_6^{2-} + 2e^- \\ 2e^- + I_2 \longrightarrow 2I^- \end{array}\right| \begin{array}{l} \times\ 1 \\ \times\ 1 \end{array}$$

$$2S_2O_3^{2-} + I_2 \longrightarrow S_4O_6^{2-} + 2I^-$$

(c)

$$\left.\begin{array}{l} H_2O + SO_3^{2-} \longrightarrow SO_4^{2-} + 2H^+ + 2e^- \\ 6e^- + 6H^+ + ClO_3^- \longrightarrow Cl^- + 3H_2O \end{array}\right| \begin{array}{l} \times\ 3 \\ \times\ 1 \end{array}$$

$$3SO_3^{2-} + ClO_3^- \longrightarrow Cl^- + 3SO_4^{2-}$$

(d)

$$\left.\begin{array}{l} Fe^{2+} \longrightarrow Fe^{3+} + e^- \\ 2e^- + 2H^+ + H_2O_2 \longrightarrow 2H_2O \end{array}\right| \begin{array}{l} \times\ 2 \\ \times\ 1 \end{array}$$

$$2H^+ + 2Fe^{2+} + H_2O_2 \longrightarrow 2Fe^{3+} + 2H_2O$$

(e)

$$\left.\begin{array}{l} 2I^- \longrightarrow I_2 + 2e^- \\ 2e^- + 2H^+ + AsO_4^{3-} \longrightarrow AsO_3^{3-} + H_2O \end{array}\right| \begin{array}{l} \times\ 1 \\ \times\ 1 \end{array}$$

$$AsO_4^{3-} + 2I^- + 2H^+ \longrightarrow I_2 + AsO_3^{3-} + H_2O$$

(f)

$$\left.\begin{array}{l} Zn \longrightarrow Zn^{2+} + 2e^- \\ 8e^- + 10H^+ + NO_3^- \longrightarrow NH_4^+ + 3H_2O \end{array}\right| \begin{array}{l} \times\ 4 \\ \times\ 1 \end{array}$$

$$4Zn + NO_3^- + 10H^+ \longrightarrow 4Zn^{2+} + NH_4^+ + 3H_2O$$

14-22 (a) $2OH^- + SnO_2^{2-} \longrightarrow SnO_3^{2-} + H_2O + 2e^-$

(b) $6e^- + 4H_2O + 2ClO_2^- \longrightarrow Cl_2 + 8OH^-$

(c) $6OH^- + Si \longrightarrow SiO_3^{2-} + 3H_2O + 4e^-$

(d) $8e^- + 6H_2O + NO_3^- \longrightarrow NH_3 + 9OH^-$

14-24 (a)

$$\left.\begin{array}{l} 8OH^- + S^{2-} \longrightarrow SO_4^{2-} + 4H_2O + 8e^- \\ I_2 + 2e^- \longrightarrow 2I^- \end{array}\right| \begin{array}{l} \times\ 1 \\ \times\ 4 \end{array}$$

$$S^{2-} + 8OH^- + 4I_2 \longrightarrow SO_4^{2-} + 8I^- + 4H_2O$$

(b) $\quad 8OH^- + I^- \longrightarrow IO_4^- + 4H_2O + 8e^-$ | x 1

$\qquad e^- + MnO_4^- \longrightarrow MnO_4^{2-}$ | x 8

$\overline{8OH^- + I^- + 8MnO_4^- \longrightarrow 8MnO_4^{2-} + IO_4^- + 4H_2O}$

(c) $\quad 2OH^- + SnO_2^{2-} \longrightarrow SnO_3^{2-} + H_2O + 2e^-$ | x 1

$\qquad 2e^- + 3H_2O + BiO_3^- \longrightarrow Bi(OH)_3 + 3OH^-$ | x 1

$\overline{2H_2O + SnO_2^{2-} + BiO_3^- \longrightarrow SnO_3^{2-} + Bi(OH)_3 + OH^-}$

(d) $32OH^- + CrI_3 \longrightarrow CrO_4^{2-} + 3IO_4^- + 16H_2O + 27e^-$ | x 2

$\qquad 2e^- + Cl_2 \longrightarrow 2Cl^-$ | x 27

$\overline{2CrI_3 + 64OH^- + 27Cl_2 \longrightarrow 2CrO_4^{2-} + 6IO_4^- + 32\,H_2O + 54Cl^-}$

14-26 (a) $\qquad H_2 \longrightarrow 2H^+ + 2e^-$ | x 2

$\qquad 4e^- + 4H^+ + O_2 \longrightarrow 2H_2O$ | x 1

$\overline{\qquad 2H_2 + O_2 \longrightarrow 2H_2O}$

$\qquad 2OH^- + H_2 \longrightarrow 2H_2O + 2e^-$ | x 2

$\qquad 4e^- + 2H_2O + O_2 \longrightarrow 4OH^-$ | x 1

$\overline{\qquad 2H_2 + O_2 \longrightarrow 2H_2O}$

(b) $\qquad H_2O_2 \longrightarrow O_2 + 2H^+ + 2e^-$ | x 1

$\qquad 2e^- + 2H^+ + H_2O_2 \longrightarrow 2H_2O$ | x 1

$\overline{\qquad 2H_2O_2 \longrightarrow O_2 + 2H_2O}$

$\qquad 2OH^- + H_2O_2 \longrightarrow O_2 + 2H_2O + 2e^-$ | x 1

$\qquad 2e^- + H_2O_2 \longrightarrow 2OH^-$ | x 1

$\overline{\qquad 2H_2O_2 \longrightarrow O_2 + 2H_2O}$

14-27 The following reactions **are** predicted to occur based on the information in Table 14-1.

(a) $2Na + 2H_2O \longrightarrow H_2 + 2NaOH$

(c) $Fe + 2H^+ \longrightarrow Fe^{2+} + H_2$

(e) $Cu + 2Ag^+ \longrightarrow Cu^{2+} + 2Ag$

(f) $2Cl_2 + 2H_2O \longrightarrow 4Cl^- + O_2 + 4H^+$

Reactions (b), (d), and (g) do not occur in the directions written.

14-29 (a) no reaction

(b) $Br_2(aq) + Sn(s) \longrightarrow SnBr_2(aq)$
(c) no reaction

(d) $O_2(g) + 4H^+(aq) + 4Br^-(aq) \longrightarrow 2Br_2(l) + 2H_2O(l)$

(e) $3Br_2(l) + 2Al(s) \longrightarrow 2AlBr_3(s)$

14-31 (c) $2F_2 + 2H_2O \longrightarrow 4F^- + 4H^+ + O_2$

(e) $Mg + 2H_2O \longrightarrow Mg(OH)_2(s) + H_2$

14-33 From Table 14-1:

$Fe + 2H_2O \longrightarrow$ no reaction

$Fe + 2H^+ \longrightarrow Fe^{2+} + H_2$

In the first reaction, $[H_3O^+] = 10^{-7}$ M, in the second reaction $[H_3O^+] = 1.00$ M.

In acid rain, the H_3O^+ concentration is much greater than in pure water. This means that the second reaction may become spontaneous.

14-36 $Zn(s)$ reacts at the anode and $MnO_2(s)$ reacts at the cathode. The total reaction is:

$Zn(s) + 2MnO_2(s) + 2H_2O \longrightarrow Zn(OH)_2(s) + 2MnO(OH)$

14-38 anode: $Cd + 2OH^- \longrightarrow Cd(OH)_2 + 2e^-$

cathode: $NiO(OH) + H_2O + e^- \longrightarrow Ni(OH)_2 + OH^-$

14-40 The spontaneous reaction is:

$Pb(NO_3)_2(aq) + Fe(s) \longrightarrow Fe(NO_3)_2(aq) + Pb(s)$

anode: $Fe \longrightarrow Fe^{2+} + 2e^-$

cathode: $Pb^{2+} + 2e^- \longrightarrow Pb$

In the anode compartment a piece of Fe is immersed in a $Fe(NO_3)_2$ solution.
In the cathode compartment a piece of Pb is immersed in a $Pb(NO_3)_2$ solution.
The electrodes are connected to the motor (or light bulb, etc.) by wires and the two compartments are connected by a salt bridge.

14-41 The Daniell cell is more powerful. The greater the separation between oxidizing and reducing agent in the Table, the more powerful is the cell.

14-43 $3Fe(s) + 2Cr^{3+}(aq) \longrightarrow 3Fe^{2+}(aq) + 2Cr(s)$

Zinc would spontaneously form a chromium coating as illustrated by the following equation:

$3Zn(s) + 2Cr^{3+}(aq) \longrightarrow 3Zn^{2+}(aq) + 2Cr(s).$

14-47 The strongest oxidizing agent is reduced the easiest. Thus the reduction of Ag^+ to Ag occurs first. This procedure can be used to purify silver.

14-48 K_3N (-3), N_2H_4 (-2), NH_2OH, (-1), N_2 (0), N_2O (+1), NO (+2), N_2O_3 (+3), N_2O_4 (+4), $Ca(NO_3)_2$ (+5)

14-49 The following reactions occur:

$Cd + NiCl_2(aq) \longrightarrow Ni + CdCl_2(aq) \quad Zn + CdCl_2(aq) \longrightarrow Cd + ZnCl_2(aq)$

These reactions indicate that Cd^{2+} is a stronger oxidizing agent than Zn^{2+}, but weaker than Ni^{2+}. It appears about the same as Fe^{2+}.

14-51 The reactions that occur are

$4Au^{3+} + 6H_2O \longrightarrow 4Au + 12H^+ + 3O_2$

$2Au + 3F_2 \longrightarrow 2AuF_3$

These reactions and the fact that Au does not react with Cl_2 ranks Au^{3+} above H^+ and Cl_2 but below F_2.

14-52

$$Zn \longrightarrow Zn^{2+} + 2e^- \qquad\qquad\qquad | \; x \, 5$$
$$\underline{10e^- + 12H^+ + 2NO_3^- \longrightarrow N_2 + 6H_2O} \; | \; x \, 1$$
$$12H^+(aq) + 5Zn(s) + 2NO_3^-(aq) \longrightarrow 5Zn^{2+}(aq) + N_2(g) + 6H_2O(l)$$

$$\textbf{g N}_2 \longrightarrow \textbf{mol N}_2 \longrightarrow \textbf{mol Zn} \longrightarrow \textbf{g Zn}$$

$$0.658 \; \text{g N}_2 \times \frac{1 \; \text{mol N}_2}{28.02 \; \text{g N}_2} \times \frac{5 \; \text{mol Zn}}{1 \; \text{mol N}_2} \times \frac{65.39 \; \text{g}}{\text{mol Zn}} = \underline{7.68 \; \text{g Zn}}.$$

14-54

$$3e^- + 4H^+ + NO_3^- \longrightarrow NO + 2H_2O \qquad | \; x \, 2$$
$$\underline{2H^+ + Cu_2O \longrightarrow 2Cu^{2+} + H_2O + 2e^-} \; | \; x \, 3$$
$$14H^+(aq) + 2NO_3^-(aq) + 3Cu_2O(s) \longrightarrow 6Cu^{2+}(aq) + 2NO(g) + 7H_2O(l)$$

$$\textbf{g Cu}_2\textbf{O} \longrightarrow \textbf{mol Cu}_2\textbf{O} \longrightarrow \textbf{mol NO} \longrightarrow \textbf{vol. NO}$$

$$10.0 \; \text{g Cu}_2\text{O} \times \frac{1 \; \text{mol Cu}_2\text{O}}{143.1 \; \text{g Cu}_2\text{O}} \times \frac{2 \; \text{mol NO}}{3 \; \text{mol Cu}_2\text{O}} \times \frac{22.4 \; \text{L(STP)}}{\text{mol NO}} = \underline{1.04 \; \text{L of NO}} \; \text{(STP)}$$

14-56

$$4OH^- + Zn \longrightarrow Zn(OH)_4^{2-} + 2e^- \qquad | \; x \, 4$$
$$\underline{8e^- + 6H_2O + NO_3^- \longrightarrow NH_3 + 9OH^-} \; | \; x \, 1$$
$$7OH^-(aq) + 4Zn(s) + 6H_2O(l) + NO_3^-(aq) \longrightarrow NH_3(g) + 4Zn(OH)_4^{2-}(aq)$$

$$\textbf{g Zn} \longrightarrow \textbf{mol Zn} \longrightarrow \textbf{mol NH}_3 \longrightarrow \textbf{vol. NH}_3$$

$$6.54 \; \text{g Zn} \times \frac{1 \; \text{mol Zn}}{65.39 \; \text{g Zn}} \times \frac{1 \; \text{mol NH}_3}{4 \; \text{mol Zn}} = 0.0250 \; \text{mol NH}_3$$

$$V = \frac{nRT}{P} = \frac{0.0250 \text{ mol} \times 0.0821\frac{L \cdot atm}{K \cdot mol} \times 300 \text{ K}}{1.25 \text{ atm}} = \underline{0.493 \text{ L}}$$

14-57 (a) $MnO_4^- + 8H^+ + 5e^- \longrightarrow Mn^{2+} + 4H_2O$ | x 1

$Fe^{2+} \longrightarrow Fe^{3+} + e^-$ | x 5

$MnO_4^- + 8H^+ + 5Fe^{2+} \longrightarrow 5Fe^{3+} + Mn^{2+} + 4H_2O$

(b) $MnO_4^- + 8H^+ + 5e^- \longrightarrow Mn^{2+} + 4H_2O$ | x 2

$2Br^- \longrightarrow Br_2 + 2e^-$ | x 5

$2MnO_4^- + 16H^+ + 10Br^- \longrightarrow 5Br_2 + 2Mn^{2+} + 8H_2O$

(c) $MnO_4^- + 8H^+ + 5e^- \longrightarrow Mn^{2+} + 4H_2O$ | x 2

$C_2O_4^{2-} \longrightarrow 2CO_2 + 2e^-$ | x 5

$2MnO_4^- + 16H^+ + 5C_2O_4^{2-} \longrightarrow 10CO_2 + 2Mn^{2+} + 8H_2O$

14-58 **g FeCl$_2$** $\longrightarrow$ **mol FeCl$_2$ (mol Fe^{2+})** $\longrightarrow$ **mol MnO$_4^-$ (mol KMnO$_4$)**
$\longrightarrow$ **vol. KMnO$_4$**

$$25.0 \text{ g FeCl}_2 \times \frac{1 \text{ mol FeCl}_2}{126.8 \text{ g FeCl}_2} \times \frac{1 \text{ mol KMnO}_4}{5 \text{ mol FeCl}_2} = 0.0394 \text{ mol KMnO}_4$$

$$V = \frac{n}{M} = \frac{0.0394 \text{ mol}}{0.220 \text{ mol/L}} = 0.179 \text{ L} = \underline{179 \text{ mL}}$$

14-59 **vol. KMnO$_4$** $\longrightarrow$ **mol KMnO$_4$(MnO$_4^-$)** $\longrightarrow$ **mol Br$^-$(KBr)** $\longrightarrow$ **vol. KBr**

$$0.125 \text{ L} \times \frac{0.220 \text{ mol KMnO}_4}{L} \times \frac{10 \text{ mol KBr}}{2 \text{ mol KMnO}_4} \times \frac{1 \text{ L KBr}}{0.450 \text{ mol KBr/L}} = 0.306 \text{ L} = \underline{306 \text{ mL}}$$

15

Reaction Rates and Equilibrium

Review of Part A *Collisions of Molecules and Reactions at Equilibrium*

OBJECTIVES AND DETAILED TABLE OF CONTENTS

15-1 How Reactions Take Place

OBJECTIVE *Describe how a chemical reaction takes place on a molecular level by using an energy diagram.*

15-1.1 Collision Theory
15-1.2 Activation Energy
15-1.3 The Course of a Reaction
15-1.4 The Heat of a Reaction

15-2 Rates of Chemical Reactions

OBJECTIVE *List the factors that affect the rate of a chemical reaction.*

15-2.1 The Magnitude of the Activation Energy
15-2.2 The Temperature
15-2.3 The Concentrations of Reactants
15-2.4 The Effect of Particle Size
15-2.5 The Presence of a Catalyst

15-3 Equilibrium and LeChâtelier's Principle

OBJECTIVE *Using LeChâtelier's principle, predict what effect changing certain conditions will have on the equilibrium point*

15-3.1 Reversible Reactions
15-3.2 Systems at Equilibrium
15-3.3 Le Châtelier's Principle
15-3.4 Changing Concentrations
15-3.5 Changing Pressure
15-3.6 Changing Temperature
15-3.7 Adding a Catalyst

SUMMARY OF PART A

Collision theory tells us that reactants are transformed into products through collisions of reactant molecules or ions with each other. All collisions between reactants do not lead to products, however, since collisions must have the proper orientation and a minimum amount of energy known as the **activation energy**. This is the minimum kinetic energy that colliding molecules must have in order for old bonds to break so that new ones can form. The lower the activation energy, the faster the reaction. At maximum impact, colliding molecules form an **activated complex**, where the bonds rearrange to form products. The activation energy is equal to the difference in potential energy between reactants and the activated complex.

Chemical reactions take place at a wide range of rates. The chemical reaction in an explosion is almost instantaneous, whereas the reaction in the aging of a fine wine takes years. The rate at which a particular reaction takes place is affected by five factors. They are as follows:

1. *The activation energy.* This depends on the particular reaction.

2. *The temperature.* The temperature relates to the average kinetic energy of the molecules. Thus, the higher the temperature, the more molecules that will have the minimum kinetic energy (the activation energy) for a reaction to occur. The rate of a reaction is also proportional to the frequency of collisions between reacting molecules. Since a higher temperature increases the average velocity of the molecules, collisions become more frequent at higher temperatures, which also increases the rate.

3. *The concentration of reactants.* Concentration also relates to the frequency of collisions. The more concentrated the reactants, the more frequent the collisions and the faster the rate of the reaction.

4. *Particle size.* In a heterogeneous reaction, the more surface area for a particular reactant, the more frequent the collisions and the faster the reaction.

5. *Catalysts.* A **catalyst** provides an alternate reaction pathway with lower activation energy which thus increases the rate of the reaction.

The hypothetical reaction discussed in this text

$$A_2(g) + B_2(g) \rightleftharpoons 2AB(g)$$

is a **reversible reaction** that reaches a **point of equilibrium** where both the forward and reverse reactions occur under the same conditions. This phenomenon is sometimes referred to as a **dynamic equilibrium** since the identity of reactants and products is constantly changing although the relative concentrations are constant.

Equilibrium phenomena have been discussed several times in previous chapters. For example, we discussed gas phase equilibrium in Chapter 8, liquid-vapor equilibrium in Chapter 11, the equilibrium between a salt and its saturated solution in Chapter 12, and finally, the equilibria of weak acids and bases in Chapter 13.

A system at equilibrium remains at equilibrium as long as there is no change in conditions. Should any condition change, such as temperature, volume of the container, pressure, or the concentration of a reactant or product, the point of equilibrium changes. According to **Le Châtelier's principle**, a system at equilibrium shifts to counteract any stress. For example, consider the commercially important process used to make ammonia for fertilizers.

$$N_2(g) + 3H_2(g) \rightleftharpoons 2NH_3(g) + heat$$

The following conditions increase the concentration of NH_3 present at equilibrium:

1. *A large concentration of N_2 and/or H_2* Increased concentration of substances on one side of an equation leads eventually to an increase on the other.

2. *Compression of the reaction mixture* This means that the reaction occurs at a higher pressure and a lower volume. A higher pressure in the container favors the side with the fewer moles of gas. In this case, four moles of reactants combine to form two moles of products. Therefore, the equilibrium shifts to the right when the reaction mixture is compressed.

3. *A low temperature* Heat can be considered as a reactant in endothermic reactions or as a product in exothermic reactions. Addition or removal of heat affects the point of equilibrium just like addition or removal of any other component. In this case, removal of heat by lowering the temperature causes the equilibrium to shift to the right.

4. *Addition of a catalyst* It takes time for equilibrium to be established (the lower the temperature, the more time it takes). A catalyst is a substance that increases the rate of the reaction so that equilibrium is achieved faster. (It does not affect the eventual distribution of reactants and products.) In the industrial process just discussed, the presence of a catalyst counteracts the decreased rate of the reaction brought on by the lower temperature. Still, the reaction is run at about 400°C.

ASSESSMENT OF OBJECTIVES

A-1 Multiple Choice

_____ 1. Which of the following is not an example of a reaction that reaches a point of equilibrium?

(a) the reaction of N_2 and H_2 to produce NH_3
(b) the ionization of a weak acid
(c) the ionization of a strong acid
(d) the formation of crystals in a saturated solution of a salt

_____ 2. Which of the following is NOT an assumption of collision theory?

(a) Molecules react through collisions with each other.
(b) Colliding molecules must have a minimum energy to react.
(c) Molecules react because they collide with the sides of the container.
(d) Colliding molecules must have the proper orientation.

_____ 3. A reaction at equilibrium

(a) goes to completion.
(b) goes so far to the right then stops.
(c) does not go to the right.
(d) goes to the right and to the left simultaneously.
(e) evolves a gaseous product.

_____ 4. When two reactants are mixed and come to a point of equilibrium, when does the rate of the reverse reaction reach a maximum?

(a) at the beginning of the reaction
(b) when the rate of the forward reaction is at a maximum
(c) at equilibrium
(d) when the rate of the forward reaction is decreasing

_____ 5. Phosphorus burns spontaneously in air, but coal must be ignited (heated) before it begins to burn. What conclusion can we draw?

(a) The combustion of phosphorus has a higher activation energy than the combustion of coal.
(b) The concentration of phosphorus is greater than that of coal.
(c) More phosphorus molecules have the proper orientation during collisions at room temperature.
(d) The combustion of coal has a higher activation energy than the combustion of phosphorus.

_____ 6. Increasing the temperature of a reaction mixture

(a) increases the average velocity of the molecules.
(b) increases the average energy of colliding molecules.
(c) both (a) and (b).
(d) increases the average mass of the colliding molecules.
(e) has little effect on the rate of a reaction.

____ 7. Increasing the concentrations of reactant molecules

   (a) increases the average velocity of the molecules.
   (b) increases the frequency of collisions between molecules.
   (c) increases the average energy of the colliding molecules.
   (d) increases the temperature.
   (e) decreases the volume of the container.

____ 8. Hot iron metal burns rapidly in pure oxygen but slowly in air. This difference implies

   (a) the activation energy is lower in pure oxygen.
   (b) the temperature is higher in pure oxygen.
   (c) the concentration of oxygen is important in the rate of the reaction.
   (d) collisions in pure oxygen are more energetic.

____ 9. Which of the following does not contribute to a fast rate of reaction?

   (a) a low activation energy
   (b) a low temperature
   (c) a high concentration of reactants
   (d) a high temperature

____ 10. Given the gaseous equilibrium: $H_2(g) + I_2(g) \rightleftharpoons 2HI(g)$, which of the following happens at equilibrium if the pressure on the system is increased at constant temperature?

   (a) shifts to the right   (b) shifts to the left   (c) no effect

A-2 Problem

1. Assume the following system at equilibrium:

    $$heat + 4HCl(g) + O_2(g) \rightleftharpoons 2Cl_2(g) + 2H_2O(g)$$

    Determine what effect the following changes will have on the amount of O_2 present at equilibrium.

    (a) heat the reaction mixture   _____

    (b) increase [HCl]   _____

    (c) decrease [Cl$_2$]   _____

    (d) add a catalyst   _____

    (e) compress the reaction
         mixture   _____

    (f) add some $H_2O(g)$ _____

    (g) increase the volume of
         the container   _____

Review of Part B   *The Quantitative Aspects of Reactions at Equilibrium*

OBJECTIVES AND DETAILED TABLE OF CONTENTS

15-4 The Equilibrium Constant

OBJECTIVE *Perform calculations involving the equilibrium constant and equilibrium concentrations.*

15-5 Equilibria of Weak Acids and Bases in Water

OBJECTIVE *Perform calculations involving equilibrium constants and pH for acid-base reactions.*

15-6 Solubility Equilibria

OBJECTIVE *Perform calculations involving solubility constants and molar solubility.*

SUMMARY OF PART B

It has long been known that reactants and products are distributed in proportions at equilibrium in a predictable quantitative manner. For the gaseous reaction

$$H_2(g) + I_2(g) \rightleftharpoons 2HI(g)$$

When the concentrations of products are written in the numerator and the concentrations of the reactants are written in the denominator all raised to the power of their coefficients, we have the following relationship.

$$K_{eq} = \frac{[HI]^2}{[H_2][I_2]} \qquad [X] = mol/L$$

K_{eq} is known as the **equilibrium constant**. Its numerical value gives some indication of the distribution of reactants and products at equilibrium *at a specified temperature*. A large value

300

tells us that products are favored, whereas a small value tells us that reactants are favored. The fraction to the right of the equal sign is known as the **equilibrium constant expression**.

In another example, consider the following hypothetical gaseous equilibrium

$$3A(g) + B(g) \rightleftharpoons C(g) + 2D(g)$$

The equilibrium constant and its expression are written as follows.

$$K_{eq} = \frac{[C][D]^2}{[A]^3[B]}$$

The value of K_{eq} at a specified temperature is determined from a measurement of the concentrations of reactants and products at equilibrium as illustrated in the following example.

Example B-1 Calculation of the Value of K_{eq}

What is the value of K_{eq} for the following reaction?

$$3A(g) + B(g) \rightleftharpoons C(g) + 2D(g)$$

If at equilibrium [A] = 0.0112, [B] = 0.0314, [C] = 0.0667, and [D] = 0.0432.

PROCEDURE

Write the equilibrium constant expression and substitute appropriate concentrations.

SOLUTION

$$K_{eq} = \frac{[C][D]^2}{[A]^3[B]} = \frac{(0.0667)(0.0432)^2}{(0.0112)^3(0.0314)} = \underline{2.82 \times 10^3}$$

Sometimes the concentrations of all species at equilibrium are implied by information about the initial and final concentrations of one reactant or product. An example follows.

Example B-2 Calculation of the Value of K_{eq}

Assume the hypothetical equilibrium $2E(g) + F(g) \rightleftharpoons G(g) + 2H(g)$. Initially, only E and F are present, both at a concentration of 1.00 mol/L. At equilibrium [F] = 0.60 mol/L. What is the value of K_{eq}?

PROCEDURE

Since [F] is known at equilibrium, the concentrations of all other species are implied by the stoichiometry of the reaction. The concentration of the F that *reacted* is

$$[F]_{reacted} = 1.00 - 0.60 = 0.40 \text{ mol/L}.$$

From the stoichiometry of the reaction, notice that for every one mole of F that reacts, one mole of G and two moles of H are formed. Therefore, at equilibrium 0.40 mol/L of F yields the following:

$$[G] = 0.40 \text{ mol/L} \qquad 0.40 \text{ mol/L F} \times \frac{2 \text{ mol/L H}}{1 \text{ mol/L F}} = 0.80 \text{ mol/L H}$$

301

Now we must find [E] at equilibrium. Again from the stoichiometry of the reaction, if 0.40 mol/L of F reacted, 0.80 mol/L of E reacted (two moles of E per mole of F). Therefore, at equilibrium

$$[E] = 1.00 - 0.80 = 0.20 \text{ mol/L}$$

SOLUTION

$$K_{eq} = \frac{[G][H]^2}{[E]^2[F]} = \frac{(0.40)(0.80)^2}{(0.20)^2(0.60)} = \underline{11}$$

When the equilibrium constant is known, the concentration of a substance can be calculated if the concentrations of all other species in the reaction are known or implied.

Example B-3 Calculation of a Component Using the Value of K_{eq}

In the hypothetical reaction discussed in problem B-2, what is the concentration of F at equilibrium if [E] = 0.40, [G] = 0.55, and [H] = 0.18?

PROCEDURE

$$K_{eq} = \frac{[G][H]^2}{[E]^2[F]} \quad \text{Solving for [F]} \quad [F] = \frac{[G][H]^2}{[E]^2 \, K_{eq}}$$

SOLUTION

$$[F] = \frac{(0.55)(0.18)^2}{(0.40)^2 \cdot 11} = \underline{0.010 \text{ mol/L}}$$

We can now discuss the equilibrium involved in weak acids and weak bases. K_a represents an **acid ionization constant** and K_b represents a **base ionization constant**. Examples of the equilibria and the equilibrium constant expressions follow.

weak acid $\quad HNO_2 + H_2O \rightleftharpoons H_3O^+ + NO_2^-$

$$K_a = \frac{[H_3O^+][NO_2^-]}{[HNO_2]}$$

weak base $\quad NH_3 + H_2O \rightleftharpoons NH_4^+ + OH^-$

$$K_b = \frac{[NH_4^+][OH^-]}{[NH_3]}$$

The values of the respective constants tell us how far to the right the equilibrium lies. The smaller the constant the more the equilibrium lies to the left, and thus the weaker the acid or base. The values of the constants are determined from experiments as illustrated in the following example.

Example B-4 Calculation of the Value of K_a

The pH of a 0.10 M solution of HNO_2 is 2.17. What is K_a for HNO_2?

PROCEDURE

$$HNO_2 + H_2O \rightleftharpoons H_3O^+ + NO_2^- \qquad K_a = \frac{[H_3O^+][NO_2^-]}{[HNO_2]}$$

pH = 2.17 or -log $[H_3O^+]$ = 2.17 $\qquad$ log $[H_3O^+]$ = -2.17 $\qquad$ $[H_3O^+] = 6.8 \times 10^{-3}$
In this solution at equilibrium
$[H_3O^+] = [NO_2^-] = 6.8 \times 10^{-3}$ $\qquad$ $[HNO_2] = 0.10 - 0.0068 \approx 0.$
$\qquad\qquad\qquad\qquad\qquad\qquad\qquad$ (0.0068 is negligible compared to 0.10)

SOLUTION

$$K_a = \frac{(6.8 \times 10^{-3})(6.8 \times 10^{-3})}{(0.10)} = \underline{4.6 \times 10^{-4}}$$

Once the constants are known, they can be used to calculate the pH of a solution of any original concentration of acid as follows.

Example B-5 Calculation of the pH of a Weak Acid Solution

What is the pH of a 0.027 M solution of HNO_2?

PROCEDURE

In this case, the unknown is $[H_3O^+]$ which we say is X. Therefore,

$[H_3O^+] = [NO_2^-] = X$ $\quad$ and $\quad$ $[HNO_2] = 0.027 - X$

(If X mol/L of H_3O^+ is formed, then X mol/L of HNO_2 reacted. Therefore, the HNO_2 left at equilibrium is the original concentration minus X.)
Since X is probably small compared to 0.027, we can attempt the following approximation.

$$[HNO_2] = 0.027 - X \approx 0.027$$

SOLUTION

$$\frac{[H_3O^+][NO_2^-]}{[HNO_2]} = \frac{X \cdot X}{0.027} = Ka = 4.6 \times 10^{-4}$$

$$X^2 = 0.12 \times 10^{-4} \qquad X = 3.5 \times 10^{-3} \qquad pH = -\log(3.5 \times 10^{-3}) = \underline{2.46}$$

[Notice X (0.0012) is less than 10% of the original concentration (0.027).]
The approximation is therefore valid.

Buffer solutions contain a weak acid (or weak base) and a salt providing its conjugate base (or acid). Calculation of the pH of buffers is simplified since one can consider the initial concentrations of acid and conjugate base to be essentially unchanged by any ionization. By rearranging the equilibrium constant expression for a weak acid and using logarithms, we can derive the **Henderson-Hasselbalch equation**, which is used specifically for buffer solutions.

$$pH = pK_a + \log \frac{[\text{base}]}{[\text{acid}]} \quad \text{or} \quad pOH = pK_b + \log \frac{[\text{acid}]}{[\text{base}]}$$

Example B-6 Calculation of the pH of a Buffer Solution

What is the pH of a solution made by dissolving 0.024 mol of hydrofluoric acid (HF) and 0.082 mol of sodium fluoride (NaF) in 2.0 L of water? For HF, $K_a = 6.7 \times 10^{-4}$.

PROCEDURE

$$HF + H_2O \rightleftharpoons H_3O^+ + F^- \qquad pK_a = -\log K_a = -\log(6.7 \times 10^{-4}) = 3.17$$

$$NaF(s) \longrightarrow Na^+(aq) + F^-(aq) \quad \text{(The } Na^+ \text{ is a spectator ion.)}$$

HF is the acid species so [acid] = 0.024 mol/2.0L
F^- is the base species so [base] = 0.082 mol/2.0L
For a buffer solution use $pH = pK_a + \log \frac{[\text{base}]}{[\text{acid}]}$

SOLUTION

$$pH = 3.17 + \log \frac{0.082 \ \cancel{\text{mol/2.0 L}}}{0.024 \ \cancel{\text{mol/2.0 L}}} = 3.17 + 0.53 = \underline{3.70}$$

The solubility of an ionic compound in water is also a dynamic equilibrium. The equilibrium constant which is known as the **solubility product (K_{sp})** is simply equal to the concentrations of the two ions each raised to a power equal to their coefficients. For example, the solubility of the compound $Ba_3(PO_4)_2$ is set up as follows.

$$Ba_3(PO_4)_2(s) \rightleftharpoons 3Ba^{2+}(aq) + 2PO_4^{3-}(aq) \qquad K_{sp} = [Ba^{2+}]^3[PO_4^{3-}]^2$$

By knowing the molar solubility, the value of K_{sp} can be calculated. On the other hand, if the value of K_{sp} is known the molar solubility can be calculated. Also, if K_{sp} and the concentration of one ion are known, the concentration of the other can be calculated. The solubility product can also be used to determine whether a precipitate will or will not form from given concentrations of the two ions. If the concentrations of ions are substituted into the equilibrium constant expression and the result is greater than the value of K_{sp} then a precipitate forms.

Example B-7 Calculation of K_{sp} from the Molar Solubility

The molar solubility of Ag_3PO_4 is 4.4×10^{-5} mol/L. What is the value of K_{sp}?

PROCEDURE

The equilibrium involved is $Ag_3PO_4(s) \rightleftharpoons 3Ag^+(aq) + PO_4^{3-}(aq)$

Notice for each mole of Ag_3PO_4 that dissolves three moles of Ag^+ and one mole of PO_4^{3-} are present in solution. Therefore, in solution

$$[Ag^+] = 3 \text{ x solubility} = 3 \text{ x } 4.4 \text{ x } 10^{-5} \text{ mol/L} = 1.32 \text{ x } 10^{-4} \text{ mol/L}$$
$$[PO_4^{3-}] = \text{solubility} = 4.4 \text{ x } 10^{-5} \text{ mol/L}$$

SOLUTION

$$K_{sp} = [Ag^+]^3[PO_4^{3-}] = [1.32 \text{ X } 10^{-4}]^3[4.4 \text{ X } 10^{-5}] = \underline{1.0 \text{ x } 10^{-17}}$$

Example B-8 Calculation of Molar Solubility from the Value of K_{sp}

What is the molar solubility of $NiCO_3$? For $NiCO_3$, $K_{sp} = 1.45 \text{ x } 10^{-7}$.

PROCEDURE

The equilibrium involved is $NiCO_3(s) \rightleftharpoons Ni^{2+}(aq) + CO_3^{2-}(aq)$

If we let X = the molar solubility, then at equilibrium

$$[Ni^{2+}] = [CO_3^{2-}] = X$$

SOLUTION

$$K_{sp} = [Ni^{2+}][CO_3^{2-}] = [X][X] = X^2 = 1.45 \text{ X } 10^{-7}$$

$$X = \underline{3.81 \text{ x } 10^{-4} \text{ mol/L}}$$

ASSESSMENT OF OBJECTIVES

B-1 Multiple Choice

_____ 1. For the hypothetical reaction A + B $\rightleftharpoons$ C + D, $K_{eq} = 10^{-5}$. At equilibrium which of the following is true?

(a) [A][B] > [C][D]   (d) [B][D] > [A][C]
(b) [C][D] > [A][B]   (e) cannot tell
(c) [C][B] > [A][D]

_____ 2. For the hypothetical reaction 2A + B $\rightleftharpoons$ C + D, what are the units of K_{eq}?

(a) mol/L   (d) L^2/mol^2
(b) L/mol   (e) other
(c) mol^2/L^2

_____ 3. Assume the hypothetical reaction $2A + B \rightleftharpoons 3C$. If one starts with 0.50 mol/L of A and the concentration of A at equilibrium is 0.20 mol/L, what is the concentration of C at equilibrium?

(a) 0.50 mol/L   (d) 0.45 mol/L
(b) 0.30 mol/L   (e) 0.20 mol/L
(c) 0.75 mol/L

_____ 4. For $HCHO_2$ (formic acid) $K_a = 1.8 \times 10^{-4}$. Which of the following statements is true?

(a) $HCHO_2$ is a strong base.
(b) $HCHO_2$ is a strong acid.
(c) $HCHO_2$ is mostly ionized in water.
(d) Most $HCHO_2$ is not ionized in water.
(e) $HCHO_2$ is a weak base.

_____ 5. Which of the following approximations is not valid?

(a) $0.10 - 0.003 \approx 0.10$ (c) $0.010 - 0.005 \approx 0.010$
(b) $1.23 - 0.002 \approx 1.23$ (d) $0.020 - 0.001 \approx 0.20$

_____ 6. Which of the changes on a system at equilibrium changes the value of K_{eq}?

(a) the pressure
(b) the volume
(c) the concentration of a reactant or product
(d) addition of a catalyst
(e) the temperature

_____ 7. Which of the following do not form a buffer solution when the two compounds are dissolved in water?

(a) HNO_3 and $Ca(NO_3)_2$ (c) H_2S and KHS
(b) CH_3NH_2 and $CH_3NH_3{}^+Br^-$ (d) $Mg(NO_2)_2$ and HNO_2

_____ 8. For $CaCrO_4$ the value of $K_{sp} = 1.0 \times 10^{-4}$. If the concentration of Ca^{2+} in a solution is equal to 1.0×10^{-3} mol/L, the concentration of $CrO_4{}^{2-}$ in that solution can be no more than

(a) 1.0×10^{-4} mol/L (c) 1.0×10^{-1} mol/L
(b) 1.0×10^{-3} mol/L (d) 1.0×10^{-5} mol/L

B-2 Problems

1. Write the equilibrium constant expression for each of the following:

(a) $P_4(g) + 6H_2(g) \rightleftharpoons 4PH_3(g)$

(b) $2NOCl(g) \rightleftharpoons 2NO(g) + Cl_2(g)$

(c) $SO_2(g) + NO_2(g) \rightleftharpoons SO_3(g) + NO(g)$

(d) $HMnO_4(aq) + H_2O \rightleftharpoons H_3O^+(aq) + MnO_4^-(aq)$

(e) $Cr_2(C_2O_4)_3(s) \rightleftharpoons 2Cr^{3+}(aq) + 3C_2O_4^{2-}(aq)$

2. Assume the following equilibrium: $2NOCl(g) \rightleftharpoons 2NO(g) + Cl_2(g)$. At a specified temperature, at equilibrium $[NO] = 1.32 \times 10^{-3}$ mol/L, $[Cl_2] = 3.76 \times 10^{-3}$ mol/L, and $[NOCl] = 0.542$ mol/L. What is the value of K_{eq}?

3. Assume the following equilibrium: $COBr_2(g) \rightleftharpoons CO(g) + Br_2(g)$. If one starts with 1.00 mole of $COBr_2$ in a 10.0-L container, it is later found that 8.50% of the $COBr_2$ dissociates (reacts) at equilibrium. What is the value of K_{eq}?

4. Assume the following equilibrium: $2NO(g) + O_2(g) \rightleftharpoons 2NO_2(g)$. If one starts with the 0.050 moles of both NO and O_2 in a 1.0-L container, the concentration of NO_2 at equilibrium is found to be 0.022 mol/L. What is K_{eq}?

5. Assume the following equilibrium: $SO_2(g) + NO_2(g) \rightleftharpoons SO_3(g) + NO(g)$. The value of K_{eq} at a specified temperature is 7.5×10^{-2}. What is $[SO_2]$ at equilibrium if $[SO_3] = [NO] = 0.045$ mol/L and $[NO_2] = 0.13$ mol/L at equilibrium?

6. Consider the same equilibrium and constant as in example 5 above. If one starts with 2.00 moles of NO_2 and some SO_2 in a 10.0-L container and no products, it is later found that 0.038 mol/L of NO_2 reacted. How many moles of SO_2 are present at equilibrium?

7. For the following equilibrium: $2CO(g) + O_2(g) \rightleftharpoons 2CO_2(g)$, $K_{eq} = 2.0 \times 10^3$ at a specified temperature. What is the concentration of CO at equilibrium if $[CO_2] = 2.0 \times 10^{-3}$ and $[O_2] = 6.7 \times 10^{-5}$ mol/L?

8.	What is K_a for the hypothetical acid HX of a 0.25 M solution of HX if the pH of the solution is 3.58?

9.	What is the pH of a 0.54 M solution of hypobromous acid (HBrO)? For HBrO, $K_a = 2.1 \times 10^{-9}$.

10.	Novocaine (Nv) is an ammonia-like nitrogen base with $K_b = 6.6 \times 10^{-6}$. What is the pH of a 0.62 M solution of novocaine that also is 0.35 M in the salt, NvH^+Cl^-?

11.	What is the value of K_{sp} for silver(I) oxalate ($Ag_2C_2O_4$) if the molar solubility of the compound equals 1.2×10^{-6} mol/L?

Chapter Summary Assessment

S-1 Matching

_____ Collision theory   _____ Catalyst

_____ Activated complex   _____ Activation energy

_____ An equilibrium constant   _____ An equilibrium constant for
expression   a partial ionization

_____ Equilibrium

(a)	$\dfrac{[NH_3]^2}{[N_2][H_2]^3}$

(b)	The minimum kinetic energy needed for reactant molecules to be transformed into products

(c)	Reactants are transformed into products by means of collisions of molecules

(d)	$K_a = 4.5 \times 10^{-4}$

(e)	A substance that shifts the point of equilibrium to the right

(f)	The reactant molecules at the instant of maximum compression

(g)   A substance that does not affect the point of equilibrium but does affect the rate of the reaction

(h)   A system at equilibrium shifts to counteract stress

(i)   A situation where the rate of the forward and reverse reactions are equal

(j)   A situation where a reaction stops before it is complete

(k) $K_b = 1.5 \times 10^{-4}$

Answers to Assessments of Objectives

A-1 Multiple Choice

1. **c**   the ionization of a strong acid (The reaction is essentially complete.)

2. **c**   Molecules react because they collide with the sides of the container. (Actually, they react because they collide with each other.)

3. **d**   A reaction at equilibrium goes to the right and to the left simultaneously.

4. **c**   at equilibrium

5. **d**   The combustion of coal has a higher activation energy than the combustion of phosphorus.

6. **c**   both (a) and (b)

7. **b**   increases the frequency of collisions between molecules

8. **c**   The concentration of oxygen is important in the rate of the reaction.

9. **b**   a low temperature

10. **c**   no effect. (There are the same number of moles of gas on each side of the equation.)

A-2 Problem

1.   (a) Heating decreases the amount of O_2 present at equilibrium.

(b) Increasing [HCl] decreases O_2.

(c) Decreasing [Cl_2] decreases O_2.

(d) Adding a catalyst has no effect.

(e) Compressing the reaction mixture decreases the amount of O_2.

(f) Adding $H_2O(g)$ increases O_2.

(g) Increasing the volume increases the amount of O_2.

B-1 Multiple Choice

1. **a**   [A][B] > [C][D]   The reactants are favored. The equilibrium lies far to the left.

2. **b**   L/mol $K_{eq} = \dfrac{\cancel{[mol/L]}\cancel{[mol/L]}}{\cancel{[mol/L]}^2[mol/L]} = \dfrac{1}{mol/L} = L/mol$

3. **d**   0.45 mol/L   If $0.50 - 0.20 = 0.30$ mol/L of A reacted,

$$0.30 \,\cancel{mol\ A} \times \frac{3\ mol\ C}{2\,\cancel{mol\ A}} = 0.45 \text{ mol A/L is formed.}$$

4. **d**   The constant indicates that the equilibrium lies to the left, which means that most $HCHO_2$ is not ionized in water.

5. **c** $0.010 - 0.005 \approx 0.010$ (Notice that the second number is 50% of the first number.)

6. **e**   the temperature   (K_{eq} is larger at higher temperatures for an endothermic reaction and smaller for an exothermic reaction.)

7. **a** HNO_3 and $Ca(NO_3)_2$ (HNO_3 is a strong acid.)

8. **c** $K_{sp} = [Ca^{2+}][CrO_4^{2-}]$ $[CrO_4^{2-}] = K_{sp}/[Ca^{2+}] = 1.0 \times 10^{-1}$ mol/L

B-2 Problems

1.   (a) $K_{eq} = \dfrac{[PH_3]^4}{[P_4][H_2]^6}$

   (b) $K_{eq} = \dfrac{[NO]^2[Cl_2]}{[NOCl]^2}$

   (c) $K_{eq} = \dfrac{[SO_3][NO]}{[SO_2][NO_2]}$

   (d) $K_a = \dfrac{[H_3O^+][MnO_4^-]}{[HMnO_4]}$

   (e) $K_{sp} = [Cr^{3+}]^2[C_2O_4^{2-}]^3$

2. $K_{eq} = \dfrac{[NO]^2[Cl_2]}{[NOCl]^2} = \dfrac{(1.32 \times 10^{-3})^2(3.76 \times 10^{-3})}{(0.542)^2} = 2.23 \times 10^{-8}$

3. The amount of $COBr_2$ that reacts is 0.0850×1.00 mol $= 0.0850$ mol. From the stoichiometry, notice that if 0.0850 moles of $COBr_2$ react, then 0.0850 moles of both CO and Br_2 are formed. The concentrations are

$$[Br_2] = [CO] = 0.0850 \text{ mol}/10.0 \text{ L} = 8.50 \times 10^{-3} \text{ mol/L}$$

The $COBr_2$ left at equilibrium is the initial amount minus the amount that reacts.

$$1.00 - 0.085 = 0.91 \text{ mol} \quad [COBr_2] = 0.91 \text{ mol}/10.0 \text{ L} = 0.091 \text{ mol/L}$$

$$K_{eq} = \frac{[CO][Br_2]}{[COBr_2]} = \frac{(8.50 \times 10^{-3})(8.50 \times 10^{-3})}{(0.091)} = \underline{7.9 \times 10^{-4}}$$

4. (a) Find the concentrations of NO and O_2 that reacted to form 0.022 mol/L of NO_2.
   (b) From the initial concentrations of NO and O_2 find the concentration of each remaining at equilibrium (the initial minus the amount that reacts).

$$0.022 \text{ mol NO}_2 \times \frac{2 \text{ mol NO}}{2 \text{ mol NO}_2} = 0.022 \text{ mol/L NO reacted}$$

$$0.022 \text{ mol NO}_2 \times \frac{1 \text{ mol O}_2}{2 \text{ mol NO}_2} = 0.011 \text{ mol/L O}_2 \text{ reacted}$$

At equilibrium: $\quad [NO] = 0.050 - 0.022 = 0.028$ mol/L
$\qquad\qquad\qquad [O_2] = 0.050 - 0.011 = 0.039$ mol/L

$$K_{eq} = \frac{[NO_2]^2}{[NO]^2[O_2]} = \frac{(0.022)^2}{(0.028)^2(0.039)} = \underline{16}$$

5. $K_{eq} = \frac{[SO_3][NO]}{[SO_2][NO_2]}$ Solving for $[SO_2]$, $[SO_2] = \frac{[SO_3][NO]}{[NO_2] \cdot K_{eq}}$

$$[SO_2] = \frac{(0.045)(0.045)}{(0.13)(7.5 \times 10^{-2})} = \underline{0.21 \text{ mol/L}}$$

6. (a) Find $[NO_2]$ at equilibrium from the initial concentration minus the concentration that reacted.
   (b) From the concentration of NO_2 that reacted find the concentration of SO_3 and NO present at equilibrium.
   (c) Substitute and solve for $[SO_2]$. Then find the total amount in 10.0 L.

$$[NO_2]_{eq} = (2.00 \text{ mol}/10.0 \text{ L}) - 0.038 \text{ mol/L} = 0.162 \text{ mol/L}$$

From the stoichiometry, $[SO_3] = [NO] = 0.038$ mol/L at equilibrium

Solving for $[SO_2]$ $\quad [SO_2] = \frac{[SO_3][NO]}{[NO_2] \cdot K_{eq}} = \frac{(0.038)(0.038)}{(0.162)(7.5 \times 10^{-2})} = 0.12$

$$0.12 \text{ mol/L} \times 10.0 \text{ L} = \underline{1.2 \text{ mol SO}_2}$$

7. Solve for [CO]

$$K_{eq} = \frac{[CO_2]^2}{[CO]^2[O_2]} \qquad [CO]^2 = \frac{[CO_2]^2}{[O_2] \cdot K_{eq}} = \frac{(2.0 \times 10^{-3})^2}{(6.7 \times 10^{-5})(2.0 \times 10^3)}$$

$$[CO]^2 = 30 \times 10^{-6} \qquad [CO] = \underline{5.5 \times 10^{-3} \text{ mol/L}}$$

8. $HX + H_2O \rightleftharpoons H_3O^+ + X^- \qquad K_a = \frac{[H_3O^+][X^-]}{[HX]}$

pH = 3.58; log $[H_3O^+]$ = -3.58 $\qquad [H_3O^+] = 2.6 \times 10^{-4}$

In this solution $[X^-] = [H_3O^+]$ and $[HX] = 0.25 - (2.6 \times 10^{-4}) \approx 0.25$

$$K_a = \frac{(2.6 \times 10^{-4})(2.6 \times 10^{-4})}{(0.25)} = \underline{2.7 \times 10^{-7}}$$

9. $HBrO + H_2O \rightleftharpoons H_3O^+ + BrO^- \qquad K_a = \frac{[H_3O^+][BrO^-]}{[HBrO]} = 2.1 \times 10^{-9}$

Let $X = [H_3O^+] = [BrO^-]$; then at equilibrium $[HBrO] = 0.54 - X \approx 0.54$
(the value of the constant tells us that X is small compared to 0.54).

$$K_a = \frac{X \cdot X}{0.54} = 2.1 \times 10^{-9} \quad X^2 = 1.13 \times 10^{-9} = 11.3 \times 10^{-10}$$

$$X = 3.4 \times 10^{-5} \quad \text{pH} = -\log(3.4 \times 10^{-5}) = \underline{4.47}$$

10. $Nv + H_2O \rightleftharpoons NvH^+ + OH^- \qquad$ This is a buffer solution.

$NvH^+Cl^-(s) \rightarrow NvH^+(aq) + Cl^-(aq) \qquad Cl^-$ is a spectator ion.
$pK_b = -\log(6.6 \times 10^{-6}) = 5.18$
Nv = base so [base] = 0.62 M; NvH^+ = acid so [acid] = 0.35 M
$pOH = pK_b + \log\frac{[acid]}{[base]} \qquad pOH = 5.18 + \log\frac{0.35 \text{ M}}{0.62 \text{ M}} = 5.18 - 0.25 = 4.93$

pH = 14.00 - 4.93 = $\underline{9.07}$

11. The equilibrium involved is $Ag_2C_2O_4(s) \rightleftharpoons 2Ag^+(aq) + C_2O_4^{2-}(aq)$

$$K_{sp} = [Ag^+]^2[C_2O_4^{2-}]$$

At equilibrium, $[Ag^+]$ = 2 x solubility and $[C_2O_4^{2-}]$ = solubility
Thus $[Ag^+] = 2 \times 1.2 \times 10^{-6} = 2.4 \times 10^{-6}$ mol/L $[C_2O_4^{2-}] = 1.2 \times 10^{-6}$ mol/L

$$K_{sp} = [2.4 \times 10^{-6}]^2[1.2 \times 10^{-6}] = \underline{6.9 \times 10^{-18}}$$

S-1 Matching

c **Collision theory** tells us that reactants are transformed into products by means of collisions.

f An **activated complex** is formed at the point of maximum compression of reactant molecules.

a An **equilibrium constant expression** is $\dfrac{[NH_3]^2}{[N_2][H_2]^3}$.

i **Equilibrium** is a situation where the rates of the forward and reverse reactions are equal.

g A **catalyst** is a substance that does not affect the point of equilibrium but does affect the rate of the reaction.

b **Activation energy** is the minimum kinetic energy needed for reactant molecules to be transformed into products.

k $K_b = 1.5 \times 10^{-4}$ is an equilibrium constant for a **partial ionization of a weak base**.

Answers and Solutions to Green Text Problems

15-1 Colliding molecules must have the proper orientation relative to each other at the time of the collision and the colliding molecules must have the minimum kinetic energy for the particular reaction.

15-3 (a) As the temperature increases, the frequency of collisions between molecules increases as well as the average energy of the collisions. Both contribute to the increased rate of reaction.
(b) The cooking of eggs initiates a chemical reaction that occurs more slowly at lower temperatures.
(c) The average energy of colliding molecules at room temperature is not sufficient to initiate a reaction between H_2 and O_2.
(d) A higher concentration of oxygen increases the rate of combustion.
(e) When a solid is finely divided, a greater surface area is available for collisions with oxygen molecules. Thus it burns faster.
(f) The souring of milk is a chemical reaction that slows as the temperature drops. It takes several days in a refrigerator.
(g) The platinum is a catalyst. Since the activation energy in the presence of a catalyst is lower, the reaction can occur at a lower temperature.

15-6 The rate of the forward reaction was at a maximum at the beginning of the reaction; the rate of the reverse reaction was at a maximum at the point of equilibrium.

15-7 In many cases, reactions do not proceed directly to the right because other products are formed between the same reactants. For example, combustion may produce carbon monoxide as well as carbon dioxide.

15-8 Products are easier to form because the activation energy for the forward reaction is less than for the reverse reaction. This is true of all exothermic reactions. The system should come to equilibrium faster starting with pure reactants.

15-10 (a) right   (e) right
         (b) left   (f) has no effect
         (c) left   (g) the yield decreases
         (d) right   (h) the yield increases but the rate of formation decreases

15-12 (a) increase   (c) decrease   (e) decrease
         (b) increase   (d) decrease   (f) no effect

15-14 (a) Since there are the same number of moles of gas on both sides of the equation, pressure (or volume) has no effect on the equilibrium.
         (b) decrease the amount of NO
         (c) decrease the amount of NO

15-15 (a) $K_{eq} = \dfrac{[COCl_2]}{[CO][Cl_2]}$ (b) $K_{eq} = \dfrac{[CO_2][H_2]^4}{[CH_4][H_2O]^2}$

        (c) $K_{eq} = \dfrac{[Cl_2]^2[H_2O]^2}{[HCl]^4[O_2]}$ (d) $K_{eq} = \dfrac{[CH_3Cl][HCl]}{[CH_4][Cl_2]}$

15-17 The large value of the equilibrium constant indicates that products are favored at equilibrium.

15-19 The intermediate size of the equilibrium constant indicates that there will be appreciable concentrations of both reactants and products at equilibrium.

15-20 $K_{eq} = \dfrac{[O_3]^2}{[O_2]^3} = \dfrac{(0.12)^2}{(0.35)^3} = \underline{0.34}$

15-21 $K_{eq} = \dfrac{[NO_2]^2}{[N_2][O_2]^2} = \dfrac{(6.20 \times 10^{-4})^2}{(1.25 \times 10^{-3})(2.50 \times 10^{-3})^2} = \underline{49.2}$

15-23 $K_{eq} = \dfrac{[CO_2][H_2]^4}{[CH_4][H_2O]^2} = \dfrac{(2.20/30.0)(4.00/30.0)^4}{(6.20/30.0)(3.00/30.0)^2} = \underline{0.0112}$

15-25 (a) $0.60 \;\cancel{\text{mol HI}} \times \dfrac{1 \text{ mol } H_2}{2 \;\cancel{\text{mol HI}}} = 0.30 \text{ mol } H_2$ $[H_2] = [I_2] = \underline{0.30 \text{ mol/L}}$

        (b) As in (a) $[H_2] = \underline{0.30 \text{ mol/L}}$ $[I_2] = 0.20 + 0.30 = \underline{0.50 \text{ mol/L}}$

        (c) $[HI] = [HI]_{initial} - [HI]_{reacted} = 0.60 - 0.20 = \underline{0.40 \text{ mol/L}}$

        $0.20 \;\cancel{\text{mol HI}} \times \dfrac{1 \text{ mol } H_2}{2 \;\cancel{\text{mol HI}}} = 0.10 \text{ mol /L } H_2$ $[I_2] = [H_2] = \underline{0.10 \text{ mol/L}}$

        (d) $K_{eq} = \dfrac{[H_2][I_2]}{[HI]^2} = \dfrac{(0.10)(0.10)}{(0.40)^2} = \underline{0.063}$

        (e) $K_r = \dfrac{1}{K_{eq}} = \dfrac{1}{0.063} = \underline{16}$

This is a smaller value than that used in Table 15-1. This indicates that the equilibrium in this problem was established at a different temperature than that of Table 15-1.

15-27 (a) $[N_2]_{reacted} = 0.50 - 0.40 = 0.10$ mol/L

0.10 ~~mol N_2~~ x $\dfrac{3 \text{ mol } H_2}{1 \text{ mol } N_2}$ = 0.30 mol/L H_2 (reacts)

$[H_2]_{eq} = 0.50 - 0.30 = \underline{0.20 \text{ mol/L}}$

0.10 ~~mol N_2~~ x $\dfrac{2 \text{ mol } NH_3}{1 \text{ mol } N_2}$ = 0.20 mol/L NH_3 formed

$[NH_3]_{eq} = 0.50 + 0.20 = \underline{0.70 \text{ mol/L}}$

(b) $K_{eq} = \dfrac{(0.70)^2}{(0.20)^3(0.40)} = \underline{150}$

15-28 (a) The concentration of O_2 that reacts is

0.25 ~~mol NH_3~~ x $\dfrac{5 \text{ mol } O_2}{4 \text{ mol } NH_3}$ = $\underline{0.31 \text{ mol/L } O_2}$

(b) $[NH_3]_{eq} = [[NH_3]_{init.} - [NH_3]_{reacted} = 1.00 - 0.25 = \underline{0.75 \text{ mol/L}}$

$[O_2]_{eq} = 1.00 - 0.31 = \underline{0.69 \text{ mol/L } O_2}$

0.25 ~~mol NH_3~~ x $\dfrac{4 \text{ mol } NO}{4 \text{ mol } NH_3}$ = $\underline{0.25 \text{ mol/L } NO}$ (formed)

0.25 ~~mol NH_3~~ x $\dfrac{6 \text{ mol } H_2O}{4 \text{ mol } NH_3}$ = 0.38 mol/L H_2O

(c) $K_{eq} = \dfrac{[NO]^4[H_2O]^6}{[NH_3]^4[O_2]^5} = \dfrac{(0.25)^4(0.38)^6}{(0.75)^4(0.69)^5}$

15-30 $K_{eq} = \dfrac{[PCl_5]}{[PCl_3][Cl_2]}$ $\qquad [PCl_5] = K_{eq}[PCl_3][Cl_2] = (0.95)(0.75)(0.40) = \underline{0.28 \text{ mol/L}}$

15-32 $[H_2O]^2 = \dfrac{[CO_2][H_2]^4}{K_{eq}[CH_4]} = \dfrac{(0.24)(0.20)^4}{0.0112(0.50)} = 6.86 \times 10^{-2}$ $\qquad [H_2O] = \underline{0.26 \text{ mol/L}}$

15-33 Let $X = [HCl] = [CH_3Cl]$ $\quad K_{eq} = \dfrac{[HCl][CH_3Cl]}{[CH_4][Cl_2]} = \dfrac{X^2}{(0.20)(0.40)} = 56$

$X^2 = 4.5$ $\quad X = \underline{2.1 \text{ mol/L}}$

15-35 (a) $K_a = \dfrac{[H_3O^+][BrO^-]}{[HBrO]}$ (d) $K_a = \dfrac{[H_3O^+][SO_3^{2-}]}{[HSO_3^-]}$

(b) $K_b = \dfrac{[NH_4^+][OH^-]}{[NH_3]}$ (e) $K_a = \dfrac{[H_3O^+][H_2PO_4^-]}{[H_3PO_4]}$

(c) $K_a = \dfrac{[H_3O^+][HSO_3^-]}{[H_2SO_3]}$ (f) $K_b = \dfrac{[(CH_3)_2NH_2^+][OH^-]}{[(CH_3)_2NH]}$

15-37 The acid, HB, is weaker because it produces a smaller hydronium ion concentration (higher pH) at the same initial concentration of acid. The stronger acid, HX, has the larger value of K_a.

15-39 $HOCN + H_2O \rightleftharpoons H_3O^+ + OCN^-$

(a) $[HOCN]_{eq} = [HOCN]_{initial} - [HOCN]_{ionized}$

From the equation $[HOCN]_{ionized} = [H_3O^+] = [OCN^-]$

$[HOCN]_{eq} = 0.20 - 0.0062 = \underline{0.19}$

(b) $K_a = \dfrac{[H_3O^+][OCN^-]}{[HOCN]} = \dfrac{(6.2 \times 10^{-3})(6.2 \times 10^{-3})}{(0.19)} = 2.0 \times 10^{-4}$

(c) $-\log(6.2 \times 10^{-3}) = \underline{2.21}$

15-40 (a) $HX + H_2O \rightleftharpoons H_3O^+ + X^-$ $K_a = \dfrac{[H_3O^+][X^-]}{[HX]}$

(b) From the equation $[H_3O^+] = [X^-] = 10.0\%$ of $[HX]_{init}$

$0.100 \times 0.58 = [H_3O^+] = [X^-]$ $[HX]_{eq} = 0.58 - 0.058 = 0.52$

(c) $K_a = \dfrac{(0.058)(0.058)}{(0.52)} = 6.5 \times 10^{-3}$

(d) $pH = -\log(5.8 \times 10^{-2}) = \underline{1.24}$

15-43 $Nv + H_2O \rightleftharpoons NvH^+ + OH^-$

$pH = 11.46$ $pOH = 14.00 - 11.46 = 2.54$ $[OH^-] = 2.88 \times 10^{-3}$

Since $[NvH^+] = [OH^-] = 2.88 \times 10^{-3}$ $[Nv]_{eq} = [Nv]_{init} - [OH^-] = 1.25 - 0.00288 = 1.25$

$K_b = \dfrac{[NvH^+][OH^-]}{[Nv]} = \dfrac{(2.88 \times 10^{-3})^2}{(1.25)} = \underline{6.6 \times 10^{-6}}$

15-44 $HC_2Cl_3O_2 + H_2O \rightleftharpoons H_3O^+ + C_2Cl_3O_2^-$

$[H_3O^+] = [HC_2Cl_3O_2]_{ionized} = [HC_2Cl_3O_2]_{initial} - [HC_2Cl_3O_2]_{eq}$
$= 0.300 - 0.277 = 0.023$ $pH = -\log(2.3 \times 10^{-2}) = \underline{1.64}$

$K_a = \dfrac{(0.023)^2}{0.277} = \underline{1.9 \times 10^{-3}}$

15-45 $HBrO + H_2O \rightleftharpoons H_3O^+ + BrO^-$

Let $X = [H_3O^+] = [BrO^-]$ Then, $[HBrO]_{eq} = [HBrO]_{init} - [HBrO]_{ionized}$

Thus, $[HBrO]_{eq} = 0.50 - X$. Since K_a is small, X is small and

$[HBrO] \approx 0.50$ $K_a = \dfrac{(X)(X)}{0.50} = 2.1 \times 10^{-9}$ $[X] = [H_3O^+] = 3.2 \times 10^{-5}$ $\underline{pH = 4.43}$

15-47 $NH_3 + H_2O \rightleftharpoons NH_4^+ + OH^-$

Let $X = [OH^-] = [NH_4^+]$ $[NH_3] = 0.55 - X \approx 0.55$

$K_b = \dfrac{[NH_4^+][OH^-]}{[NH_3]} = \dfrac{X^2}{0.55} = 1.8 \times 10^{-5}$ $X = [OH^-] = \underline{3.2 \times 10^{-3}}$

15-49 $(CH_3)_2NH + H_2O \rightleftharpoons (CH_3)_2NH_2^+ + OH^-$

Set $X = [OH^-] = [(CH_3)_2NH_2^+]$ $[(CH_3)_2NH] = 1.00 - X \approx 1.00$

$K_b = \dfrac{[(CH_3)_2NH_2^+][OH^-]}{[(CH_3)_2NH]} = \dfrac{X^2}{1.00} = 7.4 \times 10^{-4}$

$X = 2.7 \times 10^{-2}$ $pOH = 1.57$ $pH = \underline{12.43}$

15-51 $HCN + H_2O \rightleftharpoons H_3O^+ + CN^-$ $[HCN]$ = acid, $[CN^-]$ = base

$pK_a = -\log(4.0 \times 10^{-10}) = 9.40$ $pH = 9.40 + \log \dfrac{0.45 \text{ mol/2.50 L}}{0.45 \text{ mol/2.50 L}}$

$pH = pK_a = \underline{9.40}$

15-53 $HBrO + H_2O \rightleftharpoons H_3O^+ + BrO^-$

$pK_a = -\log(2.1 \times 10^{-9}) = 8.68$ $pH = 8.68 + \log \dfrac{0.20 \text{ mol/0.850 L}}{0.60 \text{ mol/0.850 L}} = 8.68 - 0.48 = \underline{8.20}$

15-56 $N_2H_4 + H_2O \rightleftharpoons N_2H_5^+ + OH^-$

$1.50 \text{ g N}_2\text{H}_4 \times \dfrac{1 \text{ mol } N_2H_4}{32.05 \text{ g N}_2\text{H}_4} = 0.0468 \text{ mol } N_2H_4$

$1.97 \text{ g N}_2\text{H}_5\text{Cl} \times \dfrac{1 \text{ mol } N_2H_5Cl}{68.51 \text{ g N}_2\text{H}_5\text{Cl}} = 0.0288 \text{ mol } N_2H_5Cl \ (N_2H_5^+)$

$pK_b = -\log(9.8 \times 10^{-7}) = 6.01$ $pOH = 6.01 + \log \dfrac{0.0288 \text{ mol/2.00 L}}{0.0468 \text{ mol/2.00 L}}$

$pOH = 6.01 - 0.21 = 5.80$ $\underline{pH = 8.20}$

15-58 When the concentration of acid and base species are equal then pH = pK_a and pOH = pK_b. If a buffer of pH = 7.50 is required then we look for an acid with $K_a = [H_3O^+] = 3.2 \times 10^{-8}$ or a base with $K_b = [OH^-] = 3.2 \times 10^{-7}$. Since $K_a = 3.2 \times 10^{-8}$ for HClO, an equimolar mixture of HClO and KClO produces the required buffer.

15-60 (a) $FeS(s) \rightleftharpoons Fe^{2+}(aq) + S^{2-}(aq)$ $\qquad K_{sp} = [Fe^{2+}][S^{2-}]$

(b) $Ag_2S(s) \rightleftharpoons 2Ag^+(aq) + S^{2-}(aq)$ $\qquad K_{sp} = [Ag^+]^2[S^{2-}]$

(c) $Zn(OH)_2(s) \rightleftharpoons Zn^{2+}(aq) + 2OH^-(aq)$ $\qquad K_{sp} = [Zn^{2+}][OH^-]^2$

5-62 $CuI(s) \rightleftharpoons Cu^+(aq) + I^-(aq)$ $\qquad K_{sp} = [Cu^+][I^-]$

At equilibrium $[Cu^+] = [I^-] = 2.2 \times 10^{-6}$ mol/L

$K_{sp} = [2.2 \times 10^{-6}][2.2 \times 10^{-6}] = \underline{4.8 \times 10^{-12}}$

15-64 $Ca(OH)_2(s) \rightleftharpoons Ca^{2+}(aq) + 2OH^-(aq)$ $\quad K_{sp} = [Ca^{2+}][OH^-]^2$

At equilibrium $[Ca^{2+}] = 1.3 \times 10^{-2}$ mol/L

$[OH^-] = 2 \times (1.3 \times 10^{-2})$mol/L $= 2.6 \times 10^{-2}$ mol/L

$K_{sp} = [1.3 \times 10^{-2}][2.6 \times 10^{-2}]^2 = \underline{8.8 \times 10^{-6}}$

15-66 $AgI(s) \rightleftharpoons Ag^+(aq) + I^-(aq)$ $\quad K_{sp} = [Ag^+][I^-] = 8.3 \times 10^{-17}$

Let x = molar solubility of AgI, then $[Ag^+] = [I^-] = x$

$[x][x] = 8.3 \times 10^{-17}$ $\quad x = \underline{9.1 \times 10^{-9}\text{ mol/L}}$

15-68 $AgBr(s) \rightleftharpoons Ag^+(aq) + Br^-(aq)$ $\quad K_{sp} = [Ag^+][Br^-] = 7.7 \times 10^{-13}$

$[7 \times 10^{-6}][8 \times 10^{-7}] = 6 \times 10^{-12}$

This is a larger number than K_{sp} <u>so a precipitate of AgBr does form.</u>

15-70 (a) $3Cl_2(g) + NH_3(g) \rightleftharpoons NCl_3(g) + 3HCl(g)$

(b) $K_{eq} = \dfrac{[NCl_3][HCl]^3}{[NH_3][Cl_2]^3}$ $\qquad$ (c) no effect $\qquad$ (d) decreases [NH₃]

(e) reactants $\qquad$ (f) $\dfrac{(2.0 \times 10^{-4})(1.0 \times 10^{-3})^3}{[NH_3](0.10)^3} = 2.4 \times 10^{-9}$ [NH₃] = $\underline{0.083\text{ mol/L}}$

15-72 (a) $2N_2O(g) \rightleftharpoons 2N_2(g) + O_2(g)$

(b) $K_{eq} = \dfrac{[N_2]^2[O_2]}{[N_2O]^2}$ $\qquad$ (c) decreases [N₂] $\qquad$ (d) decreases [N₂O]

(e) $[N_2O]_{eq} = 0.10 - (0.015 \times 0.10) = 0.10$ mol/L; $[N_2] = 0.015 \times 0.10 = 1.5 \times 10^{-3}$ mol/L;

$[O_2] = \dfrac{1.5 \times 10^{-3}}{2} = 7.5 \times 10^{-4}$ mol/L $\qquad K_{eq} = \dfrac{(1.5 \times 10^{-3})^2(7.5 \times 10^{-4})}{(0.10)^2} = \underline{1.7 \times 10^{-7}}$

15-74 (a) A solution of $NaHCO_3$ is slightly basic because the hydrolysis reaction occurs to a greater extent than the acid ionization reaction. That is, K_b is larger than K_a.

(b) It is used as an antacid to counteract excess stomach acidity.

(c) $HCl + H_2O \rightarrow H_3O^+ + Cl$ (HCl is a strong acid.)

$H_3O^+(aq) + HCO_3^-(aq) \rightarrow H_2CO_3(aq) + H_2O(l)$

$H_2CO_3(aq) \rightarrow H_2O(l) + CO_2(g)$

(d) No. The HSO_4^- does not react with H_3O^+ because it would form H_2SO_4 which is a strong acid. The molecular form of a strong acid does not exist in water.

15-76 $HC_2H_3O_2 + H_2O \rightarrow H_3O^+ + C_2H_3O_2^-$

$0.265\ L \times 0.22\ mol/L = 0.058\ mol\ HC_2H_3O_2$

$0.375\ L \times \dfrac{0.12\ \cancel{mol\ Ba(C_2H_3O_2)_2}}{L} \times \dfrac{2\ mol\ C_2H_3O_2^-}{\cancel{mol\ Ba(C_2H_3O_2)_2}} = 0.090\ mol\ C_2H_3O_2^-$

$pK_a = 4.74 \quad pH = 4.74 + \log \dfrac{0.090\ \cancel{mol/0.640\ L}}{0.058\ \cancel{mol/0.640\ L}} = 4.74 + 0.19 = \underline{4.93}$

15-77 The addition of the NaOH reacts with part of the acetic acid to produce sodium acetate. The net ionic equation for the reaction is

$HC_2H_3O_2 + OH^- \rightarrow C_2H_3O_2^- + H_2O$

0.20 mol of OH^- reacts with 0.20 mol of $HC_2H_3O_2$ to produce 0.20 mol of $C_2H_3O_2^-$. Thus $1.00 - 0.20 = 0.80$ mol of $HC_2H_3O_2$ remains and 0.20 mol of $C_2H_3O_2^-$ is formed from the addition of 0.20 mol of OH^-.

$pK_a = 4.74 \quad pH = 4.74 + \log \dfrac{0.20\ \cancel{mol/V}}{0.80\ \cancel{mol/V}} = 4.74 - 0.60 = \underline{4.14}$

15-78 $Mg(OH)_2(s) \rightleftharpoons Mg^{2+}(aq) + 2OH^-(aq) \quad K_{sp} = [Mg^{2+}][OH^-]^2$

$pH = 10.50 \quad pOH = 14.00 - 10.50 = 3.50 \quad [OH^-] = -inv \log [OH^-]$

$[OH^-] = 3.2 \times 10^{-4} \quad [Mg^{2+}] = [OH^-]/2 = 1.6 \times 10^{-4}$

$K_{sp} = [1.6 \times 10^{-4}][3.2 \times 10^{-4}]^2 = \underline{1.6 \times 10^{-11}}$

15-80 $BaSO_4(s) \rightleftharpoons Ba^{2+}(aq) + SO_4^{2-}(aq) \quad K_{sp} = [Ba^{2+}][SO_4^{2-}] = 1.1 \times 10^{-10}$

Let x = molar solubility, then $[Ba^{2+}] = [SO_4^{2-}] = x$

$[x][x] = 1.1 \times 10^{-10} \quad x = 1.0 \times 10^{-5}\ mol/L$

$1.0 \times 10^{-5}\ \cancel{mol\ BaSO_4} \times \dfrac{233.4\ g\ BaSO_4}{\cancel{mol\ BaSO_4}} = \underline{0.023\ g\ BaSO_4/L}$

16

Nuclear Chemistry

Review of Part A *Natural Occurring Radioactivity*

OBJECTIVES AND DETAILED TABLE OF CONTENTS

16-1 Radioactivity

OBJECTIVE *Write balanced nuclear equations for the five types of radiation.*

16-2 Rates of Decay of Radioactive Isotopes

OBJECTIVE *Using half-lives, estimate the time needed for various amounts of radioactive decay to occur.*

    16-2.1 Half-life
    16-2.2 Radioactive Decay Series

16-3 The Effects of Radiation

OBJECTIVE *Describe how the three natural types of radiation affect matter.*

    16-3.1 Generation of Heat
    16-3.2 Formation of Ions and Free Radicals
    16-3.3 Ionization and Types of Radiation

16-4 The Detection and Measurement of Radiation

OBJECTIVE *Describe the methods and units used to detect and measure radiation.*

    16-4.1 Film Badges and Radon Detectors
    16-4.2 Geiger and Scintillation Counters
    16-4.3 Radiation and the Human Body

SUMMARY OF PART A

The nucleus of an atom is not necessarily stable and unchanging. For over a century, it has been known that some nuclei are **radioactive isotopes**. These isotopes are unstable and disintegrate to form the nuclei of other elements and high energy particles or forms of light. Three types of natural **radiation** are illustrated by the nuclear equations below. A **nuclear equation** indicates the original nucleus or nuclei on the left and the product nuclei and particles on the right.

Alpha (α) particles

$$^{238}_{92}U \longrightarrow \; ^{234}_{90}Th \; + \; ^{4}_{2}He \quad \longleftarrow \left[\begin{array}{l} \text{He nucleus is called an alpha } (\alpha) \text{ particle. It} \\ \text{lowers the atomic number of the original} \\ \text{isotope by two and the mass number by four.} \end{array} \right]$$

Beta (β) particles

$$^{14}_{6}C \longrightarrow \; ^{14}_{7}N \; + \; ^{0}_{-1}e \quad \longleftarrow \left[\begin{array}{l} \text{An electron from the nucleus is called a} \\ \text{beta } (\beta) \text{ particle. It raises the atomic} \\ \text{number of the original isotope by one.} \end{array} \right]$$

Gamma (γ) rays

$$^{60}_{27}\text{Co} \longrightarrow ^{60}_{27}\text{Co} + \gamma \longleftarrow$$

[High energy light from the nucleus is called a gamma ray (γ). It does not affect the atomic number or mass number of the original isotope.]

In addition to these, there are two other forms of radiation that occur either rarely in nature (electron capture) or only in a few artificially produced isotopes.

Positron emission

$$^{26}_{14}\text{Si} \longrightarrow ^{26}_{13}\text{Al} + ^{0}_{+1}\text{e} \longleftarrow$$

[A positron is a positively charged electron. It lowers the atomic number of the original isotope by one.]

Electron capture

$$^{179}_{73}\text{Ta} + ^{0}_{-1}\text{e} \longrightarrow ^{179}_{72}\text{Hf} \longleftarrow$$

[When an isotope captures an orbital electron, the atomic number of the original isotope increases by one as in positron emission.]

Each radioactive isotope has a definite and unchanging rate of decay. The rate of decay of an isotope is indicated by its **half-life ($t_{1/2}$)**, which is the time required for one-half of a given sample to decay. Half-lives vary from fractions of a second to billions of years. For example, $^{238}_{92}\text{U}$ is a radioactive isotope that is still present because its half-life is about the same as the age of the Earth. This isotope is the beginning nucleus of a **radioactive decay series** that ends with a stable isotope of lead.

Radiation affects living tissue because of its ability to form ions as the particles or rays penetrate matter. Gamma radiation is considered the most dangerous type of radiation mainly because of its ability to penetrate through large quantities of matter. This destructive ability of gamma radiation can be put to constructive use when it is focused on malignant (cancer) cells, thus causing their destruction. Alpha and beta emitters are not very dangerous when outside the body but are potentially very serious when ingested. In close contact with tissue, these isotopes can cause cancer.

Radiation is detected in basically three ways. The first is **film badges**, which take advantage of the ability of radiation to penetrate and expose film. Film badges can detect radioactive dosage for those who work in the nuclear industry. More precise measurements are made with **Geiger counters**, which work by detecting the ionization caused by radiation, and **scintillation counters**, which work by the use of phosphors that emit light when they interact with radiation.

ASSESSMENT OF OBJECTIVES

A-1 Multiple Choice

____ 1. Which of the following is a beta particle?

(a) $^{-1}_{0}\text{e}$　(b) $^{0}_{1}\text{e}$　(c) $^{4}_{2}\text{He}$　(d) $^{0}_{-1}\text{e}$　(e) $^{1}_{0}\text{H}$

___ 2. If $^{234}_{92}U$ emits an alpha particle, which of the following is the product?

(a) $^{230}_{90}U$ (b) $^{234}_{91}Np$ (c) $^{230}_{90}Th$ (d) $^{232}_{90}Th$ (e) $^{238}_{94}Pu$

___ 3. If the half-life of a particular isotope is 2.0 days, how much will remain of an 8.0-g sample in 8.0 days?

(a) none   (b) 4.0 g   (c) 0.5 g   (d) 1.9 g   (e) 2.0 g

___ 4. Sodium-20 emits a positron. The isotope remaining is

(a) magnesium-20   (b) neon-21
(c) magnesium-19   (d) neon-20

___ 5. Gallium-68 is an isotope formed as a result of electron capture. The priginal isotope was

(a) germanium-68   (b) gallium-69
(c) arsenic-68   (d) germanium-69

___ 6. A radioactive isotope decays by converting a neutron into a proton. This isotope emitted which of the following?

(a) alpha particle   (b) beta particle
(c) gamma ray   (d) positron

A-2 Matching

___ Alpha particle

___ Electron capture

___ Beta particle

___ Gamma ray

___ Positron emission

___ Half-life

(a) A high-energy form of light

(b) Half of the time it takes a sample of a radioactive isotope to decay

(c) An electron absorbed from outside the nucleus

(d) A proton emitted from a nucleus

(e) A helium nucleus

(f) The time it takes half of a sample of a radioactive isotope to decay

(g) A proton with a negative charge

(h) A process whereby a nucleus gives off a particle to form a nucleus with the next lower atomic number

(i) An electron emitted from a nucleus

A-3 Problems

1. Complete the following nuclear equations. (You will need a periodic table.)

   (a) $^{135}_{53}\text{I} \longrightarrow \ ^{135}_{54}\text{Xe} \ + \ \underline{\hspace{1cm}}$

   (b) $^{232}_{90}\text{Th} \longrightarrow \ ^{228}_{88}\text{Ra} \ + \ \underline{\hspace{1cm}}$

   (c) $^{20}_{8}\text{O} \longrightarrow \underline{\hspace{1.5cm}} \ + \ ^{0}_{-1}\text{e}$

   (d) $^{80}_{37}\text{Rb} \ + \ ^{0}_{-1}\text{e} \longrightarrow \underline{\hspace{1cm}}$

   (e) $^{50}_{25}\text{Mn} \longrightarrow \underline{\hspace{1cm}} \ + \ ^{0}_{+1}\text{e}$

2. In order for a long-lived isotope to still exist in any appreciable amount, it must have a half-life around the age of the Earth. If the Earth is about 5×10^9 years old (five billion) and a certain isotope has a half-life of 1×10^9 (one billion) years, about what percent of the original amount is still present?

Review of Part B *Induced Nuclear Changes and Their Uses*

OBJECTIVES AND DETAILED TABLE OF CONTENTS

16-5 Nuclear Reactions

OBJECTIVE *Write balanced nuclear equations for the synthesis of specific isotopes by nuclear reactions.*

16-6 Applications of Radioactivity

OBJECTIVE List and describe the beneficial uses of radioactivity.

16-7 Nuclear Fission and Fusion

OBJECTIVE *Identify the differences between natural radioactivity, nuclear fission, and nuclear fusion.*

    16-7.1 The Atomic Bomb
    16-7.2 Nuclear Power
    17-7.3 Nuclear Fusion

SUMMARY OF PART B

Besides naturally occurring nuclear reactions, artificial nuclear reactions have been studied for over 70 years. The following nuclear equation illustrates how the **transmutation** of one element to another occurs from the reaction of an alpha particle and an isotope of aluminum.

$$\ce{^{27}_{13}Al} + \ce{^{4}_{2}He} \longrightarrow \ce{^{30}_{15}P} + \ce{^{1}_{0}n}$$

Notice in the previous nuclear equation that the total of the mass numbers on the left (27 + 4) equals the total of the mass numbers on the right (30 + 1). The total charge on the left (13 + 2) also equals the total charge on the right (15 + 0). Remember that the charge on the nucleus determines the identity of the element and vice versa (e.g., 15 = P). With the invention of the particle accelerator, which provides extra energy to the colliding nuclei, many new heavy elements have been artificially synthesized up to and including element #118. The particle accelerator is needed to supply the positively charged colliding nuclei with enough energy to overcome the strong force of repulsion between them. The following nuclear equation illustrates how a new element was made from naturally occurring elements by use of a particle accelerator.

$$\ce{^{238}_{92}U} + \ce{^{12}_{6}C} \longrightarrow \ce{^{244}_{98}Cf} + 6\,\ce{^{1}_{0}n}$$

There are many important uses of radioactivity. One familiar application is **carbon dating**, whereby the proportion of radioactive carbon present in a sample indicates its age. A nuclear reaction known as **neutron activation analysis** produces radioactive isotopes from stable ones. This process has many medical applications. Gamma radiation preserves food and destroys cancer cells. Positron emissions are used in an important medical test known as a **PET scan**.

In the 1930s, a new type of nuclear reaction was discovered whereby a very heavy nucleus ($\ce{^{235}_{92}U}$) was split into two similar-sized parts when it absorbed a neutron.

$$\ce{^{235}_{92}U} + \ce{^{1}_{0}n} \longrightarrow \ce{^{139}_{56}Ba} + \ce{^{94}_{36}Kr} + 3\,\ce{^{1}_{0}n} + \text{energy}$$

A nuclear reaction where a large nucleus is split into two smaller nuclei is known as **fission**. In a nuclear fission reaction, a chain reaction occurs because more neutrons are produced than consumed. Thus the reaction is self-sustaining. Also, since product nuclei weigh less than the original nuclei, the reaction produces a large amount of energy by conversion of mass ($E = mc^2$). This reaction forms the basis of the atomic bomb, but is controlled in nuclear plants for the generation of electrical power.

Another type of nuclear reaction was shown to be feasible in the late 1940s. In this case, two light nuclei undergo **fusion** to form a heavier nucleus. Since a mass loss also occurs, significant energy is released.

$$\ce{^{3}_{1}H} + \ce{^{2}_{1}H} \longrightarrow \ce{^{4}_{2}He} + \ce{^{1}_{0}n} + \text{energy}$$

This reaction is the basis of the hydrogen bomb but has not yet been controlled for use in the generation of electrical power, although much research is currently under way. The fusion reaction, however, is the original source of almost all of our energy since it is the reaction that continuously generates energy in the Sun.

ASSESSMENT OF OBJECTIVES

B-1 Multiple Choice

___ 1. Which of the following isotopes is used in carbon dating?

(a) ^{12}C (b) ^{11}C c) ^{14}C (d) ^{13}C

___ 2. When stable isotopes are neutron activated, they

(a) become alpha particle emitters.   (b) become beta particle emitters.
(c) become positron emitters.   (d) undergo fission.

___ 3. Which of the following are detected outside the body in a PET scan?

(a) positrons   (b) electrons   (c) gamma rays   (d) alpha particles

___ 4. Which of the following reactions represents a fission reaction?

(a) $^{1}_{1}H + ^{2}_{1}H \longrightarrow ^{3}_{2}He$

(b) $^{239}_{94}Pu + ^{1}_{0}n \longrightarrow ^{92}_{38}Sr + ^{144}_{56}Ba + 4^{1}_{0}n$

(c) $^{239}_{94}Pu \longrightarrow ^{235}_{92}U + ^{4}_{2}He$

(d) $^{238}_{92}U + ^{1}_{0}n \longrightarrow ^{239}_{93}Np + ^{0}_{-1}e$

(e) $^{14}_{7}N + ^{4}_{2}He \longrightarrow ^{17}_{8}O + ^{1}_{1}H$

___ 5. Which of the equations in problem 4 above represents a fusion reaction?

___ 6. Which of the following is a problem in the use of fusion power?

(a) It produces large amounts of radioactive products.
(b) It uses uranium, which is rare.
(c) It can be difficult to control.
(d) It requires a very high temperature to get started.

B-2 Problems

(a) $^{96}_{42}Mo + ^{2}_{1}H \longrightarrow \underline{\hspace{1cm}} + ^{1}_{0}n$

(b) $^{27}_{13}Al + ^{2}_{1}H \longrightarrow ^{25}_{12}Mg + \underline{\hspace{1cm}}$

(c) $^{108}_{48}Cd + ^{1}_{0}n \longrightarrow \underline{\hspace{1cm}} + \gamma$

(d) $^{235}_{92}U + ^{1}_{0}n \longrightarrow ^{90}_{38}Sr + ^{144}_{54}Xe + \underline{\hspace{1cm}}$

(e) $^{249}_{98}\text{Cf} + {}^{18}_{8}\text{O} \longrightarrow \underline{\hspace{1cm}} + 4{}^{1}_{0}\text{n}$

Chapter Summary Assessment

S-1 Matching

___ A proton	(a) $^{238}_{92}\text{U}$
___ Undergoes fission	(b) Scintillation counter
___ An alpha particle	(c) $^{24}_{11}\text{X}$
___ An isotope of sodium	(d) Beta particle
___ Detects formation of ions	(e) $^{0}_{+1}\text{e}$
___ A positron	(f) $^{1}_{1}\text{H}^{+}$
___ An electron	(g) Formation of ions
___ An effect of radiation	(h) A neutral particle with a mass of about 1 amu
	(i) High energy electromagnetic radiation
	(j) $^{235}_{92}\text{U}$
	(k) $^{23}_{12}\text{X}$
	(l) $^{4}_{2}\text{He}^{2+}$
	(m) Geiger counter

Answers to Assessments of Objectives

A-1 Multiple Choice

1. **d** $^{0}_{-1}\text{e}$

2. **c** $^{230}_{90}\text{Th}$ The mass number decreases by four, the atomic number by two. The new element with atomic number 90 is Th.

3. **c** 0.5 g

4. **d** neon-20

5. **a** germanium-68

6. **b** A beta particle increases the atomic number by one when a neutron is converted into a proton.

A-2 Matching

e   An **alpha particle is** a helium nucleus.

c   **Electron capture** is a process whereby an electron is absorbed from outside the nucleus.

i   A **beta particle** is an electron emitted from a nucleus.

a   A **gamma ray** is a high-energy form of light.

h   **Positron emission** is a process whereby a nucleus gives off a particle to form a nucleus with the next lower atomic number.

f   **Half-life** is the time it takes half of a sample of radioactive isotope to decay.

A-3 Problems

1.   (a) $_{-1}^{0}e$ (b) $_{2}^{4}He$ (c) $_{9}^{20}F$ (d) $_{36}^{60}Kr$ (e) $_{24}^{50}Cr$

2.

Time elapsed		Fraction of sample remaining	Percent of sample remaining
(beginning)	0	1	100.0%
	1×10^9 yr.	1/2	50.0%
	2×10^9 yr.	1/4	25.0%
	3×10^9 yr.	1/8	12.5%
	4×10^9 yr	1/16	6.2%
(present)	5×10^9 yr.	1/32	3.1%

B-1 Multiple Choice

1. **c** ^{14}C

2. **b**   Isotopes become neutron rich and then decay by beta emission.

3. **c**   Positrons are emitted but combine with electrons and form two gamma rays, which exit the body and are detected.

4. **b** $_{94}^{239}Pu + _{0}^{1}n \longrightarrow _{38}^{92}Sr + _{56}^{144}Ba + 4_{0}^{1}n$

5. **a** $_{1}^{1}H + _{1}^{2}H \longrightarrow _{2}^{3}He$

6. **d**   It requires a very high temperature to get started.

B-2 Problems

(a) $^{97}_{43}Tc$ (b) $^{4}_{2}He$ (c) $^{109}_{48}Cd$ (d) $2\,^{1}_{0}n$ (e) $^{263}_{106}Sg$

S-1 Matching

f   A **proton** is the same as $^{1}_{1}H^{+}$.

j $^{235}_{92}U$ undergoes **fission**.

l   An **alpha particle** can be represented as $^{4}_{2}He^{2+}$.

c   An **isotope of sodium** is $^{24}_{11}X$.

m   A **Geiger counter** detects the formation of ions.

e $^{0}_{+1}e$ is a **positron**.

d   A **beta particle** is an electron.

g   An **effect of radiation** is the formation of ions.

Answers and Solutions to Green Text Problems

16-1 (a) $^{210}_{84}Po$ (b) $^{84}_{38}Sr$ (c) $^{257}_{100}Fm$ (d) $^{206}_{82}Pb$ (e) $^{233}_{92}U$

16-3 (a) $^{4}_{2}He$ (b) $^{0}_{-1}e$ (c) $^{1}_{0}n$

16-4 $^{2}_{1}H$

16-6 (a) $^{210}_{84}Po \longrightarrow\ ^{206}_{82}Pb\ +\ ^{4}_{2}He$

(b) $^{152}_{64}Gd \longrightarrow\ ^{148}_{62}Sm\ +\ ^{4}_{2}He$

(c) $^{252}_{100}Fm \longrightarrow\ ^{248}_{98}Cf\ +\ ^{4}_{2}He$

(d) $^{266}_{109}Mt \longrightarrow\ ^{262}_{107}Bh\ +\ ^{4}_{2}He$

16-8 (a) $^{3}_{1}H \longrightarrow\ ^{3}_{2}He\ +\ ^{0}_{-1}e$

(b) $^{153}_{64}Gd \longrightarrow\ ^{153}_{65}Tb\ +\ ^{0}_{-1}e$

(c) $^{59}_{26}Fe \longrightarrow\ ^{59}_{27}Co\ +\ ^{0}_{-1}e$

(d) $^{24}_{11}Na \longrightarrow\ ^{24}_{12}Mg\ +\ ^{0}_{-1}e$

16-11 $^{51}_{25}\text{Mn} \longrightarrow {}^{51}_{24}\text{Cr} + {}^{0}_{+1}\text{e}$

16-14 $^{68}_{32}\text{Ge} + {}^{0}_{-1}\text{e} \longrightarrow {}^{68}_{31}\text{Ga}$

16-16 (a) $^{0}_{-1}\text{e}$ (b) $^{90}_{38}\text{Sr}$ (c) $^{26}_{13}\text{Al}$ (d) $^{231}_{90}\text{Th}$ (e) $^{179}_{72}\text{Hf}$ (f) $^{41}_{20}\text{Ca}$ (g) $^{210}_{81}\text{Tl}$

16-18 (a) $^{230}_{90}\text{Th} \longrightarrow {}^{226}_{88}\text{Ra} + {}^{4}_{2}\text{He}$

   (b) $^{214}_{84}\text{Po} \longrightarrow {}^{210}_{82}\text{Pb} + {}^{4}_{2}\text{He}$

   (c) $^{210}_{84}\text{Po} \longrightarrow {}^{210}_{85}\text{At} + {}^{0}_{-1}\text{e}$

   (d) $^{218}_{84}\text{Po} \longrightarrow {}^{214}_{82}\text{Pb} + {}^{4}_{2}\text{He}$

   (e) $^{14}_{6}\text{C} \longrightarrow {}^{14}_{7}\text{N} + {}^{0}_{-1}\text{e}$

   (f) $^{50}_{25}\text{Mn} \longrightarrow {}^{50}_{24}\text{Cr} + {}^{0}_{+1}\text{e}$

   (g) $^{37}_{18}\text{Ar} + {}^{0}_{-1}\text{e} \longrightarrow {}^{37}_{17}\text{Cl}$

16-19 $^{234}_{90}\text{Th}$, $^{234}_{91}\text{Pa}$, $^{234}_{92}\text{U}$, $^{230}_{90}\text{Th}$, $^{226}_{88}\text{Ra}$, $^{222}_{86}\text{Rn}$, $^{218}_{84}\text{Po}$, $^{214}_{82}\text{Pb}$, $^{214}_{83}\text{Bi}$, $^{210}_{81}\text{Tl}$, $^{210}_{82}\text{Pb}$, $^{210}_{83}\text{Bi}$, $^{210}_{84}\text{Po}$, $^{206}_{82}\text{Pb}$

16-20 $\dfrac{1}{16}$

16-22 (a) 5 mg   (b) 1.25 mg   (c) 1.25 mg

16-23 2.50 g

16-25 about 11,500 years old

16-27 The energy from the radiation causes an electron in an atom or a molecule to be expelled leaving behind a positive ion. When a molecule in a cell is ionized it is damaged and may die or mutate.

16-29 Radiation enters a chamber causing ionization. The electrons formed migrate to the central electrode (positive) and cause a burst of current which can be detected and amplified. In a scintillation counter the radiation is detected by phosphors that glow when radiation is absorbed.

16-30 (a) $^{35}_{16}\text{S}$ (b) $^{2}_{1}\text{H}$ (c) $^{30}_{15}\text{P}$ (d) 8 $^{0}_{-1}\text{e}$ (e) $^{254}_{102}\text{No}$ (f) $^{237}_{92}\text{U}$ (g) 4 $^{1}_{0}\text{n}$

16-32 one neutron

16-34 $^{208}_{82}\text{Pb}$

16-35 $^{262}_{107}\text{Bh}$

16-36 (a) $^{87}_{34}\text{Se}$ (b) $^{97}_{38}\text{Sr}$

16-38 $^{2}_{1}\text{H} + ^{2}_{1}\text{H} \longrightarrow ^{4}_{2}\text{He}$

$$\nearrow \ ^{3}_{2}\text{He} + ^{1}_{0}\text{n}$$
$$\searrow \ ^{3}_{1}\text{H} + ^{1}_{1}\text{H}$$

16-39 (a) $^{106}_{46}\text{Pd} + ^{4}_{2}\text{He} \longrightarrow ^{109}_{47}\text{Ag} + ^{1}_{1}\text{H}$

   (b) $^{266}_{109}\text{Mt} \longrightarrow ^{262}_{107}\text{Bh} + ^{4}_{2}\text{He}$

      $^{262}_{107}\text{Bh} \longrightarrow ^{258}_{105}\text{Db} + ^{4}_{2}\text{He}$

   (c) $^{212}_{83}\text{Bi} \longrightarrow ^{212}_{84}\text{Po} + ^{0}_{-1}\text{e}$

   (d) $^{60}_{30}\text{Zn} \longrightarrow ^{60}_{29}\text{Cu} + ^{0}_{+1}\text{e}$

   (e) $^{239}_{94}\text{Pu} + ^{1}_{0}\text{n} \longrightarrow ^{140}_{55}\text{Cs} + ^{97}_{39}\text{Y} + 3\,^{1}_{0}\text{n}$

   (f) $^{206}_{82}\text{Pb} + ^{54}_{24}\text{Cr} \longrightarrow ^{257}_{106}\text{Sg} + 3\,^{1}_{0}\text{n}$

   (g) $^{93}_{42}\text{Mo} + ^{0}_{-1}\text{e} \longrightarrow ^{93}_{41}\text{Nb}$

16-41 (a) $^{90}_{39}\text{Y}$ (b) 50 years

16-44 60 years is five half-lives (i.e., 60/12 = 5). 0.42 g x 2^5 = <u>13.4 g</u>

16-46 $^{209}\text{Bi} + ^{64}\text{Ni} \longrightarrow ^{272}_{111}\text{Rg} + ^{1}\text{n}$

16-48 $^{208}\text{Pb} + ^{70}\text{Zn} \longrightarrow ^{277}\text{Uub} + ^{1}\text{n}$